MW01484830

MENDEL'S LEGACY

The Origin of Classical Genetics

MENDEL'S LEGACY

The Origin of Classical Genetics

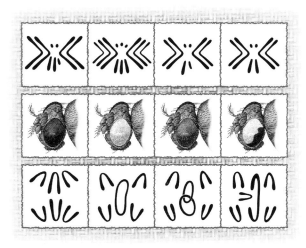

ELOF AXEL CARLSON

Stony Brook University
Stony Brook, New York

COLD SPRING HARBOR LABORATORY PRESS
Cold Spring Harbor, New York

MENDEL'S LEGACY
The Origin of Classical Genetics

Publisher	John Inglis
Acquisition Editor	Judy Cuddihy
Editorial Development Manager	Jan Argentine
Developmental Editor	Judy Cuddihy
Image and Permissions Coordinator	Nora Rice
Project Coordinators	Mary Cozza and Joan Ebert
Production Editor	Rena Steuer
Desktop Editor	Susan Schaefer
Production Manager and Interior Designer	Denise Weiss
Cover Designer	Ed Atkeson

Front cover artwork (created by Hans J. Neuhart, Electronic Illustrators Group): Top row: fruit fly chromosomes representing nondisjunction (XXY), triploid, normal male (XY), and normal female. Middle row: normal red-eyed fly, white-eyed fly, brown-eyed fly, and mosaic eye showing a patch of white and red eye. Bottom row: normal separation of chromosomes, dicentric ring chromosome, interlocking rings, and chromosome bridge and fragment from breakage-fusion-bridge cycle. The art celebrates the findings of C.B. Bridges, N. Stevens, E.B. Wilson, T.H. Morgan, H.J. Muller, A.H. Sturtevant, G.W. Beadle, and B. McClintock.

Library of Congress Cataloging-in-Publication Data

Carlson, Elof Axel.
  Mendel's legacy : The origin of classical genetics / by Elof Axel
Carlson.
      p. ; cm.
Includes bibliographical references and index.
  ISBN 0-87969-675-3 (hardcover : alk. paper)
  1. Genetics—History. 2. Genetics—Research—United States—History.
  [DNLM: 1. Genetics—history. 2. Genetic Processes. 3. Genetic
Research—history. QH 428 C284m 2004] I. Title.
QH428.C248 2004
576.5′072′073—dc22

                                   2003023295

10 9 8 7 6 5 4 3 2

All Cold Spring Harbor Laboratory Press publications may be ordered directly from Cold Spring Harbor Laboratory Press, 500 Sunnyside Blvd., Woodbury, N.Y. 11797-2924. Phone: 1-800-843-4388 in Continental U.S. and Canada. All other locations: (516) 422-4100. FAX: (516) 422-4097. E-mail: cshpress@cshl.edu. For a complete catalog of all Cold Spring Harbor Laboratory Press publications, visit our World Wide Web Site http://www.cshlpress.com/

To my twelve grandchildren

Vanessa Ariel Woodward
Natasha Nicolle Woodward
Natalie Carla Allen
Justine Allen Carlson
Caitlin Rose Allen
Miranda Lynn Woodward
Derrick Allen Carlson
Selena Michelle Woodward
Maxwell Hawkins Carlson
Jack Reuben Wiener
Hayden Anders Carlson
Owen Calder Carlson

Contents

Preface

MY MENTOR, H.J. MULLER, at one time described how he and the members of Morgan's group of fruit fly researchers struggled to work out a good portion of what we today call classical genetics. He called it "the winning of the facts." I believed then, and continue to believe now, that this was an appropriate phrase to describe a philosophy and history of science that resonates with many scientists. They experience much frustration, hesitation, false leads, and the ultimate arbiter of their work, the experimental results, to declare whether they have won or lost in their race against competitors or their contest against ignorance.

When I was a graduate student in Muller's laboratory at Indiana University (1953–1958), the term "classical genetics" was used as a derogatory reference to the work done in the first 50 years after the rediscovery of Mendel's laws. It was used by the emerging practitioners of biochemical and molecular approaches to the study of genetics. There was an irony to it, because at that time most of their methods for viruses and bacteria used the same mapping of genes, crossing of stocks, selecting of recombinants, utilization of rearranged chromosomes, agents to induce mutations, and determinations of spontaneous mutation rates that were used by those raised in fruit fly, maize, or mouse genetics. But that is the way of rivalries and the birth of new approaches to doing science. When molecular genetics had solved the immediate problems of gene structure and function, it returned once again to the organisms it had

dismissed and now it harvested nearly a century's worth of information on eukaryotic plants and animals to apply molecular techniques to once-classical organisms.

Today, classical genetics is no longer perceived as something to be shunned, but rather as something to be celebrated. The first 50 years of genetics in the twentieth century brought together the fields of breeding analysis, reproductive cell biology, cytology, and evolution. How that happened is the subject of this book. I have written it as a narrative to give readers a sense of being there and the opportunity to learn what the debates were about and to see the logic (rather than the primitiveness) of the assumptions that guided research.

I wrote this book for several reasons. No single volume has attempted to explore the way in which classical genetics emerged. There have been short histories of genetics but they are primarily based on a model that might be called a "review article" treatment.[1] The surviving ideas in such presentations are highlighted and the failures along the way are largely forgotten or omitted. The intent of these early authors was to provide readers with a chronology of the major successful events and some idea of the key intellectual issues that were prevalent at the time. The individual papers and the events that led up to them are not usually treated in these shorter surveys. Regardless of the reason that no history of classical genetics has emerged thus far, I believe that one is due. It is almost a half-century now that classical genetics has shifted without dispute into molecular genetics with the publication of the Watson–Crick model of DNA. That is a reasonable number of decades to have passed for an assessment of a period of history. I also believe that persons with both a background in writing history of science and the experience of doing that science at a professional level are rare. I consider myself fortunate to be among those with this dual interest. I believe this background to be important because every paper written by scientists reveals their scientific assumptions, stated or not. If given just the bibliography of a paper, it is not uncommon for a scientist in the field to infer who wrote that article and whose work was left out because it might present a different point of view.

Another reason for writing this work is a puzzle I hoped to explore. All of the tributaries that came together to produce classical genetics originated in Europe. Breeding analysis is chiefly associated with the work of Mendel, but other Continental and British breeders were either contemporaries or predecessors of Mendel. Cytological work, especially in German universities, was largely a European effort. Evolutionary science was developed by Lamarck in France and by Darwin and Wallace in Great Britain. A fourth tributary was reproductive biology and its ties to development that were worked out through studies of fertilization and formulations such as Weismann's theory of the germ plasm.[2] That too had a European origin. Why then were these all put together in the United States beginning with Sutton's chromosome theory of heredity in 1902?[3] I believe that the invention of the American graduate school had much to do with the shift in the way in which research was done.

Finally, I wanted to explore the contributions of plant and animal genetics to the formation of classical genetics. I was a third-generation member of Morgan's *Drosophila* school through H.J. Muller, with whom I received my Ph.D. It is easy to be so dazzled by the success of Morgan and his students that one overlooks what occurred before and what other groups had to contribute. Decades ago, my initial idea was to write a book on the fly lab. I took extensive notes and accumulated quite a bit of material from interviews that I did not use either in my early monograph, in 1966, on the history of the gene concept or in my biography of Muller (1981).[4] I abandoned that effort in favor of a broader view. Also, I feared being typecast as an apologist for the *Drosophila* group (especially Muller's perception of it); I hope my objectivity has not been blinded by my experience as his student. At least, in good conscience, I have given an honest reading of the literature in an attempt to understand the early history of classical genetics.

I offer a final comment regarding the scope of this book. This is not a history of genetics (short or otherwise). It is a history of classical genetics, however biased my view might be. What I mean is that genetics is a huge enterprise involving many aspects including its applications to agriculture, forestry, medicine, and pharmacology. Genetics has also been applied to the breeding of pets, ornamental flowers, and zoo animals. Although I suspect a substantial percent of all geneticists has been involved in applied aspects, very little of that is included in this book and I did not read all of the voluminous literature available on these subjects. I excluded this because my conception of classical genetics is that of its major principles, especially those listed in Chapter 23. But I do include chapters on classical genetics in the service of society (dealing with political influences on eugenics, the Lysenko controversy, and the radiation controversy of the Cold War) because these important chapters show that the applications of scientific knowledge are rarely value neutral. I also include a chapter on human and medical genetics because some basic principles of classical genetics emerged from them (including the Hardy–Weinberg law and the finding of balanced polymorphism). The social applications of genetics in the eugenics movement[5] and in the radiation controversy made use of classical genetics to justify some false or debatable points of view. In some instances, the outcomes were harmful to innocent people. Many geneticists who contributed important principles of classical genetics strongly believed that their findings should be applied to benefit humanity. This is a motivating principle for many scientists, including those whose laboratory work we would describe as pure science. Unfortunately, such motivations can lead to evil abuses as in Nazi race hygiene, the attempt to repress classical genetics in the Lysenko years (1935–1965), and the smearing of reputations on both sides of the radiation controversy. If there are lessons of history, these events are certainly worth describing for a younger generation of largely idealistic scientists who may have no idea of those past abuses.

I have been the beneficiary of those who allowed me to interview them or in whose classes I was a student including Alfred H. Sturtevant, Theophilus S. Painter, Edgar Altenberg, Jack Schultz, Fernandus Payne, Tracy M. Sonneborn, Ralph E. Cleland, Thomas J. King, Harold Plough, Mark Graubard, Zhores Medvedev, Bentley Glass, Irwin Herskowitz, and Willard F. Libby. I benefited from the reprint collections at the Lillie Library at the Marine Biology Laboratory of Woods Hole and especially the Davenport–Demerec reprint collection now at the Banbury Center of Cold Spring Harbor Laboratory. I also benefited from the Muller manuscripts at the Lilly Library at Indiana University, Bloomington. Valuable suggestions were made by A. Peter Gary, who read the entire manuscript. I thank Claudia Carlson for preparing 12 of the informative illustra-

tions. Over the years I have benefited from discussions with my former students, my fellow graduate students, and many colleagues and correspondents interested in the history of genetics including Harry Corwin, John Jenkins, John Southin, Ron Sederoff, Dale Grace, Robert Hendrickson, Scott Gilbert, Garland Allen, Krishna Dronamraju, Abe Krikorian, Ernest Hook, Ed Novitski, Irwin Oster, Seymour Abrahamson, Abraham Schalet, Shanta Iyengar, J. Richard Whittaker, and Ruth Cowan. I am grateful to the Cold Spring Harbor Laboratory Press for their enthusiasm for this project and the importance they see in publishing works on the history of genetics. It is a pleasure to work with editors who have literary skills and a remarkable understanding of science. I acknowledge the superb advice of John Inglis, Director of the Press; Judy Cuddihy, Developmental Editor, whose advice and careful reading were of enormous help in shaping the manuscript; Joan Ebert and Mary Cozza, Project Coordinators, for their enthusiastic organizational skills in seeing the manuscript through to production; Rena Steuer, Production Editor, for her enthusiasm and steady stream of email queries; Nora Rice, Permissions Coordinator, for her generous effort in locating hard-to-find items; Susan Schaefer, Desktop Editor; Denise Weiss, Production Manager, for her pleasing and exciting book design; and Clare Bunce, at Cold Spring Harbor Laboratory Library Archives. All of the members of the Cold Spring Harbor Laboratory Press staff contributed immensely to the evolution of this book from its incomplete first draft to its present form. I am also grateful to H.J. Muller, who in his lecture classes, relived the history of his battles in the winning of the facts as he thought aloud on the issues that occupied his career studying mutation and the gene.

ELOF AXEL CARLSON

End Notes and References

[1] Dunn L.C. 1965. *A Short History of Genetics*. McGraw-Hill, New York; and Sturtevant A.H. 1965. *A History of Genetics*. Harper and Row, New York.

Sturtevant's treatment is mixed. His account of the fly lab is very personal. He does some sleuthing and accuses de Vries of possible dishonesty in editing one of his discovery papers to include Mendel and to protect the independence of his rediscovery. Most of this small survey is written in review article fashion. Dunn's account is a more thorough coverage but it is written in a textbook or review style with very few personal reminiscences or asides.

[2] Weismann A. 1885. The continuity of the germ plasm as the foundation of a theory of heredity. In *Essays Upon Heredity and Kindred Biological Problems* (ed. E.B. Poulton, S. Schonland, and A.E. Shipley), pp. 163–236. Translated to English in 1891–1892. Clarendon Press, United Kingdom.

[3] Sutton W.S. 1903. The chromosomes in heredity. *Biological Bulletin* **4**: 231–251.

Although texts usually limit the theory to Sutton, in his time it was frequently called the Sutton-Boveri-Cannon-Wilson theory (or permutations of those four). See Chapter 9 for the history of this theory.

[4] Carlson E.A. 1966. *The Gene: A Critical History*. W.B. Saunders, Philadelphia; and Carlson E.A. 1981. *Genes, Radiation, and Society: The Life and Work of H.J. Muller*. Cornell University Press, Ithaca, New York.

[5] Carlson E.A. 2001. *The Unfit: A History of a Bad Idea*. Cold Spring Harbor Laboratory Press, Cold Spring Harbor, New York; and Kevles D.J. 1985. *In the Name of Eugenics: Genetics and the Uses of Human Heredity*. Alfred A. Knopf, New York.

Chronology of Classical Genetics

T HE FOLLOWING CHRONOLOGY COVERS THE TRIBUTARIES leading to classical genetics, as well as events after 1903 when classical genetics was clearly an interdisciplinary science, and those events leading up to the establishment of molecular genetics. I have also included events other than scientific papers and books, which establish historical background, such as the founding of institutions in which science was carried out, thus greatly facilitating scientific investigation. The dates are chiefly those of papers and books cited in the text.

1651: William Harvey identifies the egg as the basis of life (*ex ova omni*).

1665: Robert Hooke publishes *Micrographia* and describes cork as composed of myriads of "cells," the cells being empty boxes that hence account for the buoyancy of cork.

1670: Jan Swammerdam introduces preformation by observing the egg-like cocoons of insects.

1677: Antoni van Leeuwenhoek describes "animalcules" or spermatozoa in his semen.

1694: René de Graaf identifies follicles of ovaries, which he misinterprets as eggs.

1694: Rudolph Camerarius proposes sexuality in plants, the stamens and pistils corresponding to sex organs.

1741: Charles Bonnet uses aphids to support the ovist preformation theory.

1751: Carl Linnaeus argues that some species may arise from hybrid crosses.

1761: Josef Gottlieb Kölreuter notes the role of insects in carrying out pollination of flowers; he observes that tobacco (*Nicotiana*) hybrids have empty anthers.

1799: Thomas Robert Malthus publishes the first edition of his work on population, arguing that there are natural restraints on population growth.

1801: Marie Francois Xavier Bichat uses dissection without microscopy to describe organs as composed of tissues.

1801: Erasmus Darwin publishes *Zoonomia* and argues for an evolution of life based on ideas of progress, development, and metamorphosis.

1802: William Paley publishes *Natural Theology* and argues that life reflects a complex Designer (i.e., God) at work. He also provides evidence for the precision of adaptations to the environment.

1815: Jean Baptiste Lamarck proposes an evolution based on acquired characteristics.

1817: Georges Cuvier proposes a classification of higher taxa for animals and recognizes the successive replacement of species after catastrophes.

1823: Thomas Andrew Knight mates gray-coated

peas with white-coated peas and finds the F_1 generation all gray. The F_2 generation contains both gray- and white-coated peas. He does not use ratios.

1824: Joseph Jackson Lister uses minerals dissolved in quartz to produce achromatic glass lenses for telescopes and microscopes.

1830: Giovanni Battista Amici describes the pollen tube and observes its entry into the embryo sac of flowers. He constructs and uses a microscope with lenses that are free of chromatic and spherical aberration.

1830: Charles Lyell publishes *Principles of Geology* and establishes uniformitarianism as the major theory of geologists.

1833: Robert Brown describes the cell nucleus as a feature of all cells.

1835: Alexander and Karl von Humboldt establish scholarship as the chief function of the German university.

1835: Louis Agassiz observes the antiquity and abundance of fossils and attributes them to past extinctions from catastrophes. He emigrates to America and establishes comparative anatomy and the museum tradition for zoologists.

1838: Matthias J. Schleiden proposes a cell theory, arguing that plants are communities of cells.

1840: Theodor Schwann proposes that animals are also communities of cells. With Schleiden, he proposes a "free formation" of cells by crystallization.

1853: John Henry (Cardinal) Newman proposes a new Catholic university based on the liberal arts and "knowledge for its own sake."

1855: Rudolph Virchow proposes that "all cells arise from pre-existing cells" (the cell doctrine) and claims that tumors arise from single abnormal cells.

1855: Robert Chambers anonymously publishes *Vestiges of Creation*. He argues that there are discontinuities in the evolution of life.

1856: Carl Theodor von Siebold describes parthenogenesis in insect species.

1857: Joseph von Gerlach develops stain technology, leading to greatly enhanced histological studies in anatomy and pathology.

1858: Charles Darwin publishes the idea of evolution by natural selection in a joint paper with Alfred Russel Wallace for the Linnaean Society.

1860: Max Schultze coins the term "protoplasm" for the substance of the cell.

1864: Charles Darwin studies pea plants for gradations of characters using height, manner of growth, and period of maturity.

1864: Herbert Spencer proposes that physiological units for heredity are a size larger than atoms and known chemical molecules and smaller than cells.

1865: Gregor Mendel discovers the law of segregation (3:1 ratio) and the law of independent assortment (9:3:3:1 ratio) for peas.

1866–1870: Gregor Mendel confirms Karl Nägeli's idea that hawkweeds (*Hieracium*) do not follow Mendel's laws.

1866: Ernest Haeckel uses size to argue that the nucleus and not the cytoplasm must be the bearer of hereditary traits.

1868: Darwin publishes *The Variation of Animals and Plants under Domestication* and proposes a "provisional theory of pangenesis," with gemmules as circulating units of inheritance.

1868: Thomas Henry Huxley proposes a liberal education that includes the laws of nature. He favors the German model of universities for research.

1869: Thomas Henry Huxley claims that protoplasm is the basis of life.

1871: Francis Galton tests pangenesis in rabbits by circulating blood between two different colored rabbits. He finds no evidence of transmission of gemmules through blood circulation.

1872: Anton Dohrn founds the Naples Zoological Station as an international center for scientific research.

1875: Johns Hopkins University is founded and Daniel Coit Gilman becomes its first president. He creates an American Ph.D. degree, which is interdisciplinary in spirit.

1876: Oscar Hertwig describes meiosis in sea urchin eggs and notes the formation of three polar bodies and an egg.

1877: Hermann Fol describes the formation of a male pronucleus from a sperm and a female pronucleus in the egg that join in fertilization. He describes this as the "beginning of ontogeny."

1879: Walther Flemming describes mitosis in stained and living cells of salamanders.

1880: William Keith Brooks publishes *Heredity* and argues that it is a subject for scientific study. He stresses cell lineage studies for his students, rather than preservation of museum specimens.

1880: Eduard Strasburger describes mitosis in plants.

1883: Edouard van Beneden describes meiosis at the chromosomal level and notes a reduction in chromosome number.

1884: Eduard Strasburger argues that the pollen tube nucleus contains the hereditary material, because most of the pollen tube does not enter the embryo sac.

1885: August Weismann proposes a theory of the continuity of the germ plasm; it is isolated and set aside from the soma.

1886: Francis Galton argues that quantitative traits tend to "regress to the mean" from parents who show departures from the mean.

1887: The Hatch Act, funded by the U.S. Department of Agriculture, establishes experimental stations for agricultural land-grant state universities.

1888: Heinrich Wilhelm Gottfried Waldeyer names the mitotic threads and calls them chromosomes.

1888: Wilhelm Roux introduces the idea of Entwicklungsmechanik as experimental embryology applied to embryos. He argues that differentiation begins at the two-cell stage and that each blastomere has a different fate.

1888: August Weismann demonstrates that mutilations (in humans or induced in mice by cutting off their tails) do not lead to the inheritance of acquired characteristics.

1890: Hugo de Vries publishes *Intracellular Pangenesis*. In it, he argues that pangenes (Darwin's gemmules) never leave the cell and that they are the bearers of hereditary traits.

1890: August Weismann argues that meiosis must involve two divisions, one of them a reduction division, to account for biparental inheritance and to maintain a constant chromosome number.

1891: Hermann Henking finds an atypical chromosome which he calls an "X element" that may be nucleolar or chromosomal (or both) in the fire wasp, *Pyrrhocoris*.

1894: William Bateson publishes *Materials for the Study of Variation* and argues that many traits arise as sports or discontinuities.

1894: Hans Driesch finds that sea urchin blastomeres are totipotent. This launches a debate on the developmental fate of early cleavage-stage cells in embryos. His totipotent blastomeres introduce artificial cloning from embryonic parts and explain identical twins.

1895: Theodor Boveri fertilizes one species of echinoderm egg cytoplasm with sperm of a different echinoderm species. He obtains half-sized, but otherwise normal, larvae that show the traits of the sperm donor and not those of the cytoplasmic source. He thus argues that the nucleus and not the cytoplasm determines the hereditary potential of the organism.

1896: Edmund B. Wilson publishes *The Cell in Development and Inheritance*. He argues that nuclein may be the bearer of hereditary traits and the source of enzymatic activity in the cytoplasm.

1900: Thomas Harrison Montgomery, Jr. finds chromosomes paired as homologs in locust and grasshopper chromosomes during meiosis.

1900: Hugo de Vries observes Mendelism in 20 different species of plants.

1900: Carl Correns confirms Mendel's law of segregation in peas and confirms the constancy of extracted recessives for six generations.

1900: Erich Tschermak von Seysenegg confirms Mendel's law of segregation in peas.

1900: Jacques Loeb describes artificial parthenogenesis in sea urchin eggs.

1901: Clarence McClung describes the accessory chromosome (Henking's X element) as male determining.

1901: Hugo de Vries publishes *The Mutation Theory*. New species of *Oenothera* (evening primrose) arise de novo as saltations or discontinuities.

1902: William Bateson extends Mendelism to poultry.

1902: William Austin Cannon uses cotton to propose a correlation of meiotic reduction division with Mendelian segregation.

1902: William Bateson publishes *Mendel's Principles of Heredity: A Defence*. Bateson criticizes the biometric school for resisting the role of Mendelism in interpreting heredity.

1902: Theodor Boveri demonstrates that chromosomes differ in their hereditary specificities. He uses echinoderm larvae with different combinations of chromosomes to produce abnormal larvae, each abnormal in a different way.

1902–1906: William Bateson and colleagues report modified 9:3:3:1 ratios, which they describe as showing epistasis (interaction of traits). They introduce the terms allele (allelomorphs), heterozygous, and homozygous.

1903: Walter Sutton uses grasshopper chromosomes to associate both of Mendel's laws with the random meiotic alignment and separation of paired homologs.

1903: Lucien Cuénot finds Mendelian traits in mice.

1903: Wilhelm Johannsen finds bean size to be a continuous trait that he attributes to polygenic inheritance, producing bell-shaped curves. He finds no further shift in variation when selection is applied to pure (homozygous) inbred lines.

1904: Rollins A. Emerson finds epistasis in bean hybrids.

1905: Nettie M. Stevens finds two sizes of heterochromosomes in the beetle *Tenebrio*, the male having both and the female having two of the larger heterochromosomes.

1905: Edmund B. Wilson finds two sizes of idiochromosomes in his hemipteran bugs, with males producing two types of spermatozoa and females producing only one type of egg.

1906: William Bateson reports coupling and repulsion of traits.

1906: Leonard Doncaster and Gilbert H. Raynor find sex-limited traits in Lepidoptera.

1906: William E. Castle and his students use fruit flies to study inbreeding and selection effects.

1906: William Bateson introduces the term "genetics."

1907: Erwin Baur finds a semilethal mutation leading to loss of chlorophyll in snapdragons (*Antirrhinum*).

1908: William Bateson proposes a reduplication hypothesis for coupling and repulsion of traits.

1908: George H. Shull demonstrates the highly varied composition of a field of corn plants.

1908: Fernandus Payne gives Thomas Hunt Morgan a culture of fruit flies to use for genetic and evolutionary studies.

1909: Wilhelm Johannsen defines genotype as the inferred genetic composition of an individual through breeding analysis. He defines phenotype as the appearance of a character trait. He replaces gemmules, pangenes, and unit characters with the undefined term, "gene."

1909: Herman Nilsson-Ehle proposes a modified 9:3:3:1 ratio, producing a 1:4:6:4:1 ratio for quantitative traits such as coat color in cereal grains.

1909: Reginald Ruggles Gates argues that new species of *Oenothera* are chromosomal aberrations and not of major significance for the origin of most species.

1909: George H. Shull demonstrates hybrid vigor using two inbred lines of corn (maize) crossed to each other. This suggests to him that this might be used to enormously increase yields in agriculture.

1909: William E. Castle and John C. Phillips use ovarian transplants in guinea pigs to show no influence of host genotype on donor genotype.

1909: Frans Alfons Janssens publishes figures of intertwined homologous pairs of chromosomes in salamanders and calls these twists "chiasmatypie."

1910: William E. Castle and Clarence C. Little obtain a 2 yellow:1 nonyellow ratio for the yellow mouse $F_1 \times F_1$ cross and argue that this is a consequence of the lethality of the homozygous yellow.

1910: Fernandus Payne publishes a paper on fruit flies raised for 49 generations in the dark with no change in eye color or body color.

1910: Edward M. East finds that macerated tissue and juices of red tomatoes injected into nonred tomato plant ovaries or stamens have no effect on the heredity of the nonred plants obtained.

1910: Edward M. East finds quantitative inheritance in maize.

1910: Edmund B. Wilson uses the term "sex chromosomes" for idiochromosomes and heterochromosomes. Males are heterogametic and XO or XY and females are homogametic and uniformly XX.

1910: Fernandus Payne finds that reduviid bugs have multiple sex chromosomes.

1910: Thomas Hunt Morgan finds a trident pattern (with) and a darkening at the wing juncture (speck) as his first variations in fruit flies. He then finds white-eyed flies and establishes that they are sex limited and produce a modified Mendelian ratio.

1911: Bradley Moore Davis (and independently Otto Renner) shows that *Oenothera* species differ in their combinations of aberrant chromosome associations.

1911: Thomas Hunt Morgan discovers recombinant sex-limited traits (*white*, *rudimentary*, and *miniature*) and interprets these as arising from crossing-over between homologous X chromosomes during meiosis. He cites Janssens' work on chiasmatypie as the basis for his interpretation.

1912: William Ernest Castle uses hooded mice to study variable traits. He argues that they vary because of "unit-character instability" (contrary to Mendel's claims that extracted recessives breed true).

1912: Thomas Hunt Morgan describes an X-linked lethal in fruit flies.

1913: Charles Davenport applies quantitative inheritance to human skin color using racial hybrids in Jamaica.

1913: Alfred H. Sturtevant interprets Morgan's data and constructs the first chromosome map involving six traits.

1913: Alfred H. Sturtevant proposes multiple allelism to describe the relationship of white, apricot, eosin, and red eye color as X-linked traits.

1914: Hermann J. Muller publishes a criticism of William Castle's hooded rats and attributes their variation to modifier genes.

1915: Thomas Hunt Morgan and his students estimate the number and size of genes based on mapping data. They argue that fruit flies have more than 1000 genes and that the gene is too small to be seen with an optical microscope.

1916: Calvin B. Bridges publishes his analysis of nondisjunction and relates abnormal karyotypes to the phenotypes they express for white or red eye color and for the fertility or lack of fertility of the males produced.

1916: Thomas Hunt Morgan, Alfred Sturtevant, Hermann J. Muller, and Calvin B. Bridges publish *The Mechanism of Mendelian Inheritance*.

1918: Hermann J. Muller demonstrates chief genes and modifier genes in the variability of the *Beaded* wings trait.

1919: Hermann J. Muller and Edgar Altenburg report the first mutation rate for fruit flies: one lethal mutation per 1000 X chromosomes.

1922: Hermann J. Muller argues that genes have the capacity to reproduce their variations and cannot be distinguished from viruses that may be "naked genes."

1925: Alfred H. Sturtevant and Thomas Hunt Morgan discover position effect for the Bar locus, whose "reverse mutations" are actually crossovers within a duplicated gene.

1926: Hermann J. Muller proposes that the gene is the basis of life and the evolution of life is traceable to the first gene.

1927: Hermann J. Muller and Lewis J. Stadler (1928) independently find that X rays induce mutations. Muller uses fruit flies and the ClB stock to detect them and then map them. Stadler uses barley and oats.

1929: Israel I. Agol and Alexandr S. Serebrovsky propose that gene structure can be resolved through complementation analysis. Sturtevant rejects their view.

1929: Curt Stern discovers dosage compensation using the *bobbed* locus on the Y and **X** chromosomes. Muller and colleagues demonstrate (1931) that dosage compensation involves genes on the X chromosome that modify other genes. Genes that fail to respond to the modifiers are not dosage compensated.

1929: Ronald A. Fisher publishes the *Genetical Theory of Natural Selection*. His contributions to mathematical genetics, along with those of John Burdon Sanderson Haldane and Sewall Wright, establish a consistency of Mendelism with population genetics and evolution by natural selection.

1932: Hermann J. Muller proposes a functional classification of mutations as amorphs (lacking all function), hypomorphs (showing partial function), hypermorphs (having more than normal function), and neomorphs (showing a new function not seen before).

1933: Theophilus S. Painter discovers banding in giant salivary chromosomes of fruit fly larvae.

1934: Calvin B. Bridges constructs a detailed map of the salivary chromosomes correlating them to the linkage maps obtained by breeding analysis.

1935: Calvin B. Bridges and Hermann J. Muller (1936) independently discover that Bar is a duplication. Muller argues that this implies a "gene doctrine," with every gene arising from a preexisting gene to the first gene-like molecule with genic properties.

1937: Julia Bell and John Burdon Sanderson Haldane map the first human X-linked traits.

1938: Barbara McClintock discusses the breakage-fusion-bridge cycle as the mechanism producing streaks and spotting in corn kernels. Hermann J. Muller and Guido Pontecorvo use the breakage-fusion-bridge cycle to interpret embryonic death of fertilized eggs from X-ray-treated sperm (1940). Muller (1946) interprets radiation sickness at Hiroshima and Nagasaki as identical to the "radiation necrosis" described in their 1940 paper.

1940: Daniel Raffel and Hermann J. Muller use the "left-right test" to map the genes on the distal tip of the X chromosome and to identify genes as surrounded by stretches of breakable material.

1940: Clarence Peter Oliver discovers lozenge alleles that are pseudoallelic.

1941: George W. Beadle and Edward L. Tatum demonstrate biochemical pathways worked out by a study of vitamin-deficient mutations of the fungus *Neurospora*. They propose the one-gene–one-enzyme theory. With Boris Ephrussi in 1937, Beadle had earlier demonstrated that eye color in fruit flies involved biochemical pathways worked out with the use of embryonic anlage transplanted into host larvae.

1941: Edward B. Lewis argues that pseudoalleles are a consequence of tandem repeats or duplications that experienced differential mutational histories leading to cognate functions.

1941: Max Delbrück works out the life cycle of the bacteriophage, T2. With Salvador Luria, he identifies mutations of the virus and establishes it as a useful genetic organism.

1944: Oswald Avery, Maclyn McCarty, and Colin MacLeod establish DNA as the genetic material (transforming principle) in pneumococcus (*S. pneumoniae*).

1945: Erwin Schrödinger uses the work of Max Delbrück, Karl G. Zimmer, and Nikolai V. Timoféeff-Ressovsky to interpret genetic material as having a code script and as being an aperiodic crystal. His views in *What is Life?* bring several physical scientists into biology.

1946: Charlotte Auerbach reports mutations induced with mustard gas and nitrogen mustard using fruit flies. Her work, done in 1941, was censored during the war.

1946: Joshua Lederberg and Edward L. Tatum demonstrate sex and recombination of genes in bacteria. Lederberg develops tools (replica plating) for isolating bacterial mutations.

1949: Linus Pauling interprets sickle-cell anemia as a molecular disease. With Vernon Ingram (1956), he associates a specific site and amino acid with that disorder.

1952: Alfred D. Hershey and Martha Chase demonstrate that viral DNA but not viral protein enters a bacterial host cell and that both components of the virus are produced by the entering viral DNA.

1952: Guido Pontecorvo proposes that all genes are pseudoallelic because recombination occurs within genes and not just between genes.

1953: James D. Watson and Francis H.C. Crick publish the double-helix model of DNA and propose the model that will explain mutation, coding, and replication. Their work begins the era of molecular genetics.

1954: Seymour Benzer introduces the idea of genetic fine structure, with genes mapped to their presumed nucleotide sequences. He also introduces the concepts of cistrons, recons, and mutons.

1955: Milislav Demerec analyzes *S. typhimurium* and shows that both duplicated genes and intragenic recombination within them coexist, resolving the dispute between Edward B. Lewis and Guido Pontecorvo.

1957: Edward B. Lewis discovers developmental genes at the *bithorax* locus that account for insect evolution of body parts, especially wings, abdominal segments, and halteres.

What Is Classical Genetics?

Mendelism found a hearty welcome in America from two groups of biologists, those who were interested in the study of evolution from a pure science viewpoint, and those who sought more efficient ways to produce improved varieties of plants and breeds of animals.

WILLIAM E. CASTLE*

CLASSICAL GENETICS MAY BE DEFINED in several ways. Its most common usage is based on chronology and runs from 1900 (the date of the rediscovery of Mendelism by Carl Correns, Hugo de Vries, and Erich Tschermak von Seysenegg) to 1953 (the date of the introduction of the Watson–Crick model of DNA).[1] There are good reasons for using these two events to mark the beginning and end of classical genetics. The rediscovery of Mendelism quickly shifted the study of heredity-by-breeding analysis into a practical as well as theoretical science. The discovery of DNA's double-helical structure changed the way genetics was done, with a profound shift to molecular interpretations of the gene and its functions.

There are other contenders for these benchmark beginnings and ends of classical genetics. Some shifted the beginning to the interdisciplinary efforts of bringing breeding analysis to cytology (the chromosome theory of heredity in 1902–1903, especially the form proposed by Walter Sutton) or to the fusion of breeding analysis, cytology, and evolution (the work of Thomas Hunt Morgan and his students, roughly 1910–1916).[2] Others went back even earlier by referring to the cytological schools in Germany and other European laboratories that worked out fertilization, mitosis, and meiosis (mostly from the 1870s to 1890s), if not in the detail as we now know it, then at least in sufficient scope to indicate their role in heredity.

Some thought to end the study much earlier than 1953 and chose the first direct evidence for DNA as genetic material (this occurred in 1944 with the publication of the Avery, McLeod, and McCarty experiments on *Pneumococcus*)[3] or the first use of bacteriophage as genetic systems with a discernible life cycle (the work of Max Delbrück and his school, beginning in 1939).[4] However, all would agree that classical genetics was done without knowledge of the molecular basis of heredity.

*1951. The Beginnings of Mendelism in America. In *Genetics in the 20th Century* (ed. L.C. Dunn), pp. 59–76. Macmillan, New York.

What Is the Scope of Classical Genetics?

Classical genetics is important because it has had a profound effect in shaping our understanding of life. It is a field that is anathema to holistic, vitalist, mystical, and revealed models of life because its methods and findings are supremely reductionistic. It ties heredity to evolution and points the way to a molecularization of the gene, an idea that was never doubted by its main contributors. The philosophic implications of such a view of life are threatening to some and liberating to others, but all must confront the evidence for such a perception.

The practical applications of genetics are enormous. Agriculture was transformed by hybrid corn and the introduction of a variety of disease-resistant strains of crop plants. The dairy and meat industries are dependent on having students with a sound knowledge of genetics who can devise more effi-

cient ways of bringing food to populations, using methods that would have been considered impossible in Malthus' day. One could argue that our longevity and relative freedom from famine and disease are owed in large measure to the contributions of genetics to our lives. The first commercial quantities of penicillin were made possible by selection for X-ray-induced mutations that increased the yield of penicillin from molds. At the same time, classical genetics was used to justify racism and bigotry, and dubious beliefs about human behavioral inequality or failure were abused in the eugenics movement and the Holocaust. This is not unexpected in the history of science because all knowledge can be used or abused, but it is certainly important since it alerts both scientists and the public of their obligation to note potential abuses of science.

What Are the Components of Classical Genetics?

Classical genetics, as I interpret it, includes such major concepts as Mendelism, the chromosome theory of heredity, the theory of the gene, the mapping of genes to chromosomes, the genetic consequences of chromosome rearrangements, the genetics associated with polyploidy and aneuploidy, the relationship of genes to expressed characters, the genetics of traits exhibiting a continuous distribution, the study of spontaneous and induced mutations, the relationship of genetics and cytogenetics to evolution, the mathematical description of genes in populations, and the study of gene structure and function at a phenotypic level. It represents the fusion of the fields

of breeding analysis, cytology, evolution, reproductive biology, and developmental biology. Once classical genetics encountered biochemical and molecular approaches to genetics, the nature of genetics changed, and it was properly renamed molecular genetics because of its new shift to the chemical and physical properties of genes. This description of classical genetics (and my historical treatment) largely excludes biochemical genetics and molecular genetics. It also excludes most applied aspects of genetics, such as its use in the dairy and livestock industries and the widespread use of classical genetics in the production of all fruits and vegetables as well as cereal crops. It is difficult to distinguish the applied and the basic science in a consistent way. I do include such practical findings as hybrid corn, because the analysis of this phenomenon has led to major contributions of classical genetics, especially the genetic interpretation of phenotypes.

Genetics Is More Inclusive Than Classical Genetics

Genetics is a broader term than classical genetics. Classical genetics addresses major issues raised by a knowledge of biology: What are inherited traits? How are they transmitted? What is the relationship of the two parents to the heredity of their offspring? Why is there so much diversity and variation among living things? How is sex determined? If hereditary units exist, where are they to be found in a cell or organism? Which traits are acquired by training and which are essentially determined from fertilization? Are character traits simple or complex in their inheritance? Genetics necessarily includes classical genetics and uses its principles, but it also applies genetics to health, law, social problems (often erroneously and with controversy), and to the production of food and other industries.

Classical genetics has not been replaced by molecular genetics. Like the relationship of Newtonian physics to modern physics, the old is assimilated into the new. The structure of the gene at a molecular level, with its assorted regions for recognition and attachment of enzymes and other proteins, is far more complex than classical geneticists of the 1920s could have imagined.[5] But when working with diploid organisms, both molecular and classical geneticists recognize the gene as a unit of inheritance and rely on classical genetics for constructing maps of genes, combining genes to design genetic stocks, and applying tools and concepts derived from classical genetics. What makes classical genetics distinct from molecular genetics is its limitation. No matter how much these four tributaries heading to classical genetics could be exploited, they did not describe the molecular basis of heredity. This book, like the classical genetics it portrays, omits the biochemical and molecular bases of heredity.

What Are the Broad Historical Themes of the Nineteenth Century?

Classical genetics is primarily a twentieth-century science, although its roots are largely based in the nineteenth century. The issues of interest to science in the nineteenth century were considerably different from those of the twentieth. I like to interpret the life sciences in the nineteenth century as having gone through four major influences. I discuss all four in relation to the problems that they attempted to resolve and why they often failed to do so. Many of the problems associated with the development of classical genetics may be seen as responses to these older ways of interpreting life. I present these as a "big picture" or framework, so that the major ways of viewing life in the nineteenth century can be distinguished from the ways in which scientists saw the life sciences in the twentieth century.

Phase I: 1800–1830

I call this the **era of natural theology**. Its major interpreter, William Paley, interpreted life in religious terms. He believed God felt good about His creations and the world of life was in harmony. The role of science was to show the wisdom of the Creator. Life had an intelligent designer and the most that science could hope for was to learn some of that design. Opposed to Paley was the French Deist tradition that had been promoted at the end of the eighteenth century. Lamarck, in particular, was considered godless, speculative, and dangerous as a result of his ideas of an evolution of life from nonlife and for his attempt to make biology (he named that field) a secular science.

Phase II: 1830–1860

I call Phase II the **secularization of science**. The papers and presidential addresses of the British Association for the Advancement of Science demonstrate that natural theology faded away as physicists, astronomers, geologists, chemists, and biologists presented their papers. A great deal of basic and applied science emerged as the Industrial Revolution dominated early nineteenth-century society in Europe. The German university dominated scholarship in this era. Industry was well aware of its debt to scientists who described the material universe. Note that most of this secularization occurred before Darwin published his work on evolution by natural selection.

Phase III: 1860–1880

I refer to Phase III as the **Darwinian transformation of the life sciences**. The evidence for evolution was so abundant that there were few scientists willing to deny it, and religion either coexisted or was compromised in its relationship with science. At this time, variation and heredity were thought to be separate processes, the former accounting for differences within a species and the latter for the uniformity of traits associated with a species. Two types of variations were recognized: Darwinian fluctuations were subtle changes that Darwin argued would accumulate through selection over numerous generations leading to speciation. The second type was suddenly arising discontinuous variations, then called sports and today called spontaneous mutations. These were not considered of evolutionary interest (they were thought to be monstrosities that would be rapidly selected out of the population).

Phase IV: 1880–1900

I call Phase IV the **integration of the life sciences through reductionism**. This was a very rapidly developing period of the life sciences in Europe and North America. Weismann's theory of the germ plasm argued that heredity was isolated in reproductive tissue and protected from environmental influences that altered the rest of the body tissues. It crushed Lamarckian theories of heredity that saw the environment altering heredity in a direct way. Experimental science was replacing descriptive science in biology. German Entwicklungsmechanik (a form of experimental science applied to embryos) spread to North America. In general, the life sciences emulated the physical sciences. As the physical scientists became statistical, so did the life scientists, who believed that this mathematical approach would generate insights into the evolutionary process and the problem of heredity.

As we will see, the emergence of classical genetics in the twentieth century is, to a large degree, a repudiation or an extension of these different themes of the nineteenth century. These broad ideas shaped the experiments and interpretations of the first generation of geneticists. What emerged—classical genetics—was very different from most of these nineteenth-century outlooks.

End Notes and References

[1] Mendel's paper (1865), the rediscovery papers (1900), and associated papers of that period (1895–1905) are accessible in Krizenecky J. and Nemec B., eds. 1965. *Fundamenta Genetica*. Publishing House of the Czechoslovak Academy of Sciences, Prague. The double-helix papers are Watson J.D. and Crick F.H.C. 1953. Molecular structure of nucleic acids. *Nature* **171:** 737–738; and Watson J.D. and Crick F.H.C. Genetical implications of the structure of deoxyribonucleic acid. *Nature* **171:** 964–967. The way this discovery was made, from Watson's point of view, is explained in his classic work, 1968. *The Double Helix: A Personal Account of the Discovery of the Structure of DNA*. Atheneum, New York.

[2] Morgan T.H., Sturtevant A.H., Muller H.J., and Bridges C.B. 1915. *The Mechanism of Mendelian Heredity*. Henry Holt and Company, New York.

[3] Avery O.T., MacLeod C.M., and McCarty M. 1944. Studies on the chemical nature of the substance inducing transformation of *Pneumococcal* types. *Journal of Experimental Biology and Medicine* **79:** 137–158.

[4] An account of Delbrück's work, school, and influence may be read in the collection of essays by Cairns J., Stent G.S., and Watson J.D., eds. 1966. *Phage and the Origins of Molecular Biology*. Cold Spring Harbor Laboratory of Quantitative Biology, Cold Spring Harbor, New York.

This is a superb collection of papers on the history of phage genetics, with each essayist asked by the editors to be personal and reflective.

[5] Among the many good treatments of molecular genetics, I would recommend Olby R. 1974. *The Path to the Double Helix*. University of Washington, Seattle; and Judson H.F. 1979. *The Eighth Day of Creation*. Jonathan Cape, London.

Carl Linnaeus (1707–1778), professor at Lund in Sweden, showed in 1751 that new species could form from hybridization. All other species, he believed, were products of Divine Creation. Thus, Linnaeus is both a forerunner of evolution by hybridization and an icon of Creationist scientists who believe in the fixity of species. He is better known as the systematist who provided a system of classification (binomial nomenclature) by which organisms are named (using a genus and species, like that for ourselves, Homo sapiens). (Reprinted, with permission, from the Linnaean Society, London.)

Routes to Classical Genetics: Evolution

Pangenesis is a modern revival of the oldest theory of heredity, that of Democritus, according to which the sperm is secreted from all parts of the body of both sexes during copulation, and is animated by a bodily force; according to this theory also, the sperm from each part of the body reproduces the same part.

AUGUST WEISMANN*

BOTH RELIGION AND SCIENCE OFFER EXPLANA-TIONS of the origins of life. Religion does so by invoking acts of creation by an intelligent designer. The acts may be recorded in scripture or ancient texts believed by the faithful to have been inspired or dictated by God. In the Judeo-Christian-Islamic traditions, those writings are described in the book of *Genesis*. For those who take the account of God's creation literally, there are six days of effort with a day of rest after the creation of organized matter, light and dark, earth and seas, plant life (angiosperms), the astronomical universe (sun, moon, and stars), birds and sea life (including whales), and most animal life on land including humans, in roughly that sequence.[1]

Antiquarian Views of Evolution to Erasmus Darwin

Some ancient philosophers, like Lucretius and Democritus, had a vague sense of evolution and raised doubts or alternatives to the scheme of sudden creation by deities. They envisioned a random gathering of parts with a selection of those lucky progenitors with workable sets of organs. Their views were revived in the late eighteenth century by Erasmus Darwin (1731–1802) (Figure 1) in his *Zoonomia*.[2] He attempted (first in prose and then in verse form) to describe a universe coming into being without invoking miracles. He accepted notions such as the spontaneous generation of life from nonlife and a belief in a universal progress. He gave as examples of this progress the life cycles of frogs from tadpoles to adults and of insects from caterpillar stages to adult butterflies. The idea of progress

---

*1885. The continuity of the germ-plasm as the foundation of a theory of heredity. In *Essays Upon Heredity and Kindred Biological Problems* (ed. E.B. Poulton, S. Schonland, and A.E. Shipley), pp. 163–255; see p. 168. Clarendon Press, Oxford.

Figure 1. *Erasmus Darwin, the grandfather of Charles Darwin, was a writer, philosopher, and early member of the Lunar Society, a group of scientists and scholars who enjoyed the new wealth of the growing Industrial Revolution in Birmingham, England. (Courtesy of the National Library of Medicine.)*

in the late 1700s was strongly influenced by the French enlightenment movement, especially the ideas of Marie-Jean-Antoine-Nicolas Caritat, Marquis de Condorcet (1743–1794). Condorcet set no limits on progress, rationalism, or the capacity of science to change the world. He believed that the application of reason and knowledge would end sexism, slavery, political despotism, poverty, and ill-

ness. He believed that this would lead to humans regulating their own family size and keeping populations from expanding beyond their capacities to find food and housing. Condorcet's philosophy had a powerful effect on English intellectuals. Many agreed with his ideas and saw in science a superior method of explanation. Science, after all, had a string of successes in astronomy (Galileo and Kepler), biology (Harvey and Hooke), and physics (Newton) that inspired the French enlightenment.[3]

When science uses reason, observation, and experimentation instead of revelation or faith as its basis for investigating nature, it easily permits atheism or at least an erosion of popular faith based on religious tradition. Many of the scientists who made these contributions were themselves very religious and far from having an atheistic outlook. What made them good scientists was their reliance on scientific habits. They may have accepted God as a creator of the laws of science, and in their minds their inquiries were not seen as an assault on popular religion. But many in the enlightenment movement did have doubts about popular belief and sought a way out with a dualism that permitted a coexistence of two worlds that they sought to occupy. Others just dumped popular religious belief, and this threatened not only dualists, but also those who could not reconcile science with scriptural tradition.

Paley's Natural Theology Rejects Evolution

A compromise arose through the movement championed by William Paley (1743–1805).[4] He argued that a natural theology offered a role for science and that science could see the intelligent design of the Creator through God's handiwork. His famous comparison between stubbing one's toe on a stone while walking in a heath and stubbing one's toe on a dis-

carded old clock was used by generations of priests and ministers to enlighten young students about their nascent doubts. Paley's natural theology also attempted to answer some questions raised by the growing evolution movement. He explained adaptations as designs built in by the Creator for each organism to survive and flourish. His account of the eruption of teeth after breast-feeding in teething babies, the formation of milk teeth and their relation to the size of the growing jaw, and their replacement by adult teeth is meticulous in scientific description and the way in which teeth are related to human function throughout the life cycle. Charles Darwin later considered Paley his best source for studying the precision of adaptation for evolution.

Lamarck Proposes an Environmental Model of Evolution

Despite the efforts of Paley and those who tried to build a case for God as an intelligent designer, the godless doctrine of evolution imported from France continued its influence. Its chief proponent was Jean-Baptiste de Lamarck (1744–1829) (Figure 2).[5] Lamarck took an interest in science while serving in the French Army. While on the Mediterranean coast, he collected seashells and marveled at their diversity. He also took an interest in botany and began writing books (in French, since he had no formal education in Latin) on the natural plants found in his country. His support for the French Revolution gave him an opportunity to head the national museum's section on lower animals (he called these invertebrates). Lamarck's ideas were a mixture of ancient philosophy long abandoned by more modern science and empiricism of his own observations. He rigidly believed in the four-element theory of matter (air, fire, earth, and water) and attempted to interpret geological strata as decompositions or combinations of these four elements. He believed in an origin of life from spontaneous generation, but he saw a role for heredity in the process. He believed changes were slow and cumulative, with the organism responding to the changes in environment through a complex physiology of adaptation. Repeated exercise led to changes. Lack of exercising the organs led to degeneration. The idea of progress fit well his model of the inheritance of acquired characteristics. His origin of species was unusual: He believed that the genus was what the environment initially brought forth and then degraded into isolated species.[6]

Opposed to Lamarck was George (Baron) Cuvier who argued that a series of catastrophes, followed by new creations, was characteristic of the fossil record and that the Biblical account should be seen as several independent acts of creation over eras rather than as a simple week's work.

Figure 2. *Jean-Baptiste de Lamarck, a botanist turned zoologist, labeled biology as a field of science stripped of reference to God. He is better known for his theory of evolution through a hereditary mechanism called the transmission of acquired characteristics. Although this mechanism proved false, Lamarck's evolutionary ideas stimulated biological thinking throughout the nineteenth century. (Reprinted from Wheeler W.M. and Barbour T., eds. 1933. The Lamarck Manuscripts at Harvard. Harvard University Press, Cambridge, Massachusetts.)*

Chambers Proposes an Evolution of the Universe

In 1844, Robert Chambers (1802–1871) (Figure 3) anonymously offered a theory of everything that included the origin of the universe from nebular gases, the formation of the sun and the earth out of such gas clouds, the origin of life by spontaneous generation, and the evolution of that life through a combination of Lamarckism (acquired characteristics) supplemented with the belief that new species arise suddenly by new mutations (known then as sports and sometimes referred to as saltations or jumps).[7] Chambers and some of his relatives had polydactyly (extra toes and fingers). These extra dig-

its were removed by surgery, but the condition served as a constant reminder to him that the sudden appearance of traits might lead to pathologies in some instances and to new forms for evolution in other instances. His book, aptly titled *Vestiges of the Natural History of Creation*, was a best seller, but it was widely denounced as atheistic, demoting God, or casting doubt on the authenticity of the Bible. He was accused of offering the public a sterile universe void of purpose. Scientists found no evidence for his speculations and rejected them as unproven and therefore unsuitable as science.

Figure 3. *Robert Chambers was a publisher who brought articles and books to the attention of the educated public. He published his* Vestiges of the Natural History of Creation *anonymously, fearing that his business would suffer if his identify were known. The book was an attempt to explain the origin of the universe, our planet, life, and humanity using a deist's approach that believed nature's laws sufficient for explaining the material universe, including life. (Reprinted from Ward E.M. 1924.* Memories of Ninety Years. *Hutchinson, New York.)*

Darwin Proposes an Evidence-based Model of Evolution

Charles Darwin's theory of evolution by natural selection was quite different.[8] He did not arrive at his theory through speculation or philosophic principles. Indeed, when he left to sail on the *H.M.S. Beagle* as a young naturalist, he was considering a career as a minister in the Church of England (Figure 4). He was comfortable with his religious upbringing, and in his day most naturalists were ministers of the Church. They sought intellectual stimulation by studying what was widely believed to be a "bible of nature." One could learn about God by reading scripture or discover how He made the universe by studying science. By the time his voyage was over, Darwin had changed. He no longer believed in the fixity of species proposed in the later work of Linnaeus.[9] If species were naturally constrained by the space and food available to them, and if they naturally produced more progeny than could reasonably survive even in the best of circumstances, then the

Figure 4. *Charles Darwin, shown here in an 1854 portrait, was in his prime five years before he published his Origin of Species. He had avoided publishing his ideas on that topic for more than 15 years. (Courtesy of the American Museum of Natural History.)*

survivors and the losers had to differ. That difference resided in their capacity to adapt to the circumstances in which they lived. Most important, but difficult for him to solve, it implied some hereditary mechanism that brought about change in every generation.

The evidence for evolution could now be presented; 20 years of research and reflection made him certain he was right. It included 11 major findings.

1. Domestic breeds of pigeons, dogs, cats, goldfish, horses, and other animals and plants were consciously selected by breeders and unconsciously selected by farmers over centuries to establish practical breeds or strains for their human utility or pleasure.

2. Darwin used the introduction of alien species into new continents to show how rapidly a population expands and eventually crashes or levels off in response to the food, breeding territory, and other features of its host environments. His ideas on population growth and limits placed on that growth came from his reading of T.R. Malthus.[10]

3. He demonstrated a struggle for existence: Organisms vary and the disadvantaged leave no offspring or fewer offspring. Natural selection does not only act on superficial traits used in domestic selection.

4. Sexual selection leads to sexual dimorphism. It rarely leads to the death of competing males or competing females, but it can lead to a failure to reproduce based on the selection process.

5. The fossil record exists but it is spotty. Some fossils (such as the giant armadillos and sloths in parts of Latin America that he studied) show cognate forms of species that presently live in these areas.

6. There are isolating mechanisms in the evolution of new species. Among the most prominent are islands. The ancestral forms diversify in isolation, as did the birds, reptiles, and other life on the Galapagos Islands, all of which are related to forms having Ecuadorian ancestry.

7. Natural selection is independent of the idea of progress. Lower forms do not automatically become converted into higher forms. Parasites also evolved, but it is difficult to call their loss of organs innate progress.

8. Geographical distribution is consistent with an evolution of organisms over long periods of time. The plants and animals of the Galapagos Islands resemble life-forms found in Ecuador but not the life-forms on islands near Africa (e.g., the Azores), which have plants and animals resembling African forms of life rather than those of the Galapagos Islands. Yet both the Azores and the Galapagos Islands share a similar environment and habitat, since both are volcanic oceanic islands near the equator. The type of animals on oceanic islands is also important. One does not find salamanders and frogs (dependent on fresh water for their existence) on oceanic islands, most likely because they could not survive rafting on tree limbs and other flotsam lacking puddles of fresh water.

9. Comparative anatomy shows a kinship of life. The bones in a human hand, a bat's wing, the fin of a porpoise, or the foreleg of a horse show similarities in number and placement. The elephant with its short neck and the giraffe with its long neck have the same number of cervical vertebrae.

10. The existence of vestigial organs makes sense in an evolutionary model. The existence of non-functional vestigial organs suggests that they are remnants of once functional structures in their distant ancestors.

11. Geology, especially the uniformitarianism of Hutton and Lyell, suggests that there is an antiquity of the earth measured in millions of years. In uniformitarian geology, the processes we see today were operating in the past.[11]

What was missing from Darwin's theory was a source for the variations on which natural selection acted. Darwin recognized it as the Achilles' heel of his interpretation of evolution. He studied heredity by talking with breeders and reading all that he could on the subject. I suspect that he was hoping for a theory of heredity based on observation or experimentation that could help him solve some of the issues raised by natural selection but not answered by it. He could not account for atavisms, phenomena that occur when the crossing of two different breeds sometimes leads to a feral-looking wild ancestor. He had no evidence that heredity involves fluctuating minor variations that arise by chance. If they arose from environmental influence, as Lamarck had proposed, why would he need natural selection? What caused the sports (spontaneous mutations) that he and breeders described as largely pathological but are nevertheless occasionally inherited? Instead of thousand of generations of selection for short-legged sheep, one mutation in one generation produced the Ancon sheep. Darwin rejected sports as a major factor in evolution because few had a selective advantage in nature. Virtually all of the sports he had encountered from his discussions and readings were monstrosities that would not survive in nature. Where were the alleged beneficial sports?

It is not that the lack of a good genetic system prevented Darwin from publishing his theory. He felt that the public and, more important, his scientific peers would reject a theory based on speculation without supporting evidence. Darwin spent 20 years amassing that evidence and would have no doubt continued for five or more years had it not been for Alfred Russel Wallace (1823–1913), a Welsh biologist making a living as a self-supported naturalist selling specimens caught in tropical countries.[12] Wallace sent Darwin a manuscript that proposed an almost identical theory of understanding and points of argument that he himself had amassed. Darwin's friends worked out a joint presentation of their ideas in 1858 after Darwin promised that he would prepare an "abstract" of his life's work for immediate publication. He called that abstract *The Origin of Species by Natural Selection*. It was approximately 350 pages long.

Darwin's Theory of Heredity Invokes Circulating Units

Darwin felt secure that his theory would prevail because of the massive evidence that he offered to support his theory of natural selection (Figure 5), but also in his many books that followed.[13] What troubled Darwin most was his uncertainty about the heredity involved. He was not comfortable with Lamarck's theory of acquired characteristics, although he fell back on it because there was nothing else available except for the even more unsatisfactory theory of new variations arising by sports. When he did use Lamarck's theory, he tried to reduce the influence of the environment to have a less direct role than one in which specific new characters were created. Darwin hoped for a theory that would produce minor variations each generation. For a while, he believed that he had found such a model in his provisional theory of pangenesis.[14] Here, Darwin used the growing belief in the composition of living organisms from elementary units. The cell theory was well founded by the time (1868) that Darwin addressed the problem of heredity. He believed that small units, which he called gemmules, were produced by the cells and then migrated through the body, and some of them would be retained in the reproductive tissue of the gonads. Francis Galton attempted to demonstrate their existence by blood

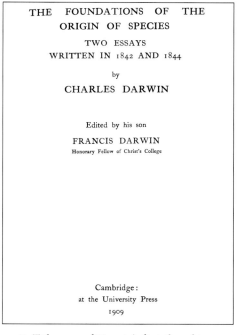

Figure 5. *Title page of Darwin's first thoughts on evolution. For the 50th anniversary of the publication of his* The Origin of Species by Natural Selection, *Darwin's son published his father's earlier sketches of that theory. This helped secure Darwin's priority over Wallace for the theory of natural selection. (Reprinted from Darwin C. 1909.* The Foundations of the Origin of Species. *Cambridge University Press, London.)*

transfusions from one strain of colored rabbit into another, but these experiments failed to show such an entry to the reproductive cells of the recipient. Pangenesis failed despite Darwin's argument that he never claimed that gemmules circulated in the blood.[15]

In the last third of the nineteenth century, Darwinism was under constant attack from evolutionary scientists. Darwin may have won the war against popular religion as an explanation for the origin of species, but he was losing the war against his fellow scientists, who were uncomfortable with an untestable mechanism of natural selection. Some retreated back to Lamarck's theory of evolution and rested their case with heredity based on the inheritance of acquired characteristics. Some believed that in some way hybridization was involved with new species formation. This was supported by Linnaeus'

original claim that hybridization led to new species and was precisely the means by which atavisms (regressions to ancestral forms) could be produced by experimental means. Some were sympathetic to Chambers' model of evolution by sport formation or at least discontinuities in evolution, rather than minuscule changes over enormous numbers of generations. Darwin's model placed evolution outside the realm of experimental science, and a new generation of scientists embracing experimental approaches was eager to find some inroad into experimental evolution, a field named by hope and not reality.

As the nineteenth century drew to a close, an air of uncertainty and anger existed as loyalists sought to justify Darwinian gradualism through mathematical inference from the statistical study of variable traits. Its critics continued to collect instances of discontinuities in evolution as the answer to a failed model.

End Notes and References

[1] In Darwin's day, Christian theology was flexible on how the book of *Genesis* should be interpreted. Inerrancy of the Bible was a later development in Protestant thought. What is called the fundamentalist movement had its origins in a set of beliefs endorsed by evangelical churches in Minnesota and New York in the early 1900s. See Numbers R. 1993. *The Creationists*. University of California Press, Berkeley, for an account of this history and its relation to the Scopes evolution trial of 1925 and the rise of Creation Science in the last half of the twentieth century.

[2] Darwin E. 1801. *Zoonomia*, 3rd edition. Johnson, London.

[3] Condorcet's views (and their relevance for the twenty-first century) are discussed in Wilson E.O. 1998. *Consilience: The Unity of Knowledge*. Knopf, New York.

Condorcet was a mathematician and active in the Royal Academy of Sciences. He completed his last work, on the idea of unlimited progress, while hiding from the extremists of the French Revolution. He died one day after his arrest, the causes unknown.

[4] Paley W. 1802. *Natural Theology*. Wilks and Taylor, London.

[5] Lamarck's life is described in L.J. Jordonova's 1984 biography, *Lamarck*. Oxford University Press, United Kingdom.

There is a long history of Lamarckism in the life sciences, some based on faulty experiments and some based on fraud. There is no convincing evidence that acquired characteristics are inherited. The chief difficulty of that doctrine would be in finding an environment that resequenced the amino acids of proteins and then drove the genetic code in reverse back to the DNA of the gametes.

[6] Unfortunately, Lamarck had an embittered outlook on life because of the deaths of his wives and the physical and mental illnesses of his children. He became blind in his old age and harshly engaged his critics in polemic battles. When he died, his death was not lamented and his eulogy was delayed in publication by some 30 years because his chief critic, Georges Cuvier, wrote it as a

scathing denunciation of a warped mind, intended to warn young scientists of the dangers of unbridled speculation.

[7] Chambers R. 1843–1846. *Vestiges of the Natural History of Creation*. Reissued 1994. Chicago University Press.

[8] There are several good biographies of Darwin including Browne J. 1995. *Darwin Voyaging*. Knopf, New York. This book covers his early years through the *Beagle* circumnavigation and its reception. Also highly informative is Brent P. 1981. *Charles Darwin: A Man of Enlarged Curiosity*. Harper and Row, New York.

[9] Darwin accepted an antiquity of the earth that was measured not in days, but in millions of years. He saw evidence of past changes that were dramatic and suggestive of an evolution of life. He did not want to repeat Chambers' mistake by proposing an evolution without a mechanism to account for it. The idea of natural selection came to Darwin a few years after the return of the *Beagle*. He had read Malthus' essay on population for relaxation and instead found a germ of an idea that he could apply.

[10] Malthus T.R., *An Essay on Population*, went through eight editions. The first edition, including contemporary works that lead up to it and respond to it, is in Appleman P., ed. 1975. The *Norton Critical Reader: Malthus*. Norton, New York.

[11] Lyell C. 1830–1833 *Principles of Geology*, John Murray, London, was published in three volumes.

For a biography of Lyell and his influence on Darwin, see Wilson L.G. 1972. *Charles Lyell: The Years to 1841: The Revolution in Geology*. Yale University Press, New Haven.

[12] A.R. Wallace's 1858 essay, "On the tendency of species to form varieties," was published by the Linnaean Society.

In 1908, Wallace was still alive and attended the Linnaean Society celebration for the joint papers that he and Darwin had written. He was very modest in his appraisal of the worth of his own contribution, considering it proportional to Darwin's time in preparing it and his own, "as 20 years is to one week." See *The Darwin-Wallace Celebration*, p. 7. Linnaean Society, London. Both Darwin's and Wallace's 1858 papers are reprinted in the celebration booklet.

[13] Darwin's original plan, I believe, was to have a multivolume set of books on his theory of natural selection appear at once, with every possible line of evidence given in massive detail. He knew that there would be opposition to so radical a theory, and he hoped that the sheer weight of evidence would convince his potential critics that this was not written out of speculative whim.

[14] Darwin C. 1868. *The Variation of Animals and Plants under Domestication*, two volumes. John Murray, London (1905 popular edition).

[15] Galton F. 1871. Experiments in pangenesis, by breeding from rabbits of a pure variety into those whose circulation blood taken from other varieties had previously been largely transfused. *Proceedings of the Royal Society (Biology)* **19**: 393–404.

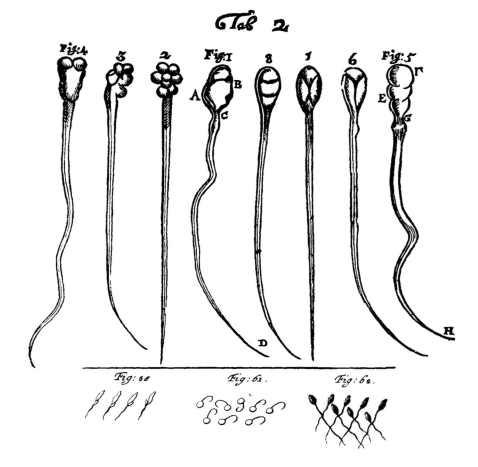

Some perceived Antoni van Leeuwenhoek's drawings of spermatozoa (drawings 1–4 are of humans; 5–8 are of dogs) as parasites or commensals in the seminal fluid and other saw them as agents of fertilization. Their existence as reproductive cells was not recognized in the seventeenth century, and this issue was not clarified until the 1870s. (Reprinted from Hughes A. 1959. A History of Cytology. Abelard-Schuman, London.)

Routes to Classical Genetics: Cytology

It is only in the very lowest orders of plants, in some algae and fungi for instance, which consist of a single cell, that we can speak of an individual in this sense. But every plant developed in any higher degree, is an aggregate of fully individualized, independent separate beings, even the cells themselves.

MATHIAS J. SCHLEIDEN*

WITH RARE EXCEPTIONS, most scientific ideas arise out of an earlier period of confusion, contradiction, and partial truths. This is well illustrated in the route that led to our present idea of the cell as the unit of life, arising from cell division by preexisting cells and carried out by mechanisms that we call mitosis in somatic cells and meiosis in reproductive cells. We think of the cell as a relatively large cytoplasmic mass composed of organelles in a matrix carrying out metabolism directed by genes in chromosomes. We primarily assign these chromosomes to the nucleus of the cell. All of these ideas, which are so familiar to those who teach biology and even to students after a good high school biology course, arose neither in logical sequence nor in fully integrated form on specific dates that we can celebrate. Every review article on the history of these ideas might bring out several dozen names from the seventeenth to the late nineteenth centuries with disputed priorities.

The Introduction of the Microscope Opens a New World

The use of optical lenses for enlarging objects seen at a distance (telescopes) and nearby (microscopes) began in the late 1500s. Both of those inventions were stimulated by the use of corrective lenses for poor eyesight, a practice that dated to thirteenth-century lens grinders. News of the telescope developed by Hans and Zacharias Janssens in the Netherlands came to Galileo's attention; he constructed his own telescope and was the first to apply it to astronomy in 1609 with spectacular results.[1] The Janssens also made the first compound microscopes, and these instruments began making their way through Europe in the 1620s.

---

*1838. Contributions to Phytogenesis, pp. 231–268. In Schwann T. 1839. *Microscopical Researches into the Accordance in the Structure and Growth of Animals and Plants.* Translated from the German by Henry Smith. Sydenham Society, London, 1847.

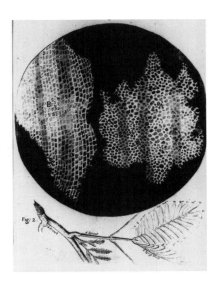

Figure 1. *The first portrayal of cells. (Left) Hooke's portrayal of cork as a myriad of boxes (or a honeycomb lattice), which he called cells, appeared in 1665. He had no theory of their function other than the buoyancy they provided for the cork he studied. Note that the cells could be interpreted as either individual units or a lattice-like structure. Until the 1830s, when Schleiden and Schwann emphasized communities of cells, both interpretations were in vogue. (Right) Hooke's compound microscope (1665). ([Left] Courtesy of the National Library of Medicine. [Right] Reprinted from Hughes A. 1959. A History of Cytology. Abelard-Schuman, London.)*

William Harvey used lenses in 1628 to study the beating heart of insects, corroborating his theory that the heart is a pump that circulates blood.

We must, however, properly celebrate Robert Hooke (1635–1702) for popularizing microscopy studies.[2] His work, *Micrographia* (1665), is a brilliant collection of engravings of the many objects he observed. One of the most important was his analysis of a piece of cork (Figure 1). He used a "sharp pen-knife" to cut a thin slice and found the section to be porous. He noted that "...it had very little solid substance" and compared the substance to a honeycomb. He also stated that "...these pores, or cells, were not very deep but consisted of a great many little boxes, separated out of

one continued long pore, by certain diaphragms... ." He concluded that he had found in these cells the explanation for cork's lightness, namely, "...a very small quantity of a solid body, extended into exceeding large dimensions."[3] Hooke thus gave us a name, the cell, and for the next century or more would persist a misleading belief that the cell wall and the cell were synonymous and provided the unique anatomic features of plants. Hooke's cell was a container for air or fluid. But Hooke was also enamored by the potential of microscopy for science. He claimed that "there is nothing so small as to escape our inquiry."[4] Most of Hooke's observations were of objects magnified about 30 times. He had not discovered the microbial world.

This microbial world was discovered by Antoni van Leeuwenhoek (Figure 2) (see also Chapter 4), Hooke's contemporary in Holland. Although van Leeuwenhoek used more efficient microscopes capable of magnifying 200–300 diameters, he referred to his specimens as "animalcules." He did not propose a cellular composition of life.

Figure 2. *Antoni van Leeuwenhoek did not provide a cell theory, and most of his observations were of what today would be called unicellular protozoa and algae. His work revealed how extensive the world was when revealed by microscopy. (Reprinted from Stubbe H. 1963. Kurze geschichte der genetik bis zur wiederentdeckung der vererbungsregeln Gregor Mendels. G. Fischer, Jena.)*

Histology Begins as an Anatomical Science

The study of tissues is called histology. Yet the founder of that field, Marie Francois Xavier Bichat (1771–1802) (Figure 3), did not use a microscope for his work (he considered it a toy and unworthy of science). He was a gifted anatomist and during the French Revolution carried out his work using very thin needles to separate organs into sheets of membranes and fibers. He concluded that organs were composed of a collection of materials that he called tissues. He also found relatively few tissue types in any one organ system or in the entire organism. He called these mucous, serous, and fibrous tissues. Bichat died of an infection at a young age, a common hazard for those practicing surgery and dissection in an era before a germ theory or effective treatments for infections existed.[5]

Jean-Baptiste Pierre Antoine de Lamarck, Chevalier de Monet (1744–1829), better known to us as Lamarck, for his contributions to evolution as a theory to account for the history of life on earth, and

Figure 3. *Marie Francois Xavier Bichat introduced the concept of tissues as the basic components of organs. (Reprinted from Bichat F.X. 1859. La Vie et la Mort. Librairie Victor Masson, Paris, France.)*

for coining the term biology (thereby showing the unity of plant and animal life), was stimulated by the discovery of tissues and combined that with Hooke's concept of cells. He argued that "...no body can possess life if its components are not a cellular tissue, or formed by cellular tissue. Thus every living body is essentially a mass of cellular tissue in which more or less of the complex fluids move more or less rapidly; so that if the body is very simple, that is without special organs, it appears homogeneous and presents nothing but cellular tissue containing fluids which move about within it slowly; but if its organization is complex, all its organs without exception, are enveloped in cellular tissue, and even were essentially formed of it."[6] It would take a charitable reading to see Lamarck's interpretation as a full-fledged cell theory because he saw the tissues (Bichat's contribution) as central to the formation of organ systems, and he recognized that these tissues are sheets, fabrics, or other structures with a cellular pattern. What was not clear to him or to his contemporaries in the coming decades was how those cells form and their relationship to tissues. Some saw the tissues as a latticework or a web and not as a brickwork of individual units.

Cell Theories Proposed in the 1830s Are Flawed

Between 1800 and 1838, many plant anatomists and some animal anatomists recognized the cellular composition of organs and organisms. There were competing names for cells in use—"utricles," "globules," "bladders," and "vesicles"—reflecting the shapes in the minds of the investigators. They also saw tubules, rings, and fibrous structures that they did not consider cellular. Some of these formulations were hard to distinguish from what later was credited to Mathias Schleiden and Theodor Schwann. C. Wolff, C.F. Brisseau-Mirbel, R.J.H. Dutrochet, and F.J.F. Meyen are among many of these contemporaries who became the losers in historical remembrance to Schleiden and Schwann.[7] Some who have delved into this history are almost ferocious in their anger regarding the fact that Schleiden and Schwann are recognized as the founders of cell theory and they assign not much more to them other than plagiarism or errors. Why is this so? I believe it has a lot to do with the personality of Schleiden and his influence on microscopic anatomy. Schleiden was a disturbing personality who had started in law, failed in that career, attempted suicide (he lodged a bullet in his brain), and then after recovery decided to become a physician. He wasn't successful at that either and switched to botany for which he gained fame as a gifted writer on plant biology (Figure 4). He appealed to the public with book titles such as *The Plant: A Biography*, and he forced his contemporaries to take notice of his ideas through his inflammatory writing style. Schwann was very different and saw the goodness in people. He admired Schleiden, whom he had met accidentally at a railroad station where he learned that they both had a passion for microscopic anatomy.[8]

The Schleiden–Schwann theory of cells was seriously flawed, but it included an important feature that was not stressed by their predecessors and contemporaries. Schleiden and Schwann believed that organisms were communities of cells. They also believed that the cell had a dual role: It had its own existence and it could enlarge and grow. In addition, it had a cooperative role in forming a tissue with specific functions. Schleiden and Schwann both knew that there had been two competing interpretations of cells. One group believed that cells arose as a result of some sort

Figure 4. Although Mathias Schleiden is best known for his contributions to cell theory, he popularized the study of botany and provided new insight into the composition of organisms. (Reprinted from Stubbe H. 1963. Kurze geschichte der genetik bis zur wiederentdeckung der vererbungsregeln Gregor Mendels. G. Fischer, Jena.)

of yet unknown process of cell division. Others believed that they arose by enlarging from smaller units within the cell. By then, two consistent structures were cited as sources of cell growth. One was the nucleus, first described by Robert Brown in 1831 (Figure 5).[9] The second was the nucleolus, sometimes seen within nuclei. We should pause when reflecting on this era and its issues: At this time, there was no stain technology nor effective instruments to section and lay out tissue in neat slices embedded in paraffin and no good-quality lenses free of chromatic and spherical aberrations. Most of those developments did not take place until the 1850s.

Schwann perceived cell growth from a physiological point of view (Figure 6). He had coined the term "metabolism," and he was interested in what would someday become biochemistry. He believed it pre-

sented a good analogy to crystallization. Well-structured crystals arose from a shapeless saturated fluid. If chemicals could do this and if life were chemical, why shouldn't cells be some sort of complex crystals coming out of saturated "life stuff?"[10] For Schleiden, there were "minute granules": Starch and other products could be seen in plant cells, especially in seeds. If cells were condensates, this would be consistent with a material structure of the universe, an idea that got a strong boost in 1810 with Dalton's atomic theory accounting for the property of molecules by the combining properties of atoms. Both Schleiden and Schwann rejected the much-admired view that many prominent biologists of that era favored—a holistic approach to life with ties to an earlier "bible of nature," in which the study of living things would be rewarded by seeing how the Creator used a common plan for life and put limits on what biologists could infer.

Schleiden's entry into this contentious field stimulated his critics to study plant and animal anatomy

Figure 5. Robert Brown noted the existence of a nucleus in cells, an observation that inspired Schleiden and Schwann to infer that these were crystalline condensates en route to becoming cells. (Reprinted from Locy W.A. 1925. The Growth of Biology. Henry Holt and Company, New York.)

Figure 6. *Theodor Schwann was an early advocate for what later became the field of biochemistry. He related life to crystal formation and rejected vitalist interpretations of life. With Schleiden, he is best known for the cell theory. (Courtesy of the National Library of Medicine.)*

in greater detail. The Schleiden–Schwann theory, which they called the "free formation of cells," claimed that cells arose from a chaotic or formless fluid or that cells arose within cells as a result of enlargement from nuclei or even the small granules themselves. As they grew, they eroded the mother cell's walls and expanded (Figure 7). Their cell theory was based on a foam-like model that was similar to the formation of soap bubbles. By the 1850s, few still believed this. One major critic was Robert Remak (1815–1867), who proposed in 1849 that cell division in embryos was discontinuous and associated with events in the cell nucleus. Remak's views were extended by Rudolph Virchow (1821–1902) (Figures 8 and 9) using his own studies of tumors. He saw each cancer as having a single-celled origin with the

Figure 7. *Schleiden's and Schwann's portrayal of cells. (Left) Theodor Schwann's portrayal of cartilaginous cells (upper left) showing free formation of new cells in the periphery. Other tissues are also represented by their cell types. (Right) To the left and middle is Matthias Schleiden's portrayal of plant parenchymal cells. Note that in the upper left, Hooke's empty cells are represented. Instead of air, sap or fluid was thought to flow through them. (Reprinted from Schwann T. 1847. Microscopical researches into the accordance in the structure and growth of animals and plants. Sydenham Society, London, United Kingdom.)*

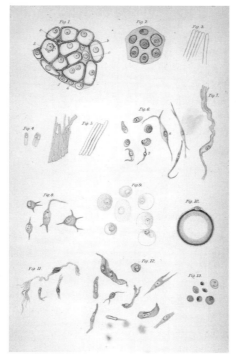

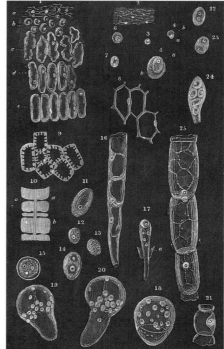

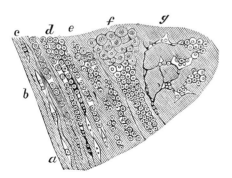

Figure 8. *Rudolph Virchow's illustration of breast tumor tissue. Virchow's belief that tumors arose from single malignant cells gave rise to his cell doctrine. Cells did not crystallize; they could trace their origins to a single cell. (Reprinted from Tyson J. 1878. The Cell Doctrine: Its History and Present State for the Use of Students in Medicine and Dentistry, 2nd edition. Lindsay and Blakiston, Philadelphia, Pennsylvania.)*

Figure 9. *As a founder of the field of cellular pathology, Rudolf Virchow transformed medical diagnosis. His cancer studies led to his formulation of the cell doctrine that later led to the discovery of its mechanism—mitosis. (Courtesy of the National Library of Medicine.)*

initial tumor cell proliferating from that. He extended that observation to embryos and those adult tissues in which cell divisions could be observed. What he saw convinced him that there was no crystallization from a liquid nor simple enlargement of cells from smaller units within the cell. Instead, there was a continuity of cells, with "each cell arising from a preexisting cell," the familiar phrase taught to our students.[11] To distinguish Virchow's contributions, biologists have called his interpretation the "cell doctrine," and they salvaged the concepts of a community of cells and the dual roles of cells from the Schleiden–Schwann theory.

The Contents of Cells Assume Importance

The cell also underwent a transformation from a lattice or collection of empty or filled boxes serving as containers, and the cell walls lost their initial structural interest for their functional contents, which became known as protoplasm in the 1860s, from Max Schultze's characterization, and later as a result of T.H. Huxley's popularization of protoplasm as the basis of life.[12] This concept endowed the cell with a potential heredity. Protoplasm was identified with life itself. The fleshing out of the cell required new technologies that included the elimination of chromatic and spherical aberrations of lenses. The lenses before the 1850s would break up white light into rings of color (chromatic aberration) and into multiple levels of focus (spherical aberration). These created fuzzy images, difficult to interpret as artifact or reality. That problem was solved by Joseph Jackson Lister, who used different salts dissolved in the quartz used for glassmaking, and these fused sandwiches of glasses of different densities compensated for the two aberrations.[13] A second innovation, beginning in the late 1850s, was Joseph von Gerlach's discovery that staining required heating the specimen. This came about accidentally because von Gerlach was not a tidy investigator. Instead of cleaning up before going home, he left some slides on a hot plate, and when he returned in the morning, he decided to look at them before he washed them. To his delight, the stains that he had tried so hard to use without success were now well penetrated into the tissues, and as a result, he could readily distinguish nucleus from cytoplasm as well as the boundaries of each cell.[14]

The microscopists in the 1850s to 1870s worked out what is known as cellular histology, the composition of tissues. Each tissue type required its own hard-won stain or preservative to result in the features being as best preserved and free of distortion as possible. Muscle cells, nerve cells, and assorted sheets of lining cells of the blood vessels, guts, and other organ systems could now be used for diagnosis in healthy and pathological specimens. This rapidly became a medical field. Most of the tissues in animals are nondividing, and thus the nucleus is intact and has a diffuse network of lint-like material seen in the stained state—this was called chromatin. Today we identify chromatin as the intact DNA-bearing threads of the unwound chromosomes associated with histone and other proteins.

The Analysis of Mitosis Shifts Attention to the Chromosomes

Histology became cytology when the interest shifted from the composition of tissues to the composition of cells. Until the 1870s, there were no effective techniques to reveal the contents of cells other than large objects such as nuclei. The emergence of stain technology changed that, especially as synthetic analine dyes became available. Also required was favorable material such as embryos or tissues likely to need constant replacement: skin, the lining of guts, or the marrow that forms blood cells. In 1879, Walther Flemming described and named the process of mitosis (Figure 10),[15] a remarkable reconstruction because Flemming would have had to look at a field of dividing cells in different stages of cell division and figure out a sequence from these. He was lucky to find a living system to guide him. He used salamander tail fins to see live cells divide and he used their gills for his stained specimens. What he saw in the live tissue were the dissolving of the nucleus and the shadowy formation of masses of threads, which

Figure 10. *Walther Flemming used salamander tissue both in vivo and stained to demonstrate that mitosis is a precise process of chromosome distribution with identifiable stages that follow in sequence. (Courtesy of the National Library of Medicine.)*

separated; what he saw in the stained cells was more orderly—a formation of doubled threads in the chromatin. He inferred how they separated as cell division progressed.

The term "chromosome," which would not appear to replace the term "threads" (or chromatin threads or loops) for another ten years, was named by Heinrich Waldeyer in 1888. These threads riveted Flemming's attention: "...one longitudinal half of each thread could move into one half of the nuclear figure, the other longitudinal half of the thread into the other half of the nuclear figure, in other words, each into a future daughter nucleus."[16] Flemming incorrectly guessed when the threads became double

(Figure 11). To him it looked like the early stage of mitosis (what is today called prophase) and that error would persist until the mid-twentieth century. There was no clear idea in his mind of a constancy of chromosome number, or that chromosomes varied in size, or that each individual chromosome made precise copies. But he did suspect that the longitudinal separation and prior copying had some underlying hereditary significance. In fact, Flemming was a very modest man in proposing a hereditary role for his theory of cell division by mitosis: "I have presented this hypothesis here because the longitudinal splitting of the thread seems to me too noteworthy not to try to provide some thought concerning it. For the

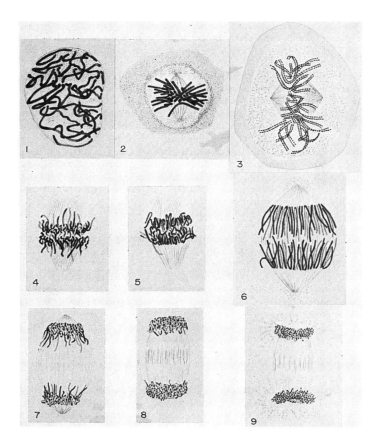

Figure 11. *Walther Flemming's portrayal of mitosis. The concept of a constant distinct number of chromosomes that retained their individuality was not yet observed when the stages of mitosis were worked out. Some believed the early prophase involved a chromatin mass consisting of one long thread (a spireme) that broke up into smaller chromosomes. The term "chromosome" had not yet been coined when Flemming worked out mitosis. (Reprinted from Flemming W. 1879. Contributions to the knowledge of the cell and its life appearance. Arkiv für Mikroskopische Anatomie* **16**: *302–406.)*

moment I throw this out purely as a possibility without in any way insisting upon it."[17] Ten years earlier, Ernest Haeckel, among others, proposed the idea that the nucleus must contain the substance of heredity because an egg is immensely larger and the sperm seemed to be almost entirely composed of its nucleus.[18] Since it was well known that males and females contributed equally to the appearance of their children, Haeckel argued that the nucleus must be the site of hereditary substances. Flemming was now able to penetrate the fuzzy lattice of chromatin and show that it consisted of threads whose doubling and distribution must somehow be tied to the problem of heredity.

The Analysis of Meiosis Strengthens the Hereditary Role of the Nucleus

What we call meiosis, the process by which eggs or sperm are formed, was initiated earlier than the study of mitosis. In 1876, Oscar Hertwig (Figures 12 and 13) studied the sea urchin *Toxopneustes lividus*. Sea urchins were easy to collect and maintain in saltwater tanks, and they could be milked of their eggs or sperm just by squeezing them. Hertwig noticed that when he placed sperm in the fluid surrounding the eggs, a sperm touched the egg cell surface and soon was replaced by a clear, growing "light spot" that

Figure 12. *Oscar Hertwig. (Reprinted from Stubbe H. 1963. Kurze geschichte der genetik bis zur wiederentdeckung der vererbungsregeln Gregor Mendels. G. Fischer, Jena.)*

became the size of the egg nucleus. He called this light spot the "sperm nucleus" and noticed that the two nuclei came together and fused. The cell with the fused nucleus then became the progenitor of cell divisions and an embryo that formed a new sea urchin. "We have now recognized that the most significant occurrence in fertilization is the fusion of the two cell nuclei."[19] Fertilization to us is obviously a union of one sperm and one egg, but that was not really known then. At best, it was a speculation until Hertwig's analysis. "So the important fact is deduced that the single nucleus which appears immediately before cleavage in the egg cell and around which the yolk platelets are arranged in rays, results from the copulation of two nuclei."[20] Hermann Fol confirmed Hertwig's theory the next year. But unlike Hertwig, Fol actually observed the penetration of the single sperm into the egg and the swelling up of the sperm head to form a nucleus. He called the egg nucleus the female pronucleus and the sperm nucleus the male pronucleus and argued that they were pronuclei because they fused to form the nucleus used by the fertilized egg to initiate cleavage and begin the embryonic journey. He also showed that if two or more sperm entered, the resulting fusion of nuclei "only produced deformed larvae."[21] This reinforced Flemming's tentative theory of a hereditary role for the chromatin threads, and it stimulated Theodor

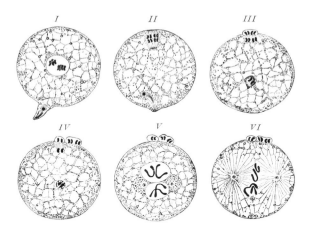

Figure 13. *In depicting meiosis, Hertwig described the formation of three polar bodies and an egg. Note the entry of the sperm in I, the sperm pronucleus in III, and the syngamy, or union of the male and female pronuclei, in V and VI. Hertwig used the nonsegmented worm Ascaris to work out meiosis and the relation of the polar bodies to the process. (Reprinted from Hertwig O. 1903. La Cellule. C. Naud, ed., Paris.)*

Figure 14. *Eduard Strasburger independently described mitosis, using plant cells and fertilization as a union of two nuclei. (Reprinted from The Darwin-Wallace celebration. 1908. Linnaean Society, London.)*

Boveri's much more elaborate experiments to demonstrate the significance of Fol's observations. Fol concluded from the pathology of polyspermy that only one sperm and one egg are both sufficient and necessary to establish a normal offspring.[22]

A confirmation of Fol's work came from botanist Eduard Strasburger in 1884 (Figure 14). He used orchids, and in *Orchis latiforia* he followed the formation of the pollen tube nuclei as they divided in the pollen tube.[23] He noted that the pollen nuclei were only a minuscule portion of the mass of the pollen tube, but it was the pollen tube nucleus that entered the egg and fertilized the egg nucleus. From this, Strasburger concluded that the cytoplasm of the male element (pollen tube cytoplasm) is minimal or nonexistent in carrying the hereditary traits from the father to the mother. Hertwig agreed. He demonstrated that in animal eggs, the difference between sperm volume and egg volume can be about 1 in 100

million. Yet the nuclear material, even in the most extreme of large eggs, is the same when the pronuclei are formed and come into contact. Hertwig inferred that the nucleus is the source of what Nägeli called the idioplasm and what we would today call the genes. "Inasmuch now as the transfer of the characteristics of the father to animals derived from eggs is necessarily linked with fertilization, the further evident consequence may be drawn, the nuclear substance is at the same time the bearer of the properties which are transferred from the parent to their progeny. Thus if it is carried further, the theory of fertilization, in the form presented by me, also includes a theory of inheritance."[24]

Although Fol, Strasburger, and Hertwig made their arguments on the fusion of nuclei, they did not describe the events occurring at a level of chromatid threads or what Waldeyer called chromosomes.[25] Edouard van Beneden in 1883 was the first to call attention to that possibility using the nematode worm, *Ascaris megalocephala*.[26] It had a small chromosome number (the diploid or adult tissue has four

Figure 15. Ascaris *meiosis and fertilization shown enlarged from Figure 13, bottom row. Ascaris has a chromosome number of four. This made the reductions of number from four to two readily visible in the four products of those divisions. (Reprinted from Hertwig O. 1903. La Cellule. C. Naud, ed., Paris.)*

chromosomes and the sperm or egg each have two). van Beneden noted that two threads from the male pronucleus join two threads from the female pronucleus to produce a set of four threads (Figure 15). Moreover, these four threads (he called them loops) split longitudinally and each of the first two cleavage cells get four chromosomes, two derived from the fertilizing maternal nucleus and two from the fertilizing paternal nucleus. In his first paper on this process of fertilization at the chromosome level, van Beneden made an error. He assumed that the maternal and paternal chromosomes stayed together and each nucleus thereafter consisted of separate paternal or maternal chromosomes. In the formation of eggs, he assumed that the paternal set of chromosomes were extruded into the polar bodies. This error was not resolved until the early twentieth century when the chromosome theory of heredity was formally proposed by E.B. Wilson's school. Despite this error, two years later, in 1885, Oscar Hertwig made an important inference that "the nucleus is the substance which not only fertilizes but also transmits the hereditary characteristics and as such corresponds with Nägeli's idioplasm."[27]

At the end of the nineteenth century and just before the rediscovery of Mendel's work on breeding analysis as a tool for understanding heredity, cytology was on the verge of bringing about a fusion of the study of chromosomes with the study of heredity. It was not yet a chromosomal theory of heredity, and it had little predictive value other than the logic of paternal and maternal contributions to heredity through their gametes. It did not see breeding analysis as a possible basis for that connection. What it did do, however, was to create in the minds of cytologists and (to a lesser extent) some breeders that there was a possible connection to be made.

End Notes and References

[1] Sobel D. 2000. *Galileo's Daughter*. Penguin Books, New York.

This book gives an excellent overview of Galileo's career and troubles with church authorities and his fellow astronomers. The author also describes how he came to construct his own telescope.

[2] Hooke R. 1665. *Micrographia*.

I had the rare good fortune, when working on the Muller biography in the Lilly Library vault in Bloomington, Indiana, to hold the first edition in my hands and leaf through the engravings. The book is quite large and the illustrations stunning. I used the section on the cellular composition of cork ("Of the schematisme or texture of Cork, and the cells and pores of some other such frothy bodies") in Carlson E.A., ed. 1967. *Modern Biology*, pp. 19–21. George Braziller, New York.

[3] Hooke R. "Schematisme..." In Carlson E.A., ed. 1967. op. cit. p. 20.

[4] Hooke R. Cited in Woodruff L.L. 1940. Microscopy before the nineteenth century. *Biological Symposia* **1**: 5–36. Quote is on p. 10.

[5] Bichat F.X. 1801–1802. *Anatomie Descriptive*. Gabon, Paris.

A brief account of Bichat's life and contributions to tissue theory may be found in Nordenskiold E. 1935. *History of Biology*. Tudor, New York.

[6] Lamarck J.B. 1809. *Philosophie Zoologique*. Dentu, Paris. Cited in Karling J.S. 1940. Schleiden's contribution to the cell theory. *Biological Symposia* **I**: 37–57; see pp. 41–42.

[7] 1940. *Biological Symposia*, volume I. Cattell, Lancaster, Pennsylvania.

Four review articles are devoted to the history of cell theory to celebrate the role cell theory has had in modern biology. In addition to the papers of Woodruff and Karling, the other two are by Mayer J. The cell theory—its past, present, and future (pp. 1–4); and Conklin E.G. Predecessors of Schleiden and Schwann (pp. 58–66).

[8] Nordenskiold E. 1935. op. cit. provides a brief account of Schleiden's life.

[9] Brown R. 1833. Observations on the organs and mode of fecundation in *Orchidae* and *Asclepiadae*. *Transactions of the Linnaean Society* **16**: 685–745.

Brown made his original fame as a traveler, like Darwin, serving as a naturalist on a late eighteenth-century expedition to Australia. His reputation for classifying the materials lead to jobs as curator and librarian for the major gardens in Great Britain. He is also known for his observation of Brownian movement (successfully interpreted by Einstein in 1905).

[10] 1847. The Schleiden and Schwann papers, *Microscopical Researches into the Accordance in the Structure and Growth of Animals and Plants* (T. Schwann) and *Contributions to Phytogenesis* (M.J. Schleiden), were translated into English for the Sydenham Society and edited by Henry Smith.

[11] Virchow R. 1855. Cellular Pathologie. *Virchow's Archiv Für Pathologische Anatomie und Physiologie* **8**: 1.

This was expanded into a book of the same title, *Cellular Pathologie*, in 1858. No recent English biography exists for this very significant scholar, whose contributions to medicine, anthropology, public health, and politics helped shape many of the applications of biology to social life throughout the world.

[12] Huxley T.H. 1869. On the physical basis of life. *The Fortnightly*.

Huxley popularized the term "protoplasm," which was used earlier by H. von Mohl (1846) and M. Schultze (1860). Huxley portrayed it as slime that covered the ocean floors and saw it as "disembodied life" organized into cells. Spiritualists quickly took to the idea of disembodied life, much to Huxley's chagrin. Huxley put his popular scientific essays into the 1870 book *Lay Sermons, Addresses, and Reviews* (American edition. 1882. Longmans, New York).

[13] For a brief history of achromatic lenses and the role of Joseph Jackson Lister, see Williams H.S. 1900. *The Story of Nineteenth-Century Science*, pp. 327–329. Harper and Brothers, New York.

Lister did his work during 1824–1830 using flint glass and crown glass as the compositions of his achromatic lenses. Lister was the father of the more eminent Joseph Lister, the founder of antiseptic surgery, who used poultices of carbolic acid in 1865 to treat compound fractures.

[14] von Gerlach's contributions may be read in Conn H.J. 1933. *A History of Staining*. Book Service of the Biological Stain Commission, Geneva, New York.

Anyone making slides goes through an elaborate process, as I learned after taking a course in stain technology as an undergraduate. Each of the following steps was the work of largely forgotten individuals known only to historians interested in the field of microscopy or histology: A specimen is placed in a preservative such as picric acid and then, after a suitable time (usually 30 minutes to several hours), in alcohol to remove the water. This might require transfer to several jars for a clean, water-free job. It is then placed in a solution called xylol, which is a fat solvent and removes the alcohol. The specimen is then placed in melted paraffin, which permeates the tissue. The paraffin is allowed to harden and a portion containing the specimen is trimmed, mounted on a microtome, and sliced into strips of paraffin with serial sections of the specimen. These strips are mounted on a slide (egg albumen is usually used to hold the paraffin on the glass slide). The slide is then moved through a series of jars (called Coplon jars) to extract the paraffin with xylol, the xylol with alcohol, and the alcohol with water, and then to place the slides in the proper dyes (favorites of histologists were hemotoxylin and eosin, the former staining the nucleus a deep purple and the latter staining the cytoplasm a pastel pink). The time involved to make such a slide thus lasted approximately 12 to 24 hours. As much as 20 different jars were used. After the slide was stained and dried, a bit of balsam was applied to it and a cover slip was placed over it to render it permanent. This can require a day or more to dry out.

[15] Flemming W. 1880. Beitrage zur kentniss der Zelle und ihre Lebenserscheinungen. *Archiv für Mickroscopische Anatomie* **16**: 302–406. In English, "Contributions to the knowledge of the cell and its life appearance." Voeller B. 1968. *The Chromosome Theory of Inheritance*, pp. 43–47. Appleton-Century-Crofts, New York.

[16] Flemming W. 1880. op. cit. p. 45.

[17] Flemming W. 1880. op. cit. p. 47.

[18] Haeckel E. 1866. *Generelle Morphologie*. G. Fischer, Jena.

[19] Hertwig O. 1876. Contribution to knowledge of the formation, fertilization, and division of the egg. *Morpholisches Jahrbuch* **1**: 347–434. In Voeller B. 1968. op. cit. p. 4.

[20] Hertwig O. In Voeller B. 1968. op. cit. p. 7.

[21] Fol H. 1877. On the beginning of ontogeny in various animals. *Archives de Zoologie Experimentale et Generale* **6**: 145–169. In Voeller B. 1968. op. cit. p. 10.

[22] Note that this was an important shift from the belief, seemingly confirmed by Kölreuter's experiments, that several pollen (and by implication, sperm) were needed to fertilize an egg.

[23] Strasburger E. 1884. *New investigations on the course of fertilization in the phanerogams as basis for a theory of inheritance* (Neue Unterschungen über den Befruchtungsvorgangbei den Phaneogamen als Grundlage für eine Theorie der Zeugung) G. Fischer, Jena. In Voeller B. 1968. op. cit. p. 22.

[24] Strasburger E. In Voeller B. 1968. op. cit. p. 23.

[25] Waldeyer H. 1888. Über Karyokinese und ihre Beziehungen zu den Befruchtungsvorgangen. *Arkiv für Mikroscopische Anatomie* **32**: 122.

[26] van Beneden E. 1883. Researches on the maturation of the egg and fertilization. *Archives de Biologie* **4**: 265–640. In Voeller B. 1968. op. cit. p. 55.

[27] Hertwig O. 1885. The problem of fertilization and isotropy of the egg, a theory of inheritance. *Jenaische Zeitschrift* **18**: 276–318. In Voeller B. 1968. op. cit. p. 33.

Nägeli used the term "idioplasm" to represent his inferred units of heredity. They were similar to Weismann's germ plasm with its assumed units and Hugo de Vries' pangenes, which were intracellular. Putting together units of inheritance, chromosomes, meiosis, and fertilization did not occur in the nineteenth century but the components were all there.

Routes to Classical Genetics: Embryology and Reproduction

In nature there is no generation but only propagation, the growth of parts. Thus original sin is explained, for all men were constrained in the organs of Adam and Eve. When their stock of eggs is finished, the human race will cease to be.

JAN SWAMMERDAM*

OF THE FOUR ROUTES TO THE FORMATION of classical genetics, the embryological route had the longest sustained experimental history and the most questions raised as the nineteenth century came to an end.[1] The major issues, directly or indirectly, were related to heredity, although much of the experimentation and findings seemed to be independent of such a connection. Some of the issues remained unresolved for more than 100 years and most of them helped shape the thinking and experimental design of the experimentalists in the nineteenth and early twentieth centuries. These issues might be called the fertilization problem, the preformation-epigenesis debates, the problems of parthenogenesis and organ regeneration, and the status of somatic influence on germinal tissue. Resolution of these issues was well under way as the nineteenth century came to an end and that resolution was made possible by the introduction of stain technology, effective microscopy, and the rise of Entwicklungsmechanik, the reductionist approach to experimentation applied chiefly to invertebrates, ironically, by vitalists and mechanists alike. That reference to vitalists and mechanists was also significant throughout the lengthy history of embryology. The findings were filtered through the viewpoint of atheists, dualists, and deists for one school of thinking and the viewpoint of those who saw in living phenomena the signature of God.

---

*ca. 1675. Cited in Needham J. 1959. *A History of Embryology*, p. 170. Abelard-Schuman, New York.

The Fertilization Problem Shifts from Noncellular to Cellular Interpretations

At the time that William Harvey (1578–1657) (Figure 1) was a physician to King Charles, Aristotelian views on biology were still dominant in the education of medical students. Aristotle believed that reproduction involved a material substance, menstrual blood, provided by the female in her womb, and some animating principle (it was not known then if this was a material substance) present in the semen.[2] This male principle allegedly introduced form to the chaotic coagulum and produced the embryo. In the 1650s, Harvey had access to quite a few deer which were killed in the weekly hunts that King Charles enjoyed for his health. Harvey opened the uteri of female deer and showed

Figure 1. *William Harvey is best known for his experiments demonstrating blood circulation with the heart as a pump. He was also a contributor to embryology and rejected Aristotelian theories of development. He argued instead that all life has its origins from eggs. (Reprinted from Stubbe H. 1963. Kurze geschichte der genetik bis zur wiederentdeckung der vererbungsregeln Gregor Mendels. G. Fischer, Jena.)*

that those females that were isolated from males and guaranteed to be virgin did not show such a coagulum and neither did females who were recently mated. He had the rare courage to challenge the revered authority of Aristotle, and he convinced King Charles, who often watched Harvey at work and enjoyed learning from him new insights into science. He did not convince most of his fellow physicians who, like Harvey, had no immediate alternative to account for the origin of the embryo. Harvey was impressed by the study of poultry, snakes, fish, frogs, and other animals that reproduced by forming eggs. He retained the possibility of an Aristotelian activation by the male, yet it was not menstrual blood but eggs in his model that were actually fertilized (Figure 2). There was no concept of cells in this fertilization process, but Harvey correctly inferred that these eggs must come from the ovaries (which were later called the female testicles) or from the uterine lining. Harvey then made a generalization, *ex ovo omnia* (all life arises from eggs), which was the new theory of reproduction that he offered in his *Exercitationes de Generatione Animalium* (*Experiments on the Generation of Animals*) in 1651.[3] He probably borrowed the phrase *ex ovo omnia* from Ovid, whose work in *Metamorphoses* depicted, in metaphorical terms, the changing nature of the material world with its numerous transformations, real and mythical. It should be noted that spermatozoa would not be described for another 35 years and their function not established for another 200 years.[4]

One of the implications in the Aristotelian model was that something formless (the coagulum) is given form by an external animating event, the presence of semen, then seen as a fluid or as an "aura seminalis,"

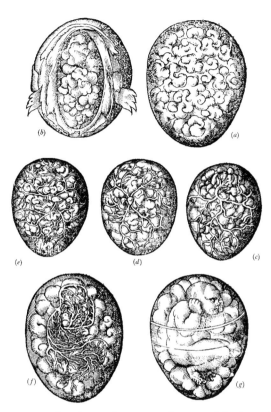

Figure 2. *The Aristotelian coagulum model. Aristotle believed that menstrual blood (a coagulum) was vitalized by the semen (not then known to contain sperm). This illustration is the imagined sequence of epigenetic development of the coagulum. Harvey shifted the process to the egg, although he was also unaware of the cellular process involved. (Reprinted from Needham J. 1959. A History of Embryology. Abelard-Schuman, New York [Courtesy of Needham Research Institute].)*

something akin to odors and emanations from matter (like magnetism). This suggested two ideas. The first was that development was inherently epigenetic. In epigenetic development, organs are not preformed; they gradually arise from earlier stages that do not hint at the future shape of the organs. The second

implication was that semen might not be the only animating factor for generating life.[5] Many physicians and scholars believed in spontaneous generation, in which unknown environmental influences caused decomposing matter to reanimate with an emergence of lower forms of life such as flies from rotting meat. A third consequence of this theory of fertilization, invoking an "aura seminalis" or other vitalizing principle on a formless matter or even an egg, made epigenesis a vitalist doctrine. Something living or spirit-like is impressed on the coagulum or egg in either Aristotle's or Harvey's interpretation.

About 21 years after Harvey's study of reproduction, René de Graaf observed what he thought were eggs, but which were actually follicles on the surface of mammalian ovaries.[6] That same year (1677), Antoni van Leeuwenhoek identified spermatozoa (animalcules) in the milt of cod and pike that were spread over the roe (eggs) of these fish. He also saw these spermatozoa in human (presumably his own) sperm.[7] But van Leeuwenhoek had also seen animalcules in hay infusions, well water, stagnant water, and even the saliva of his own mouth. Those who read about his numerous observations of a microscopic universe did not doubt the existence of such animalcules; they just did not know their significance. Some believed that these were harmless commensals, like the animalcules in hay infusions. Some believed that they, and not an animating principle, were agents of fertilization for the semen, which would place fertilization and epigenesis on a material basis instead of a vitalist basis. As the seventeenth century ended and the eighteenth century began, the arguments over fertilization spilled over into development itself. The debate was put to rest in the mid-nineteenth century when fertilization was demonstrated to be the union of a sperm and an egg. Giovanni Battista Amici made a similar observation in plants when he followed the path of pollen to the plant ovary (Figure 3).

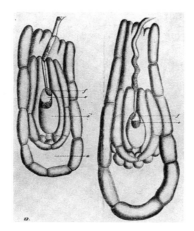

Figure 3. *Amici's illustration of the pollen tube entering the embryo sac to bring about fertilization. (Reprinted from Olby R. 1966.* Origins of Mendelism. *Constable, London. Originally from* Annales des Science Naturelles, Botanique. *Sér: 3,7 [1847].)*

The Preformation–Epigenesis Debates Represent a Struggle to Reject Vitalism

Jan Swammerdam, another member of a prominent school of Dutch scientists in the 1670s, was a gifted anatomist who could work on small specimens and reveal their anatomy with microinjections of pigments to astound those who looked at his preserved curios. One of these was the carefully folded silkworm moth he dissected from a cocoon.[8] If the cocoon was an egg (enlarged, of course) with a fully formed adult moth about to emerge, perhaps it enlarged just as the egg did. Swammerdam proposed a theory of preformation, or growth and development by enlargement. "In nature," he proposed, "there is no generation but only propagation, the growth of parts. Thus, original sin is explained, for all men were contained in the organs of Adam and Eve. When their stock of eggs is finished, the human race will cease to be."[9] This was one way out of the vitalist issue that dogged epigenesis. If the material universe works on God's scientific laws and not God's miraculous intervention in everyday affairs, then preformation would be a mechanist interpretation of life. At the same time, Swammerdam could not be accused of atheism. He provided what would later be called a deist interpretation, but in his own

day, he was supportive of the movement that sought God's handiwork through a bible of nature.

This preformation model got a boost in the eighteenth century when the Swiss naturalist Charles Bonnet (1720–1793) (Figure 4) studied aphids in 1741.[10] He found aphids within the abdomens of

Figure 4. *Charles Bonnet promoted the belief that mechanical enlargement was the basis of development. This view avoided the reliance on a supernatural process that was then assigned to the older model of epigenesis. (Courtesy of the National Library of Medicine.)*

female aphids, and they gave birth to these aphids that enlarged after birth. Here was preformation one could see with one's own eyes. To Bonnet, it was not a theory but a reality. This touched off a lengthy and hot debate among ovists (as Bonnet and his supporters were called) and epigenesists. Joining in the fray were those who ridiculed the idea of preformation by sending in hoax parodies of the ovist position. One, using the pseudonym Delanpatius, drew an imagined homunculus in a sperm and suggested preformation was not in the egg but in the sperm. Some actually believed such homunculi existed, either because they did not get Delanpatius' joke and took it as factual, or because they imagined that such forms could be seen in the sperm head. Niklaas Hartsoecker's elegant little man sitting on his haunches inside a sperm head was not seen under the microscope at all. Hartsoecker drew it as an example of what would be seen if he were to search for such homunculi (Figure 5).[11]

There was an immediate problem with the spermist point of view. It was known from van Leeuwenhoek's work that the number of spermatozoa in an ejaculate is measured in the hundreds of millions. If fertilization involved a little man or woman in the sperm, this would imply the following. Sex determination would be made through the sperm as it entered its union with an egg. Heredity would be easy to attribute to the paternal homunculus and difficult to interpret through the female egg, which lacked a corresponding being. Most disturbing to those who saw the living world as a bible of nature was that the vast amount of

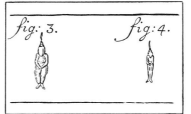
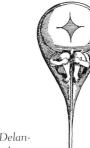

Figure 5. (Left) *The homunculus. Delanpatius (a pseudonym) provided the hoax drawing of homunculi in sperm as a way to tweak preformationists. (Right) Similarly, Hartsoecker drew an imagined homunculus as it should appear if a good microscope were used. Hartsoecker's homunculus was also a taunting parody. Neither of these fake illustrations worked; instead, they led to a school of spermists that assumed that such homunculi were proven and to textbook writers who actually believed that these persons looking under the microscope drew what they imagined. (Reprinted from Needham J. 1959. A History of Embryology. Abelard-Schuman, New York.)*

sperm produced was doomed to damnation as unbaptized souls, assuming that each preformed being was a future person as Swammerdam claimed. The debate revolved around the theological issue of what constitutes living material, most holding out for a vitalist interpretation of a God-given vitality that could not be reduced to some mechanical and material model of life. It also involved the scientific issue of what constituted embryonic development—enlargement of preformed entities or some sort of elaborate epigenetic construction whose mechanism was largely unknown.

Regeneration Presupposes a Hereditary Role for Cells in Wounded Tissue

Arthropods are well known for regenerating lost limbs. This is especially so for crustaceans (lobsters and crabs), who have adapted to this danger by generating a new limb whenever a limb is lost.[12] It is difficult to justify preformation through regeneration, although it is easy to accept epigenesis as the basis for the formation of the new limb, which begins to differentiate from a disorganized mass of cells, called the blastema, at the site of the stump of the old limb. At the same time, regeneration raises a problem. What is the nature of the hereditary material that permits a stump to form a new limb in a new non-embryonic context? It was regeneration of limbs that made Darwin sympathetic to the theory of pangenesis.[13] By invoking gemmules flowing everywhere, he could see them assembled at the wound site and then directing the cells to form a new limb. As his critics pointed out, Darwin failed to explain why a similar blastema and new limb do not occur in mammals and many other organisms. Although regeneration, when first described and later interpreted, seemed to be remote from the question of heredity, by the end of the nineteenth century, heredity was very much a key issue.

Parthenogenesis Creates Confusion on the Role of Reproductive Cells

In 1856, C.T. von Siebold, and two years later, Rudolph Leuckart, confirmed that parthenogenesis was characteristic of several insects.[14] This included bees and aphids. Parthenogenesis was troubling because nothing consistent seemed to emerge from its study. Its role was that of a spoiler of theories. In aphids, it made the semen unnecessary to produce offspring. Thus, the semen offered neither a preformed being (spermist preformation) nor a vitalizing role (epigenetic activation) for the formation of offspring. In the aphids, the parthenogenetic offspring were females. This made it seem logical that whatever was essentially female was reflected by the unfertilized egg. That was not the case for bees, because the females were products of fertilization and males (drones) were derived from unfertilized eggs.

Weismann, however, saw in parthenogenesis a clue to the nature of polar bodies and the meiotic process (Figure 6).[15] In most species that reproduced sexually, two rounds of division produced polar bodies. Polar bodies were believed to be asymmetric cell divisions in egg formation that Weismann interpreted as repositories for the elimination of chromosomes. Many theories of polar bodies existed. One was the false idea (proposed by Edouard van Beneden, among others) that male heredity was discarded from the egg and that the sperm then brought

Figure 6. *A prolific writer, August Weismann stimulated thinking in the life sciences in the last half of the nineteenth century. He disproved the theory of acquired characteristics and promoted a theory of the germ plasm in which reproductive cells are set aside and protected from the environment. Both of his major insights contributed to twentieth-century theories of evolution and heredity. (Courtesy of the National Library of Medicine.)*

in a new male heredity.[16] Weismann argued against this and favored the belief that the polar body permitted a reduction in chromosome number. Fertilization would restore the chromosome number because a comparable meiotic process occurs in males, albeit without polar body formation (Figure 7). Weismann believed that the asymmetry was not a problem because retention of the cytoplasmic mass by the egg was incidental to the more important role

of the meiotic division in reducing chromosome numbers by half. What puzzled Weismann was why there were two meiotic divisions instead of one. He came up with the idea that there is a doubling of chromosomes as meiosis begins and that the two divisions are needed to end up with one half of the predoubled state of the egg as it is about to enter meiosis. For Weismann, this pushed reduction division to the second meiotic division, which might

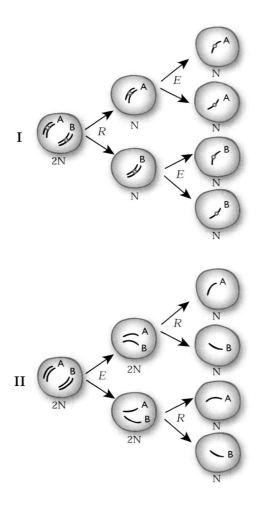

Figure 7. *When does reduction division occur? August Weismann proposed the concept of reduction division in 1890. He argued that there had to be a mechanism that reduced the hereditary content during the production of reproductive cells so that the hereditary instructions for each species would be preserved. Fol and van Beneden soon applied this to the chromosomes. In the classical presentation of the twentieth century (top) we recognize a chromosome entering meiosis as having two chromatids per chromosome, held together by a centromere. The two homologs, A and B, come together during metaphase alignment, and reduction is associated with the first meiotic division. One cell has chromosome A and the other has chromosome B. Equation division follows to separate the two chromatids of each chromosome. Thus, the four final germ cells are in a ratio of 2A to 2B. In the model proposed by Weismann, Fol, and van Beneden (bottom), the existence of centromeres is not yet recognized, and the two threads are duplicated and separated from each of A and B (it was not yet recognized that they pair until the work of Montgomery and McClung in the late 1890s). Thus, the first meiotic division is equational with each chromosome receiving an A and B thread. Reduction occurs in the second meiotic division, resulting in an identical 2A to 2B ratio of gametes, but with a very different distribution. Confusion regarding this mechanism of meiosis was abundant at the turn of the century and it did not become clear and consistent until 1905. Clarification required three concepts: The recognition that chromosomes could have one or two chromatids depending on their stage in the cell cycle, pairing of homologs occurred, and chromosomes are genetically distinctive (a correlation with uniquely different-sized chromosomes came with McClung's use of grasshoppers). (Figure drawn by Claudia Carlson.)*

seem puzzling to contemporary students and scientists who have been taught that the reduction in chromosome number occurs in the first meiotic division.[17] Neither Weismann nor anyone else knew in the 1880s and 1890s, when he formulated his theories, that homologous chromosomes existed and that these chromosomes had two chromatids per chromosome attached by a centromere when they entered meiosis and paired. That observation did not emerge until 1898–1902 when McClung, Montgomery, and Sutton began to piece together the story of meiosis in relation to heredity.[18]

Weismann showed that in several species (ostracods, daphnia, and some insects), the parthenogenetic egg only produces one polar body. This was consistent with his belief that the reduction of chromosome number by half was the essential feature of meiosis. If the egg is doubled in number before entering meiosis in the parthenogenetic mother, it will end up producing one fully reduced egg and a discarded single polar body. Weismann also proposed a broad theory of the role of sexuality in the life of the individual. He argued that meiosis resulted in a mixing of paternal and maternal heredities in an individual (he called this mixing of parental genetic components "amphimixis"). The amphimixis, he argued, was essential for Darwinian evolution acting on numerous variations in the population, the diversity stemming from amphimixis in the production of sperm and eggs.[19]

Weismann's Theory of the Germ Plasm Has a Pervasive Influence

Weismann's greatest contribution was the theory of the germ plasm. He studied alleged cases of Lamarckian inheritance and found them to be anecdotal or deficient.[20] He did his own experiments, cutting off tails of mice for six or seven generations and he found no shrinkage in tail size among the mice after breeding each generation with amputated tails. Mutilations, at least, did not show an inheritance of acquired characteristics, although a gullible public still believed that dueling scars and similar mutilations produced slash-like birthmarks in children. Weismann scoffed and countered that some three millennia of circumcision among Jews had not shrunk a Jewish baby boy's foreskin by even a millimeter! He also showed that the same was true for foot-binding among the Chinese, as well as the variety of lip and ear piercings for plug insertion over numerous past generations that had no hereditary outcome. These rituals must be performed anew each generation. With this as his argument, Weismann proposed that the reproductive tissue (which he called germ plasm) is set aside early in development, and the bulk of the body is composed of different tissue that he called the soma. Changes in the soma are not transmitted to the germ plasm.[21] Thus, genetic variations are not a consequence of exposure to good or bad environments; genetically transmissible variations arise from unknown causes independent of the environment. Somatic variations occur in response to the environment, but they have no effect on the germ cells. This was a pessimistic doctrine to Lamarckian biologists like Luther Burbank, who believed that plants (and people) could have their heredity trained by the environment. But it was also an important insight that had major influence throughout the twentieth century and still does into the twenty first.

End Notes and References

[1] An excellent source is Needham J. 1959. *A History of Embryology,* 2nd edition. Abelard-Schuman, New York.

This was originally part of a 1931 treatise on biochemical embryology written by Needham.

[2] Aristotle. 1937. *De Generatione Animalium.* Loeb Classical Library, Heinemann, London.

[3] Harvey W. 1651. *Exercitationes de Generatione Animalium.* Elzevir, Amsterdam. Cited in Needham Joseph. 1959. op. cit. p. 146.

Harvey was already famous for his work on the circulation of blood by the heart in 1628.

[4] Spermatozoa were first described by Antoni van Leeuwenhoek. An excellent account of his life and contributions may be found in Dobell C. 1932. *Antoni van Leeuwenhoek and his "Little Animals."* Baler and Danielson, London.

van Leeuwenhoek estimated that fish spermatozoa exceeded 13 billion, his estimated population of people on the earth.

[5] A considerable body of folk beliefs exists about the role of semen (or other agents) in fertilization. One good source is Thomson S. 1955–1958. *Index of Folk Motifs.* Indiana University, Bloomington, Indiana.

In some African societies (for example, the Yoruba in Nigeria), twins were believed to the product of a double fertilization, one by the human partner and the other by a deity, resulting in one twin being a demigod.

[6] de Graaf R. 1677. *Opera Omni.* Leiden. See Needham's account of de Graaf's observation of ovarian follicles in 1672. op. cit. p. 163.

[7] van Leeuwenhoek A. In Hughes A. 1959. *A History of Cytology*, p. 150. Abeland-Schuman, New York.

[8] Swammerdam J. 1672. *Biblia Naturae.* Leiden. See Needham's account. 1959. op. cit. p. 170.

[9] Swammerdam J. In Needham J. 1959. op. cit. p. 170.

[10] Bonnet C. 1745. *Traité de Insectologie.* See Needham J. 1959. op. cit. pp. 207–211.

[11] The myth of the microscopic sperm homunculus is described by Cole F.J. 1930. *Early Histories of Sexual Generation.* Oxford University Press, United Kingdom.

[12] A good description of the problems and phenomena of regeneration may be read in Morgan T.H. 1901. *Regeneration.* Macmillan, New York.

[13] Darwin C. 1868. *The Variation of Animals and Plants under Domestication*, volume 2 (1905 popular edition, John Murray, London).

On p. 443, Darwin states, "In the case of those animals which may be bisected or chopped into pieces, and of which every fragment will reproduce the whole, the power of re-growth must be diffused throughout the whole body."

[14] Leuckart R. 1853. Zeugung in R. Wagner's *Handworterbuch der Physiologie.* Cited in Weismann A. 1891. *Essays Upon Heredity and Kindred Biological Problems* (ed. E.B. Poulton, S. Schonland, and A.E. Shipley), volume I, p. 231. Oxford University Press, United Kingdom.

Leuckart was the first to stimulate an unfertilized frog egg and demonstrate that it entered early cleavage stages. He was August Weismann's mentor when Weismann was a student.

[15] Weismann A. 1887. On the number of polar bodies and their significance in heredity. In Poulton E.B. et al., eds. 1891. op. cit. pp. 343–396.

[16] van Beneden E. 1883. Researches on the maturation of the egg and fertilization. *Archives de Biologie* **4:** 265–640. In Voeller B. 1968. *The Chromosome Theory of Inheritance*, pp. 54–66. Appelton-Century-Crofts, New York.

[17] Weismann A. 1887. op. cit. pp. 343–396.

[18] For details on the coming together of the chromosome theory of heredity, see Chapter 10.

[19] Weismann A. 1891. Amphimixis or the essential meaning of conjugation and sexual reproduction. In Poulton E.B. et al., eds. *Essays Upon Heredity and Kindred Biological Problems*, volume II, pp. 99–222. Oxford University Press, United Kingdom.

[20] Weismann A. 1888. The supposed transmission of mutilations In Poulton E.B. et al., eds. 1891, volume I, op. cit. pp. 431–461.

[21] Weismann A. 1885. The continuity of the germ-plasm as the foundation of a theory of heredity. In Poulton E.B. et al., eds. op. cit. pp. 163–256.

Routes to Classical Genetics: Breeding and Hybrid Formation

It is impossible to doubt that there are new species created by hybrid generation.

CARL LINNAEUS*

Who can say how far human perseverance may not be able to penetrate into the mechanism of the brain, and to reveal a connected structure and a common principle in its countless elements? But surely this work will be most materially assisted by the simultaneous investigation of the structure and function of the nervous system in the lower forms of life—in the polypes and jellyfish, worms, and crustacea. In the same way we should not abandon the hopes of arriving at a satisfactory knowledge of the processes of heredity if we consider the simplest processes of the lower animals as well as the more complex processes met within the higher forms.

AUGUST WEISMANN†

PLANTS AND ANIMALS HAVE BEEN DOMESTICATED for at least 10 and probably 20 millennia.[1] Much of this interest was for food, and approximately 15,000 years ago, it led to the first agrarian communities. Cats and dogs as pets and ornamental flowers for decoration around the homestead are also ancient. The oldest known such use may be the wild flowers accompanying the burials of Neanderthals in the Middle East. We know of virtually no scientific studies of breeding these plants and animals until relatively recent times. Those who carried out the breeding used a vague hereditary principle—like breeds like. Most breeders also knew to cull their breeding stock and toss out (or eat) their imperfect specimens, saving their very best for the next year's harvest or for the next generation of a breed. Fortunately, the "like-breeds-like" guideline often works, although the process may be slow to bring about improvements.

---

*1751. *Plantae Hybridae*. Cited in Roberts H.F. 1921. *Plant Hybridization Before Mendel*, pp. 23–24. Hafner, New York.
†1883. On Heredity in *Essays Upon Heredity and Kindred Biological Problems* (ed. E.B. Poulton, S. Schonland, and A.E. Shipley), pp. 69–106; see p. 72. Clarendon Press, Oxford. 1891, authorized translation.

The Start of Scientific Breeding
Identifies Pollination as a Sexual Act

Scientific studies of breeding and the nature of hybrids took place in the seventeenth century. It was known since antiquity that the male and the female each play a part in producing a new generation. Exactly what was produced by each sex was largely unknown because there was no cell theory until the 1830s and no concept of fertilization by individual gametes until the 1870s. The first awareness that plants had a sex life was proposed by Rudolph Camerarius (1665–1721) in 1694–1696 (Figure 1).[2] At the time, he was a professor in Tübingen. He proposed that pollination was a sexual act and that some plants have separate sexes (these are called dioecious). Hemp (*Cannabis* or marijuana) is one such plant that he studied. He showed that if he cut off the anthers of male plants, they could not fertilize

Figure 1. *The discovery of sex in plants by Rudolph Camerarius was regarded as scandalous in his time. Linnaeus confirmed Camerarius' theory and greatly extended it with the experimental production of hybrids. (Reprinted from Stubbe H. 1963. Kurze geschichte der genetik bis zur wiederentdeckung der vererbungsregeln Gregor Mendels. G. Fischer, Jena.)*

female plants. Camerarius correctly identified the pollen of the anthers as the male element and the seed-producing components in female plants (stigma, style, and ovary) as the female apparatus.

Some 50 years later, Carl Linnaeus (1707–1778) endorsed the findings of Camerarius and many others who had confirmed this sexual theory of flowering plants, but he greatly extended it (Figure 2). Linnaeus recognized that the pollen enters the stigma and descends down the style, but he did not see or report a pollen tube; that was first identified by Giovanni Battista Amici (1786–1863) in 1823.[3] Linnaeus used emasculation of flowers of *Mirabilis longiflora* (the genus popularly known as "four o'clocks") to confirm sterility. But if he pollinated an emasculated *M. longiflora* with pollen from another *M. longiflora* plant, the seeds were normal in size, number, and capacity to germinate. When he used a different species, *Mirabilis jalapa*, the *M. jalapa* pollen that was introduced to emasculated *M. longiflora* flowers produced ovules that aborted with a few deformed nongerminating seeds or no seed at all. In 1751, Linnaeus published *Plantae Hybridae*, a treatise on the use of hybridization to produce new species.[4] Some species, like mules, were sterile; others produced fertile hybrids and still other interspecific crosses yielded partially fertile offspring. From those that he observed, Linnaeus made a hasty conclusion perhaps based on an initial or careless impression of his hybrids. He thought that the outer traits of the hybrid (color and leaf characteristics) were largely derived from the male (pollen) source and the inner attributes or tissues (venation and tubular arrays) were derived from the female (ovule) source. Most of his work was done on the goat's beard, *Tragopogon*, using *T. pratense* and *T. porrifolius*

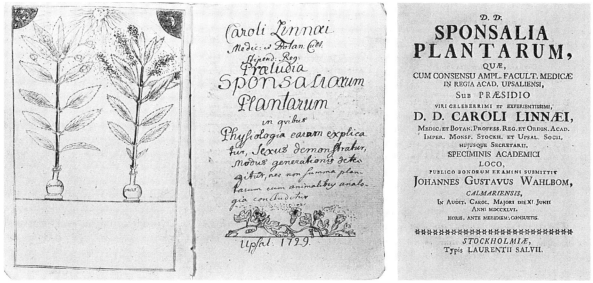

Figure 2. *Carl Linnaeus classified plants by sex organs. In 1729, Linnaeus published his classification of plants using the sex organs (pistils and stamens) because they were the basis of generation of new plants (he considered the petals the "bridal bed" of the sex act). Shown here is Linnaeus' own sketch for his title page of his first attempt at classification. (Reprinted, with permission, from Blunt W. 1971.* The Compleat Naturalist: A life of Linnaeus. *Viking Press, New York. Curtis Brown Group Ltd., London on behalf of the Estate of Wilfrid Blunt. © Wilfrid Blunt 1971.)*

reciprocal crosses. Linnaeus argued that "it is impossible to doubt that there are new species created by hybrid generation."[5] Although we credit Linnaeus in his later taxonomic work with a doctrine of fixed species, he was more open to the possibility of speciation through hybridization, an idea that echoed throughout the nineteenth century and into the early twentieth century among plant and animal breeders. It was also an idea that appealed to Darwin.

The most extensive plant breeding for analysis of hybridization was actually done by a contemporary of Linnaeus, Joseph Gottlieb Kölreuter (1733–1806) (Figure 3).[6] Kölreuter carried out 136 experiments in his garden and published the results in 1761–1766 when he was a professor at Karlsruhe. Most of these hybridization experiments were with the tobacco

Figure 3. *Joseph Kölreuter, whose work on hybrids Mendel carefully read. (Reprinted from Olby R. 1966.* Origins of Mendelism. *Constable, London.)*

species, *Nicotiana paniculata* and *Nicotiana rustica.* Kölreuter used a microscope to study the process of pollination. He erred in believing that the fluids of the pollen and stigma were the active sexual agents. He also noted that if he used fewer than ten pollen grains, the flowers did not produce seeds, and thus he added a second error to his analysis: He believed that many pollen were involved in fertilization. This actually went along with his fluid model because the amount of fluid on a stigma was far greater than the fluid he believed was exuded from the pollen (his observed exudations may actually have been the early stages of pollen tube formation, which he misinterpreted as an expressed droplet of fluid). What Kölreuter did not know was that not all pollen grains are fertile and not all pollen tubes manage to reach the ovary. He also recognized that pollination was not limited to the distribution of pollen by the wind. He was the first to recognize the role of insects in bringing pollen to flowers.

Kölreuter believed that hybridization was an unnatural state and that it had to be brought about artificially. In his cross of *N. paniculata* and *N. rustica,* the F_1 hybrid was intermediate for the traits of the two species except for the anthers that were shriveled, the pollen that was "rubbed to pieces," and many of the anthers that were "empty husks." He had found a basis for the infertility of hybrids and

claimed that "this plant is thus in the real sense a true, and as far as it is known to me, the first botanical mule which has been produced by art."[7] He noted that he had expected 50,000 seeds from his crosses and got none. Yet the plant anatomy of the hybrid was otherwise healthy. It was the fertility that was pathological, and this he attributed to the upsetting of a natural law to keep the species separate. In his words, the sterility reflects the abhorrence of hybrids between two species, "which indeed were not destined for each other by the wise Creator."[8] The belief in a harmony of nature was and still is a recurrent theme in the history of the life sciences.

Kölreuter later rejected the idea that the sexual component of the pollen or female apparatus in a flower was its fluid. He argued instead that the fluid was just a conducting medium. He applied the fluid scraped off the stigma of *N. rustica* and showed that it had no effect when mixed with *N. paniculata* pollen and placed on the shaved stigma of *N. paniculata.* No *N. rustica* traits appeared in the F_1 generation. Reciprocal crosses also showed no influence of the *N. paniculata* stigma's fluid. Only the pollen contributed to hybridization when placed on the female apparatus. He rejected the erroneous theory of Linnaeus that internal tissue is maternal and exterior tissue is paternal. None of his hybrids with tobacco showed such a hybrid outcome.

Forerunners of Mendel Do Not Connect
Their Findings into a Usable Theory

Among Mendel's forerunners were many breeders who used peas. Thomas Andrew Knight (1759–1838) was a highly regarded plant breeder, chiefly interested in edible fruits.[9] He found peas more convenient for

experimentation. When, in 1823, he crossed gray coats on peas with white coats on peas, the offspring were gray (with some variability in their intensity compared to the parental gray). In the next generation, he obtained some seeds from the hybrids that had gray coat color and some that had white. He did not keep a count of the amount of each class and thus, although he found some sort of dominance in the F_1 generation and some sort of segregation in the F_2 generation, he did not have the same insights into what his findings meant, as did Mendel. Similarly, Charles Naudin (1815–1899) carried out experiments in France in 1864 shortly before Mendel published his own results. He confirmed Knight's findings but did not keep counts of his offspring in the F_2 generation, referring instead to the results as a "disordered variation."[10] Darwin cultivated "at the same time forty-one English and French varieties. They differed greatly in height—namely from between 6 and 12 inches to 8 feet—in manner of growth, and in period of maturity." Darwin did not breed color varieties or contrasting traits the way Mendel did, and much of his cultivation was to see how much stability of types he got from growing them together. He noted that while most varities self-fertilized, occasional bees pushed their way in and necessarily had to pollinate the next flowers they visited. Darwin chose the three traits he listed because they were of importance to farmers in the past as well as in the present, but he felt that all the traits were highly modifiable: "We may therefore infer that most plants might be made, through long continued selection, to yield races as different from one another in any character as they now are in those parts for which they are valued and cultivated."[11] In fact, the weakness in the experiments (including Darwin's) on hybridization in peas that set the work apart from Mendel's is one of an underlying assumption. These experiments usually looked at traits of an *entire species* represented in a single specimen used in a cross, whereas Mendel was looking at a *single trait* in a single specimen while ignoring all other characteristics. Their interest was in a totality of effect of hybridization whether for evolutionary purposes (as motivated Darwin), for the species problem (as motivated Linnaeus and those inspired by him), or for the "harmony of nature" problem (as motivated Kölreuter). Although all observed or carried out crosses, did three-generation studies, and had reasonably abundant seeds to harvest in their gardens, the influence of these larger perceptions of biology shaped the way in which they designed their experiments, selected their data, and interpreted their findings. Mendel's approach was motivated by his atomistic model of heredity, using fundamental units related to contrasting characters whose fates (and numerical ratios) could be relentlessly pursued.

Mendel Contributes to an Understanding of Heredity

Gregor Johann Mendel (1822–1884) is properly acknowledged as the founder of the field of genetics (Figure 4).[12] Certainly, breeding analysis as we practice it today stems from his experiments published in 1865. Mendel was unusual in that he did not have a middle-class or privileged upbringing. He grew up in what a century later would be called the Sudentenland of western Czechoslovakia. His parents were peasants; however, Mendel was lucky because an uncle helped found an elementary school that he attended as a child. He was recognized as talented and a quick learner, and he was encouraged to go to the equivalent of a high school. That was not free; his sister sacrificed part of her dowry so that he could attend. Once again, Mendel's talent struck his teachers as unusual, but no additional money from

Versuche über Pflanzen-Hybriden.

Von

Gregor Mendel.

(Vorgelegt in den Sitzungen vom 8. Februar und 8. März 1865.)

Einleitende Bemerkungen.

Künstliche Befruchtungen, welche an Zierpflanzen desshalb vorgenommen wurden, um neue Farben-Varianten zu erzielen, waren die Veranlassung zu den Versuchen, die her besprochen werden sollen. Die auffallende Regelmässigkeit, mit welcher dieselben Hybridformen immer wiederkehrten, so oft die Befruchtung zwischen gleichen Arten geschah, gab die Anregung zu weiteren Experimenten, deren Aufgabe es war, die Entwicklung der Hybriden in ihren Nachkommen zu verfolgen.

Dieser Aufgabe haben sorgfältige Beobachter, wie Kölreuter, Gärtner, Herbert, Lecocq, Wichura u. a. einen Theil ihres Lebens mit unermüdlicher Ausdauer geopfert. Namentlich hat Gärtner in seinem Werke „die Bastarderzeugung im Pflanzenreiche" sehr schätzbare Beobachtungen niedergelegt, und in neuester Zeit wurden von Wichura gründliche Untersuchungen über die Bastarde der Weiden veröffentlicht. Wenn es noch nicht gelungen ist, ein allgemein giltiges Gesetz für die Bildung und Entwicklung der Hybriden aufzustellen, so kann das Niemanden Wunder nehmen, der den Umfang der Aufgabe kennt und die Schwierigkeiten zu würdigen weiss, mit denen Versuche dieser Art zu kämpfen haben. Eine endgiltige Entscheidung kann erst dann erfolgen, bis Detail-Versuche aus den verschiedensten Pflanzen-Familien vorliegen. Wer die Ar-

Figure 4. (Above) Portrait of Mendel. (Right) First page of Mendel's paper on peas, 1865, as published in the Brünn Scientific Society Proceedings. (Facsimile from J. Cramer, Weinheim.) Mendel's experiments on breeding analysis of peas were not appreciated for 35 years. After 1900, his work was confirmed in thousands of experiments involving plants and animals. (Reprinted from Mendel G. 1866. Versuche über pflanzen-hybriden. Verhandlung des Naturforschenden Vereines in Brünn. 4: 3–47 [Courtesy of Austrian Press and Information Service].)

his family was available and no scholarships existed in those days. Instead, Mendel made a sacrifice. He agreed to be considered for the priesthood and was accepted into a teaching order of Augustinian priests in Brünn (now called Brno). They sent him to the University of Vienna for two years. He failed to complete a degree but earned enough credits to teach physics and natural sciences in his monastery. He was never certified because he had some unusual neurosis that led him to have mental breakdowns

during examinations. Despite these formal failures, he was active as a college student and presented some minor papers to a student science society. He enjoyed courses in mathematics and physics and took courses in biology from one of Schleiden's students.

After his return to the monastery, Mendel began an eight-year program of breeding. He liked horticulture and was much taken by the variety of flowering plants that his father kept on their farm. He chose peas as an organism to use for breeding studies because its anatomy made it difficult to encounter wind-borne pollination. Mendel's motivation was a curiosity about hybrids and their composite features. He read the work of his predecessors (Linnaeus, Kölreuter, Gärtner, Knight, and Goss) and devised a more effective strategy for experimentation. If I were to credit him for a particular discovery or contribution to science, I would single out his conception of breeding analysis. Mendel's scientific outlook was that of an ultimate reductionist. Where his predecessors looked for holistic or cosmic significance to the meaning of hybrids, Mendel decided to strip hybrids of their mystical appeal and tease apart the components of the two parents in the hybrid by focusing on one trait at a time. He also used an approach common to some of the most successful scientists, for example, Thomas Hunt Morgan (who used fruit flies) or Seymour Benzer (who used bacteriophages); he swept all the difficult, complex, and messy problems under a mental rug and started his research with the simplest components available to him.[13]

Mendel wrote to seed suppliers and obtained 34 varieties of pea plants. He selected 22 of these for their constancy of individual traits and then trimmed this down to seven traits that he studied thoroughly. What they had in common was unambiguous contrast. These traits included the shape of peas (round or wrinkled), the color of the peas (yellow or green), the color of the coat around the pea (gray or white), the shape of the pods (inflated or constricted), the color of the pod (green or yellow), the location of flowers (axial or terminal), and the length of the stem (about six feet [2 m] or six inches [15 cm]). His scientific insights shine through his first paper, as he begins with an assessment of prior studies: "Among all the numerous experiments made, not one has been carried out to such an extent and in such a way as to make it possible to determine the number of different forms which the offspring of hybrids appear, or to arrange these forms with certainty according to their separate generations, or definitely to ascertain their statistical relations."[14] Note the combination of scientific skills he will employ: He will focus on individual traits. He will follow each trait across several generations. He will count every offspring in each generation. He will apply a statistical standard to the numerical results. It looks obvious to us now, of course, but not so obvious to his peers or to those who preceded him.

Mendel did reciprocal crosses on all seven of these traits. He noted from this that the pollen contributed as equally as the ovules to the heredity of the hybrid. He noted a simple dominance in each one of these contrasting pairs of traits. The dominant form in the hybrid could not be distinguished in appearance from the parental pure breeding dominant type. The extracted recessive in the F_2 generation bred true and was indistinguishable in appearance from the parental recessive. He also noted a consistent ratio of 3 dominant to 1 recessive. He saw that these F_2 pea traits (color and shape) were distributed randomly in the pods. Each pod held about six to nine peas, and in the F_2 for the dominant trait, it was not uncommon to have a pod filled with only the dominant. But for the recessive trait, this was very rare. He found none that were all recessive in a pod with six to nine peas. What Mendel realized was that statistically this would require a one-fourth to the sixth power, or 1 in 4,096 pods, for his best chance of finding such a pod. The ratios he obtained in the F_2 plants or the distribution of dominant to recessive traits were close to 3.0 to 1.0, varying from 2.81 to 3.15 to 1. Mendel also demonstrated that F_2 dominant peas were of two different genetic constitutions. Two-thirds were hybrid like the F_1 and one-third were dominants that bred true and did not produce recessive progeny (Figure 5).

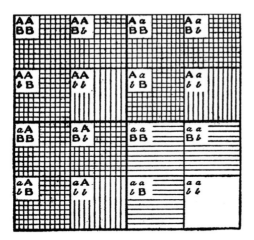

Figure 5. *Ronald Punnett devised the Punnett square for his book* Mendelism, *one of the first texts in genetics. It continues to be an effective way to teach basic hybrid and dihybrid outcomes. (Reprinted from Punnett R.C. 1909.* Mendelism. *Wilshire, New York.)*

Mendel also applied his analysis to two contrasting traits at a time and came up with a second F_2 ratio symbolically represented as 9**AB**:3**Ab**:3**aB**:1**ab**, where 9/16 are **AB** peas showing both dominant traits, 3/16 each show one or the other of the dominant traits (i.e., **Ab** or **aB**), and 1/16 are recessive **ab** peas. Mendel extended these dihybrid results to future possibilities. He had found in his preliminary crosses, not only in peas but in other flowering plants, that some traits are complex for their color inheritance; but he argued that there was no reason that they could not be reduced to their independent components by his breeding analysis approach. He concluded his paper with a cognate plant, the bean, *Phaseolus vulgaris*, and showed that in hybrids involving three traits (pod color, pod shape, and height of stem) a similar outcome was evident, with his principles confirmed.[15]

Doubts about Mendel's Integrity Are Exaggerated

Almost 80 years later, a somewhat dyspeptic but brilliant statistician, Ronald Aylmer Fisher (1890–1962), analyzed Mendel's results and found some problems with the data.[16] Fisher was responsible for modern statistical theory, including sampling distribution, randomization likelihood, and analysis of variance tests. He also published extensively on evolution and genetics. At issue in his critique of Mendel's work were missing classes of plants that should have been produced if Mendel's stated methods were strictly followed. These are called null sets, and they should arise in small samples (subsets of ten or less of a dominant phenotype) from self-fertilized heterozygous peas, all showing the dominant trait. Mendel excluded them but did not give a reason for doing so, leading Fisher to surmise that the data were improved in some way. Also, the 3:1 ratio, although it did vary among Mendel's numerous reported cases, collectively hovered too close to a 3:1 ratio to have actually occurred by chance. Fisher did not want to accuse Mendel of improving (or topping off) the data, so he blamed a mythical gardener helping him. Since that time, Mendel's reputation in thousands of high school and college classes has been smeared as that of a fraud. We do not have access to Mendel's original notes because Mendel's last years as head of his monastery were embroiled in a feud with the Emperor Franz Josef who wanted to tax the monasteries. Mendel refused on the grounds of an earlier separation of church and state agreement that he felt still applied. The fight got so ugly that his own Bishop censored him, preferring to pay symbolic payment rather than compromise the functioning of the church. Mendel was even dispossessed, removed bodily from his office, and his office was padlocked. When Mendel died, all of his papers were burned to prevent embarrassment to the monastery.

It is difficult for me to believe the accusation that Mendel faked his data. Foremost of the good reasons for believing this is not so was his effort to get Carl Nägeli to repeat his work (Figure 6). He sent packets of peas with predicted ratios for Nägeli to plant.[17] Few cheats would take the risk of being unmasked by sending evidence of their work with predicted outcomes. The published papers also reveal many of the problems that he faced. He had to protect the plants from accidental fertilization, and he tried to reduce that by emasculating some of the plants after successfully hand-pollinating them. He also tried to cover many of his plants with small sacs to keep insects from entering. That was effective for flying insects but not for *Bruchus pisi*, the pea weevil that he wrote about at the University of Vienna that could have crawled through the tied bags and spread pollen about. Mendel may indeed have encountered the statistically improbable "null sets" among his crosses and excluded them as contaminants. Anyone who reads Mendel's paper will be impressed by the immense amount of work that went into it. It is almost impossible to fake the results in such amazing detail when there was no effective model he could use to generate a fabricated set of hundreds of experiments over eight years.

It is sad that Mendel was dogged by misfortune most of his life. His correspondence with Nägeli, on the recommendation of his teacher from Vienna, turned out to be a disaster. Nägeli wrote back and sent packets of seeds of the hawkweed, *Hieracium*. Mendel dutifully bred these and confirmed that Nägeli's hawkweeds did not show 3:1 ratios. He collected local varieties and species of *Hieracium* and

Figure 6. *Carl Nägeli rejected Mendel's findings because they were not universal. His own hawkweeds did contradict Mendel's model of heredity, but neither he nor Mendel knew that the hawkweeds used an unusual mode of parthenogenetic reproduction and were thus an exception. (Reprinted from Stubbe H. 1963. Kurze geschichte der genetik bis zur wiederentdeckung der vererbungsregeln Gregor Mendels. G. Fischer, Jena.)*

found the same thing that Nägeli had told him would be found. The results suggested a maternal inheritance in some reciprocal crosses and irregular ratios in other crosses. Different species gave different results. This was very discouraging for Mendel, a person not noted for ego strength, as his response to his university testing would suggest. Mendel dropped plant breeding after publishing a second article, which was not quite a renunciation of his earlier results but had much the same effect—Mendelism was not a universal finding and thus not very interesting to pursue.

What went wrong? Neither Nägeli nor Mendel knew that in *Hieracium* many species showed a type of parthenogenetic development (apomixis or a cloning of the maternal tissue) that required the stimulation of the embryo sac by the pollen tube but not the entry of the fertilizing pollen nuclei. The tubes touched the membrane of the embryo sac but no penetration occurred. The maternal tissue began developmental activity from the stimulus of the pollen tube's contact with the cell wall of the embryo sac. In about 3% of the pollinations, an actual penetration and fertilization took place, thus resulting in the funny ratios. But the *Hieracium* story would not be worked out until the early 1900s, some 20 or more years after Mendel died.[18]

It was also sad that for 35 years, Mendel's results were available in print but remained largely unread and dismissed by those who read them as vaguely Pythagorean or mystical in assigning to a messy diversity of life some numerological fixed ratios. I do not believe at all that had Darwin read Mendel's paper he would have seen a connection to his theory of natural selection as pertinent to heredity or evolution any more than Pearson, Weldon, and other leaders of the biometric movement in Great Britain saw them in 1900, when Mendel's paper was resurrected. Sports and discontinuities then did not seem to be the stuff of heredity or evolution. The one happy outcome of Mendel's work is its belated recognition. There is no Nägeliism, Pearsonism, or Weldonism in our genetic vocabulary, but there is a lasting Mendelism, besmirched or not.

End Notes and References

[1] Roberts H.F. 1929. *Plant Hybridization Before Mendel* (facsimile reprinted 1965, Hafner, New York). Chapter 1, pp. 1–6.

Roberts discusses Egyptian and Mesopotamian practices. In antiquity the most common belief may have been like-for-like inheritance, but numerous forms of visual impressions and other forms of acquired characteristics were common. Roberts was a botanist at the University of Manitoba in Canada, and his book is the first history of genetics, particularly valuable because he quotes extensively from the original sources to give representative thinking to scientists long forgotten.

[2] Camerarius R. 1696. *De Sexu Plantarum Epistola*. Cited in Roberts H.F. 1929. op. cit. p. 29.

[3] Amici G.B. 1823. Cited in Wilson E.B. 1896. *The Cell in Development and Inheritance*, p. 162. Macmillan, New York.

[4] Linnaeus C. 1751. *Plantae Hybridae*. See Roberts H.F. 1929. op. cit. pp. 15–29.

[5] Linnaeus C. 1751. In Roberts H.F. 1929. op. cit. p. 23.

[6] Kölreuter J.G. 1893. *Vorlaufige Nachrict von einigen das Geschlect der Pflanzen betreffenden Versuchen und Beobachtungen, nehbst Fortsetzungen 1, 2, und 3 (1761–1766)*, W. Pfeffer. In Ostwald's *Klassiker der exakten Wissenschaften No. 41*, Leipzig. Cited (p. 61) and discussed (p. 34) in Roberts H.F. 1929. op. cit. See also Olby R. 1966. *The Origins of Mendelism*. Constable, London. Olby greatly adds to Roberts' account.

[7] Kölreuter J. 1761. In Roberts H.F. 1929. op. cit. p. 43.

[8] Kölreuter J. 1761. In Roberts H.F. 1929. op cit. p. 45.

It is of interest that the word hybrid is derived from the Greek word *hubris*—the act of imbalance that sets off tragic consequences.

[9] Knight T.A. 1823. *Transactions of the Horticultural Society of London* **5**: 377–380. See Roberts H.F. 1929. op. cit. pp. 85–93.

Some remarks are made on the supposed influence of the pollen in cross-breeding, the color of the seed coats of plants, and the qualities of their fruits. Knight published over 100 papers on his plant experiments.

[10] Naudin C. 1865. De l'hybidité consideré comme cause de variabilité dans les vegetaux. *Annales des Science*

Naturelles, Botanique 5me Serie **3**: 153–163. In Roberts H.F. 1929. op. cit. p. 132.

[11] Darwin C. 1868. *The Variation of Animals and Plants under Domestication*, two volumes. John Murray, London.

I used the popular edition of 1905, same publisher. Darwin's statement on his 41 varieties of peas is in volume 1, p. 404. His comment on the effects of selection is in volume 2, p. 254.

[12] Mendel's life is well described in Iltis H. 1924. *Life of Mendel.* Translated into English 1932. Norton, New York; and Orel V. 1996. *Gregor Mendel, The First Geneticist.* Oxford University Press, United Kingdom.

Orel greatly supplements Iltis' work, relying on the resources and contributions to the Mendel Museum at the monastery in Brno at which he was the director.

[13] At one meeting I attended in the mid-1950s, Benzer was being questioned by a critic of his experimental methods. He countered that some people try to start at a simple level and build up to greater complexity, in the process showing how things come to pass. Others, he claimed, try to start with the most complex problems and then throw clouds of confusion on the problem without solving it. He said he called the first type of scientist "a clarifier" and the second type of scientist "a turbidifier." He concluded his remarks with the retort, "You, sir, are a turbidifier."

[14] Mendel G. 1866. Experiments in plant hybridization. *Verhandlung des Naturforschenden Vereines in Brünn* **4**: 3–47. Translated into English by C.T. Dreury for the Royal Horticultural Society, 1901. In Bateson W. 1909. *Mendel's Principles of Heredity*, pp. 317–361. Cambridge University Press, United Kingdom. Quote is on p. 318.

[15] Mendel G. 1866. In Bateson W. 1909. op. cit. pp. 347–348.

[16] Fisher R.A. 1936. Has Mendel's work been rediscovered? *Annals of Science* **I**: 115–137.

Fisher actually hedged his accusations by allowing that Mendel had not published all of his data or that he polished his presentation as a good teacher often does to get a point across without blowing students away with an avalanche of detail and asides. According to Orel, Fisher, while still an undergraduate in 1911, found Mendel's data statistically inconsistent. But in his 1936 paper, he did not claim fraud. In 1955, when he commented on the importance of Mendel's paper, he introduced the idea of a "gardener's assistant" as deceiving Mendel. At issue were samples of ten that Mendel had used from each plant. He should have classified a sample of ten dominant peas from a plant as all homozygous dominant about 5% of the time, because that was the probability of having none of them heterozygous. His results suggest that he never encountered any exception to having at least one of the ten heterozygous, and that is where Fisher felt that some sort of fraud or unreported error had to exist. My own feeling is that Mendel threw out these sets of ten as contaminants. A cognate difficulty he cites in his letter to Nägeli (undated, 1866) is that "I tried to combine *H. pilosella* with *pratense, praealtum*, and *Auriciula*; and *H. murorum* with *umbellatum* and *pratense*, and I did obtain viable seeds; however, I fear that in spite of all precautions, self-fertilization did occur." Page 2 in *The Birth of Genetics*, supplement to *Genetics*, 1950, by Brooklyn Botanic Garden, New York, printed in Menasha, Wisconsin. Mendel also stated to Nägeli on December 31, 1866 that he obtained the seeds he was sending for confirmation under difficult circumstances, and that "Furthermore the seeds were at a time when the *Bruchus pisi* was already rampant, and I cannot acquit the beetle of possibly transferring pollen." See *The Birth of Genetics*, op. cit. p. 6.

[17] The Nägeli-Mendel letters were translated into English in *The Birth of Genetics*. 1950. op. cit.

[18] Ostenfeld C.H. 1904. Zur Kenntnis der Apogamie in der Gattung *Hieracium. Berichte der Deutschen Botanischen Gesellschaft* **22**: 537–541.

Apomixis or apogamy is the maternal production of viable eggs without fertilization. It leads to maternal inheritance.

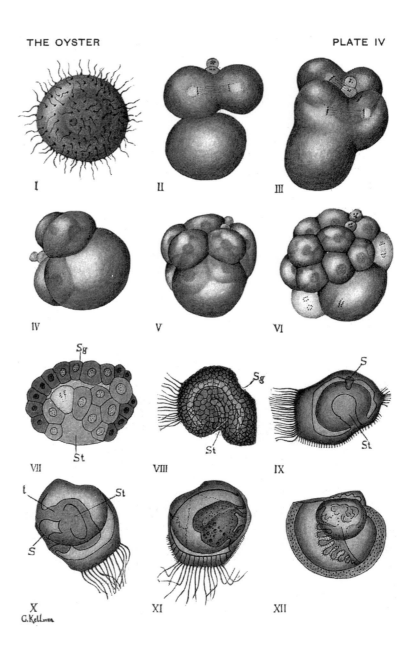

Oysters as described by W.K. Brooks. Brooks was a mentor to T.H. Morgan, E.B. Wilson, and W. Bateson, three prominent contributors to classical genetics. He moved away from preserved specimens and insisted his students study living organisms beginning from the time they are fertilized eggs. (Reprinted from Brooks W.K. 1905. The Oyster. Johns Hopkins University Press, Baltimore, Maryland.)

The Rise of the American University

It was bad enough to invite Huxley. It were better to have asked God to be present. It would have been absurd to ask them both.

ANONYMOUS*

Baltimore afforded an opportunity to develop a private endowment free from ecclesiastical or political control, where from the beginning, the old and the new, the humanities and the sciences, theory and practice, could be generously promoted.

DANIEL COIT GILMAN[†]

It was as if I had entered a new world with new outlooks on nature, a new respect for exact science, new determination to contribute to the best of my ability to "the increase and diffusion of knowledge among men."

E.G. CONKLIN[‡]

THE SOURCES OF CLASSICAL GENETICS were all developed in Europe—cell biology, evolution, experimental embryology, cytology, and reproductive biology. They emerged throughout the nineteenth century sometimes as independent fields (especially the nonmicroscopic studies of hybrids and heredity) and sometimes as cognate fields (especially embryology, with its dual interests in cell biology and reproduction). What held them together was a hope for their relationship to evolution, the dominant theoretical concept of the life sciences in the nineteenth century, first through Lamarck and then through Darwin.

North Americans read about the new biology, but until 1875, there was little movement to compete with scientists abroad or to displace the prevailing traditions of collecting, preserving, and classifying specimens and displaying them in museum collections,

*1876. Citation of a preacher's reaction to Huxley's inaugural address of September 12, 1876 at Johns Hopkins University. In Hawkins H. 1960. *Pioneer: A History of the Johns Hopkins University 1874–1889.* Cornell University Press, Ithaca, New York.

†1906. *The Launching of a University,* p. 4. Dodd, Mead, and Company, New York.

‡1958. Cited as Conklin's impressions of Johns Hopkins University. In Harvey E.N. *Biographical Memoir of the National Academy of Sciences* **31:** 54–91; see p. 61.

often affiliated with colleges or universities. The most prominent American biologist was Harvard's Louis Agassiz, a Swiss immigrant, who popularized nature studies and who looked on each new species that he described as a thought of God. Agassiz's biology meshed well with a pious tradition of American colleges and universities. The secularization of European science, begun in the 1830s, had made little headway in North America, where spawning new religions was more common than spawning new sciences.

Historical Background to Higher Education in the Nineteenth Century

The status of higher education in the nineteenth century from 1800 to 1875 might be briefly summarized for comparison as follows:

- In the United States, there were no universities as we think of them today. The Ph.D. degree did not exist and research was not of major interest to the colleges. Religious grounding and good citizenship were the major goals and responsibility of the faculty. Latin, Greek, logic, and mathematics were part of the required curriculum for all students (all of them male, with few exceptions). Applied fields (art, music, law, medicine, business, engineering, and teaching) were taught in trade schools and not in the colleges.

- In Great Britain, the dominant model was the Oxford–Cambridge tradition of the liberal arts; there were no Ph.D. degrees. The B.A. (later an M.A. was added) sufficed for a select few who held the status of fellows, allowing them to continue a subsidized research career (usually associated with their remaining unmarried). The major role of the faculty was to teach. Science was rarely studied at the university and most of the sciences were assigned to trade schools.

- In Europe, German universities were recognized as centers for scholarship, and professors had research as their major duty. The Ph.D. was the research degree (but in the life sciences, research was frequently carried out in the medical schools and students did a dissertation for the M.D. degree regardless of the organism chosen for their dissertation project). Other continental European nations largely followed the German model, in which a single professor headed a department. All others in the department were his assistants or lacked professorial status. As in the United States and Great Britain, with few exceptions, Continental colleges and universities were populated by men.

Medical education was quite varied. The United States had no medical schools equivalent to those in Great Britain and continental Europe. Much of medicine in the United States involved quackery, apprenticeship, or limited programs (some involved as little as one year of study for the M.D.). Until the Flexner reports of the 1890s, no standardized curriculum existed. The American Medical Association was not founded until 1876. For talented Americans desiring a medical education in those days, a medical degree from Edinburgh, Paris, Vienna, or other Continental school was considered essential. It should be kept in mind that as citizens of colonies of Great Britain, talented North Americans were expected to go to Great Britain for their more sophisticated education. Those lacking that talent were likely to go to inferior American medical schools, many of them arising and disappearing within a few years. By contrast, many of the great medical schools in Europe had been around since the thirteenth century.

Medical research was encouraged in the major European medical schools. Edinburgh pioneered studies in surgery and public health. Paris was famous for its physiological approaches to both disease and normal life functions (Claude Bernard's influence was particularly admired). German medical research shifted from anatomical studies to histological and cytological studies. The development of pathology as a science was chiefly a German medical school enterprise.

This brief look at medical schools suggests an intimate relationship between medicine and the life sciences in German and Continental universities. In the United States, historical circumstances made North America dependent on European schools well into the last quarter of the nineteenth century. I believe that the separation of medicine and the life sciences resulted in a different education from 1875 on in the United States. Life sciences were free from the traditions of medical education, and thus, more time and opportunity for novel approaches became available in the new American biological graduate programs. This, of course, required a graduate school worthy of the name, and in 1875, this school came into existence through the model provided at Johns Hopkins University by its president, Daniel Coit Gilman.

The American model caused all areas of scholarship to flourish, but the life sciences were meteoric in their rise to prominence. This too requires some historical perspective. The physical sciences—astronomy, physics, and chemistry—had a long history of success in Europe, in and out of the universities. Unlike medicine, which has a necessary immersion in anatomy and physiology and a focus on the human body, the physical sciences shared basic principles rooted in mathematics and (after 1810) atomic theory. This gave European scientists in those fields an enormous advantage in facilities for research and the traditions for doing that research. The new American model was not dramatically different from that of the European universities for doing research in the physical sciences. What was distinctive was the formation of a field of life sciences independent of a medical school home and faculty. For the new American medical schools after 1875, this worked out well. Their focus was on faculty research for the medical and human concerns of pathology, diagnosis, and treatment. They also chose to do away with the dissertation for the M.D. degree and focus instead on the competence for practice of each M.D. receiving that degree. See Table 1 for the educational backgrounds of the major contributors to classical genetics.

No Real Higher Education Exists for Americans Before 1876

Until 1876, higher education in the United States was virtually nonexistent. At a meeting of the National Teachers Association in Trenton, New Jersey, in 1869, one officer lamented, "We have, as yet, no near approach to a real university in America."[1] The quote reflects two issues: There was a good model of the university but it was to be found chiefly in continental Europe. In the United States, not even Harvard or Yale (established in colonial days) could boast of a graduate program that emphasized the nurturing of scholars or a research mission. It was not that the United States lacked an interest

Table 1. *Education of major contributors to the formation of classical genetics.*

Name	Education	Major workplaces	Comments
Charles Darwin (1809–1882)	Edinburgh 1825 (no degree); Cambridge 1831 (B.A.)	Independent scholar	At Edinburgh, Darwin started medical school but quit; he studied for the clergy at Cambridge. His voyage around the world as a naturalist shaped his career. He was independently wealthy.
Carl Nägeli (1817–1891)	Medical studies in Zurich (1839); Geneva 1840 (Ph.D.); Jena 1842 (second dissertation)	Zurich; Freiburg; Munich	Started medical school, shifted to botany and Hegelian philosophy, and shifted again to studying hybrids.
Gregor Mendel (1822–1884)	Vienna (no degree); self-taught	Brünn monastery	Flunked out from Vienna twice; taught physics in high school.
Francis Galton (1822–1912)	Self-educated; some college	Independent scholar	Dropped out of medical school at Cambridge; inherited father's wealth; explorer, psychologist, and meteorologist; shifted to heredity in the 1860s.
Thomas Henry Huxley (1825–1895)	Self-taught; apprentice; London 1842 (M.D.)	London School of Mines	Lower middle-class background; voyage as surgeon launched his career; chief advocate of Darwinism after 1858.
August Weismann (1834–1914)	Göttingen 1856 (M.D.)	Freiburg	Started with insect embryology; blindness shifted him from cytology to theoretical biology in midcareer.
Ernest Haeckel (1834–1919)	Vienna; Würzburg; Berlin 1857 (M.D.)	Jena	Considered art or philosophy as a career; switched to science through Johannes Muller.
Anton Dohrn (1840–1909)	Breslau 1865 (Ph.D.)	Naples Station	Started medical studies and switched to science through Haeckel; independently wealthy; founded and built Naples Zoological Station in 1872. This became a mecca for experimental life sciences.
Walther Flemming (1843–1925)	Göttingen; Tübingen; Berlin; Rostock (M.D.)	Prague; Konigsberg; Kiel	Early work in histology; developed stains and techniques for new field of cytology.
Eduard Strasburger (1844–1912)	Paris; Bonn; Jena (Ph.D.)	Bonn	Resisted parental influence to go to medical school; chose botany and cytology.
Hermann Fol (1845–1892)	Jena; Heidelberg 1869 (M.D.)	Geneva	Independently wealthy; chose cell research over practice; drowned at sea in a storm.
Hugo de Vries (1848–1935)	Leiden 1866 (Ph.D.); Würzburg; Halle	Amsterdam	Early research and training in plant physiology; later research in plant breeding.
Oscar Hertwig (1849–1920)	Jena 1872 (M.D.)	Jena; Berlin	Shared much of his early research work with brother Richard, both in cell biology.
Wilhelm Roux (1850–1924)	Jena 1878 (M.D.)	Innsbruck; Halle; Breslau	Founded field of Entwicklungsmechanik (developmental mechanics or experimental embryology); shifted from anatomy to embryology.
Edmund Beecher Wilson (1856–1939)	Johns Hopkins 1881 (Ph.D.)	Williams; MIT; Bryn Mawr; Columbia	Considered music, chose science; shifted from embryology to cell biology.

Name	Education	Location	Comments
Wilhelm Johannsen (1857–1927)	Self-taught; apprenticed in pharmacy	Copenhagen	No university training; shifted to botany, also as apprentice, to Pfeffer.
Hermann Henking (1858–1942)	Göttingen 1882 (Ph.D.)	Göttingen	Left cytology and academics in 1892 for lifelong work in fisheries.
Jacques Loeb (1859–1924)	Würzburg 1884 (**M.D.**)	Bryn Mawr; Chicago; Rockefeller	Married American and emigrated because of anti-Semitism; shifted to experimental studies after stay at Naples Station.
William Bateson (1861–1926)	Cambridge 1882 (B.A.); Johns Hopkins 1883–1884 (no degree)	Cambridge; John Innes	England did not have a Ph.D. program in the nineteenth century, but they added one later to keep Americans from going to Germany. Bateson spent two summers with Brooks and shifted from embryology to genetics (in both plants and animals).
Theodor Boveri (1862–1915)	Munich 1885 (Ph.D.)	Würzburg	Studied anatomy with Kupfer; switched to cell biology with Hertwig.
Thomas Hunt Morgan (1866–1945)	Johns Hopkins 1890 (Ph.D.)	Bryn Mawr; Columbia; Caltech	Early research in experimental embryology; later research in genetics.
William Ernest Castle (1867–1962)	Harvard 1895 (Ph.D.)	Knox; Harvard; Berkeley	Started in classics; shifted from experimental morphology to genetics.
Hans Driesch (1867–1941)	Jena 1889 (Ph.D.)	Heidelberg; Cologne; Leipzig	Independently wealthy; doctorate with Weismann but research done out of pocket mostly at Naples Station. Later professional career as a philosopher; dropped laboratory science.
Clarence Erwin McClung (1870–1946)	Kansas 1902 (Ph.D.)	Kansas; Pennsylvania	Trained in pharmacy, then entomology, then cytology; later career as administrator.
Bradley Moore Davis (1871–1957)	Harvard 1895 (Ph.D.)	Chicago; Pennsylvania; Michigan	Initial training in cryptogamic botany; shifted to *Oenothera* cytology.
Thomas Harrison Montgomery (1873–1912)	Berlin 1894 (Ph.D.)	Pennsylvania; Texas	American with B.A. from Pennsylvania; graduate school in Germany.
Reginald Ruggles Gates (1882–1962)	Chicago 1908 (Ph.D.)	McGill, Chicago; University College, London	Canadian undergraduate; shifted to *Oenothera* cytology at Marine Biological Laboratory; left U.S. in 1912 for England; switched to eugenics and anthropology in later career.

The major contributors are listed in chronological order by birth year. The M.D. degree is indicated in bold. Some started out in medical school, but switched to a Ph.D. program. This is indicated in the comments listed for each subject. Note that the earliest (those born in the 1820s) were largely self-taught. Formal graduate programs as we think of them today did not emerge in the German universities until the 1830s. Also in Germany, the life sciences were clearly part of the medical sciences for the greater part of the nineteenth century. In the 1890s, both Continental and U.S. graduate life sciences programs were almost exclusively associated with the Ph.D. The transition for this change is marked by two events: the Naples Station, which encouraged independent research in zoology, and the Johns Hopkins Ph.D. program, which was quickly adopted by other leading American universities.

in education; it had a long tradition of fostering the education of its citizens. In fact, in 1785, the Continental Congress set aside 640 acres per township of public land to be used for public schools for the 13 states. When the nation was founded in 1786, Congress saw a role for higher education in America and provided public land for constructing seminaries, colleges, and universities: "Religion, morality, and knowledge being necessary to good government and the happiness of mankind, schools and the means of education shall forever be encouraged."[2] In 1787, Congress gave its newest state, Ohio (still known as the Western Territory), 46,080 acres of land to establish a university. After 1803, each new state was provided that stipulated amount of land as part of its admission to the union.

During the Civil War, President Lincoln supported an extension of federal land and support for education through the Morrill Act of 1862. Its intent was the support of the agricultural and mechanical arts "without excluding other scientific and classical studies."[3] Its purpose was to "promote the liberal and practical education of the industrial classes in the liberal pursuits and professions in life."[4] These became known as "land-grant colleges" to their supporters and "cow colleges" to their detractors. Congress amended the act several times in cost-sharing efforts to increase their role in American education. In 1887, an important addition came from the Hatch Act, which established experimental stations for plant and animal research. These were funded by and affiliated with the U.S. Department of Agriculture, but the staffing, along with local tenure and academic policies, was provided by the universities. These became important settings for the development of classical genetics in the United States.

The intent of Congress was reflected by American higher education until the 1870s. Congress provided land to the new states but did not dictate a model for higher education. It had chosen, despite pressure from Thomas Jefferson when the nation was established, not to have a national university. Several colleges had been established in the colonies, many of them through royal or religious patronage, and Jefferson's opponents feared a uniform standard imposed by a federal government. Young men (few women went to college, and most colleges were sexually segregated), who did not choose to become preachers, went to private colleges where they received a Christian education reflecting the denomination that founded the college. The role of the colleges (even those that were nondenominational) was one of creating moral citizens. The students who received their degrees largely entered the world of business and became civic leaders, owing their social success to the contacts made in these colleges. Their education was in the liberal arts, moral philosophy, and the theological basis of their religion. They learned Latin and Greek and some mathematics to discipline their minds.[5] All students took the same curriculum. Students who chose law, medicine, music, engineering, art, or teaching did not attend these schools—they went to trade schools. Medical schools varied in quality and duration. David Starr Jordan (1851–1931), one of the prominent scholars of late nineteenth-century American biology, did not receive a Ph.D. (it did not exist in most universities) and chose instead to attend a one-year medical college in Indianapolis to obtain a doctorate. Jordan called it the "dark ages" of American medical education.[6] Many of the schools were based on quackery. Students often got law degrees by apprenticing themselves to lawyers, musicians went to conservatories, artists went to art institutes, and teachers went to normal schools. Scholarship and research were not requirements for the faculty at these schools, whether devoted to the liberal arts or the practical professions.

German Universities Provide a Scholarly Role

By contrast, the German universities were admired for their scholarship. That tradition began in the early 1800s through the influence of the Humboldt brothers. Alexander von Humboldt (1769–1859) (Figure 1) and Karl Wilhelm von Humboldt (1767–1835) were children of a prominent military

Figure 1. *Alexander von Humboldt devoted his life to science, traveled extensively, and made contributions to geology, astronomy, ecology, and meteorology. He and his brother Karl Wilhelm were major critics of higher education and their views led to strong support for the emerging German model of scholarship associated with the Ph.D. degree. von Humboldt inspired the careers of Louis Agassiz and Charles Darwin. (von Humboldt in Venezuela by F.G. Weitsch, 1806. National Galerie, Berlin.)*

family (and King's chamberlain) in Prussia. Alexander was a naturalist, world explorer, and diplomat. During his long life, he had a hand in popularizing mountain climbing (his descriptions of the mountains in Tenerife and his endurance in climbing Mount Chimborazo in the northern Andes were sensational), founding the field of ecology (he noted the succession of vegetations as he ascended mountains), making geography a science, studying astronomy (transits of Mercury and meteor showers), and helping to found meteorological stations so that weather forecasting would become a science. His ambitious project of writing an encyclopedia of knowledge, *Cosmos: Sketch of a Physical World Description* (1845–1862), was intended to unify the natural sciences by recording what was known about the earth, thereby creating "a philosophy of nature." Humboldt had studied at the University of Göttingen, one of the leading German universities at that time. *Cosmos* stimulated hundreds of scholars for several generations.[7] Humboldt was a foe of ignorance, bigotry, slavery, and oppression of the laboring class. He and his brother set a standard of objective scholarship that they hoped would be followed in the Continental universities. Karl Wilhelm von Humboldt was a philologist, poet, translator, and diplomat. He studied the Basque language and noted its unique attributes and independent origin from surrounding Spanish culture.

German universities became the model of scholarship, especially from the 1850s on. Their roots can be traced to the changing role of the universities from the scholasticism of the Middle Ages, where theology was supreme, to the Renaissance, where both Catholic and Protestant (after the reformation) universities debated the influence of humanism and

Renaissance scholarship. For the most part, that effort was defeated by the sectarianism of Protestantism and the Counter-Reformation of the Catholic universities. Even a Protestant university, such as the University of Edinburgh, held on to scholastic traditions and did not replace the Ptolemaic view of the Universe with the Copernican model, until Isaac Newton's *Principia* was widely accepted in Europe. Major changes occurred in the 1700s as more European colleges and universities were either taken over or founded by royal patronage and greater freedom for curriculum development and scholarship became possible. The original nine colleges in the United States during the colonial era were based on the Oxford–Cambridge traditions, including Harvard, William and Mary, Yale, College of New Jersey (later Princeton), King's College (later Columbia), College and Academy of Philadelphia (later the University of Pennsylvania), Rhode Island University, Queen's College (later Rutgers), and Dartmouth. These colleges were intended to provide religious ministers and classically informed citizens who would play a leadership part in governance. These roles were continued after the American Revolutionary War and did not substantially change until the mid- to late nineteenth century.[8]

The German model evolved in the nineteenth century into a collection of prominent professors, each having the control of a department in which no one else held his title (note that female professors did not exist in that sexist era). It made the title one of marked national esteem, and international recognition was required (as well as considerable astute campaigning) for a prominent scholar to be appointed for such an opening. The dominance of the professor had a liability: The professor might be so sure of his scholarly theories and findings that he could exclude competitive ideas and stifle originality that did not resonate with his scholarly vision. However, it also had an important benefit. Research was the mission of the professor, and encouraging that enthusiasm in his graduate students was an obligation. Academic freedom in the German university was not so much the choice of a field by a professor to explore or teach, but the right of a student to attend courses in any German university if he was a student in any one of them.[9] The student determined the curriculum and was given freedom to put together that combination of courses and independent research that satisfied him. Attending lectures was not mandatory. Only when the student was ready would oral and written examinations of a grueling nature be provided as part of the passage to the Ph.D.

English Universities Have a Liberal Arts Tradition

In contrast, Great Britain used a different model in which teaching, not research, was its major focus. The teaching was almost exclusively based on the sectarian-dominated idea of the university. The Protestant Reformation led to a replacement of Catholic by Protestant interests, rather than to a secular university emphasizing independent scholar-

ship. By the mid-nineteenth century, demands for reform of higher education in Great Britain erupted, leading to debates among three major groups. One group represented the traditionalists who wanted to keep the Oxford–Cambridge model intact with its emphasis on Greek, Latin, moral philosophy, and those disciplines that honed the mind (logic and mathematics). A second group represented the more practical "nouveau riche" created by the industrial revolution and strongly influenced by utilitarian thinkers (but denounced in such novels as Charles Dickens' *Hard Times*).[10] The utilitarians sought practical courses that would feed the trades and give a secular rather than sectarian slant to higher education. The most prominent spokesman for the utilitarian approach was Thomas Henry Huxley (1825–1895) (Figure 2). The third group was led by John Henry (Cardinal) Newman (1801–1890) (Figure 3) who favored a liberal arts education that was neither sectarian nor practical. His model was outlined in *The Idea of a University* (1853).

Newman was rector-elect of a new Catholic university in Dublin. In 1852, he gave public lectures outlining his vision for that university. In response to his own question "...what is the end of a university education...?" Newman answered with the memorable phrase, "...knowledge is capable of being its own end ... knowledge is its own reward."[11] He went on to acknowledge that a role for university courses would be "...that of training good members of socie-

Figure 3. *John Henry (Cardinal) Newman advocated a liberal arts education "for its own sake." His views, when incorporated into the American university, were modified to satisfy the practical demands of students and their parents. (Courtesy of the National Portrait Library Gallery, London.)*

ty." He rejected the idea of the university as a training place for scholars, leaders, heroes, and geniuses (although they do end up having their share of such people) and instead argued that "...a university training is the great ordinary means to a great but ordinary end; it aims at raising the intellectual tone of society, at cultivating the public mind, at purifying the national taste, at supplying true principles to popular enthusiasm and fixed aims to popular aspiration, at giving enlargement and sobriety to the ideas of the age, at facilitating the exercise of political power, and refining the discourse of private life."[12] Newman's model student was the refined gentleman, the person who could be counted on to be patriotic, to help colonize the Queen's empire, and who retained his charm and poise while using his cultivated mind to be a good citizen: "He knows when to speak and when to be silent; he is able to converse, he is able to listen; he can ask a question pertinently, and gain a lesson seasonably."[13] In Newman's university, there would be no majors and no degrees in law, medicine, engineering, fine arts, music, business, or education; those would be suitable trade school enterprises but

Figure 2. *Thomas Henry Huxley, seen here at the height of his career as an evolutionist, biologist, and teacher, reshaped the way liberal arts education was taught around the world. (Courtesy of the National Library of Medicine.)*

not the business of a liberal arts education. Newman's concept of knowledge for its own sake was influential among educators but rarely put to practice. Hardly a university in the world then or now provides an education "for its own sake." It became one of those academic pieties that idealists cite but both college faculties and the parents of students, for the most part, utterly oppose. Both students and their parents want some practical outcome from education and without acknowledging it, their values are utilitarian.

Thomas Huxley's approach was quite different. He did not come from wealth and he had to struggle out of lower middle-class poverty to obtain an education. His lack of contacts and standing for many years kept him from university appointments, and he had to settle for teaching in trade schools. He was an outstanding naturalist who served as a naval physician on a trip to Australia and the archipelago of islands stretching across it in the Pacific. He published voluminously on the invertebrates he collected at sea and shifted to being a full-time naturalist. By the time Darwin's theory of evolution by natural selection had appeared in 1858, Huxley was already sold on the idea of evolution and the antiquity of the earth's life on it, but he was still not committed to natural selection as its mechanism. He was, like Darwin, an agnostic and kept religion out of his science.[14] He was also a spectacular lecturer who drew crowds when he gave public lectures, and he recognized the value that new knowledge gives not only to a field of scholarship, but to the world and the shaping of our worldview. He summed up these ideas in an 1868 lecture given to the South London Working-Men's College and entitled his essay "A Liberal Education and Where to Find It."

Huxley saw education as both practical and liberating. He claimed that "...education is the instruction of the intellect in the laws of nature; under which name I include not merely things and their forces, but men and their ways; and the fashioning of the affections and of the will into an earnest and loving desire to move in harmony with these laws."[15] His analogy reflected the impact of the industrial revolution on European society in his own time and the dominant role of science in shaping it: "That man, I think has had a liberal education who has been so trained in youth that his body is the ready servant of his will, and does so with ease and pleasure all the work, as a mechanism, it is capable of; whose intellect is a clear cold, logic engine, with all its parts of equal strength, and in smooth working order; ready, like a steam engine, to be turned to any kind of work, and spin the gossamers as well as forge the anchors of the mind; whose mind is stored with a knowledge of the great and fundamental truths of nature and of the laws of her operations; one who, no stunted ascetic, is full of life and fire, but whose passions are trained to come to heel by a vigorous will, the servant of a tender conscience; who has learned to love all beauty, to hate all vileness, and to respect others as himself."[16] Huxley argued that producing new knowledge was an essential role for the faculty and that British universities were shameful in their lack of such efforts: "...why, a third-rate, poverty-stricken German university turns out more produce of that kind in one year, than our vast and wealthy foundations elaborate in ten."[17] He deplored the British model of higher education, which he associated with "boarding schools for youths" or "clerical seminaries." Instead, the universities of Germany, and he hoped, the future colleges of Britain, would be "institutions for the higher culture of men, in which the theological faculty is of no more importance or prominence than the rest, and which are truly universities, since they strive to represent and embody the totality of human knowledge, and to find room for all forms of intellectual activity." [18]

The New American University Emerges

The model of American life sciences, until the founding of Johns Hopkins University in 1876, was based on the museum model. Universities liked to collect specimens and illustrate diversity, especially the living forms found in North and South America. Darwinism was still a new and alien concept, introduced less than one generation earlier. Experimental science was not a tradition in American life sciences. The existence of national or international research institutes such as the Naples Zoological Station was also very new. Anton Dohrn had founded the Naples Station in 1872 (Figure 4). It would be many more years before Woods Hole (1888) would be founded on the Naples model.[19]

In America, the demand for higher education was voiced by those who had to go abroad to study and return to practice their scholarly fields. It was hard to do so in American colleges because it was not part of their mission statements. In 1872, Harvard offered undergraduate electives not taken by undergraduates for its own graduate program. It did not initiate new graduate level courses. The Ph.D. was rarely given, the first Ph.D. in America being conferred in 1860 at Yale. Henry Philip Tappen, the president of the University of Michigan, was one of the first educators to advocate a university based on the European model. It was a hard sell, and there was no strong demand for scholars among practical people indiffer-

Figure 4. *Anton Dohrn (left) and the Naples Research Station (right). The Naples Marine Biological Station was founded by Anton Dohrn and demonstrated the international character of science. Both the Woods Hole Marine Biological Laboratory and the Cold Spring Harbor Laboratory were inspired by its contributions to the life sciences. Dohrn used his private wealth to found and build an international station for experimental zoology, relying on the Continental model of Entwicklungsmechanik. Scholars rented table space and the facility provided the equipment and supplies needed to carry out research. (Reprinted from [left] Groeben C., ed. 1982. Charles Darwin, 1809–1882, Anton Dohrn 1840–1909: Correspondence. Napoli Macchiaroli, Italy. [Right] Kofoid C.A. 1910. The Biological Stations of Europe. United States Bureau of Education, Washington D.C.)*

ent to the liberal arts or pious people threatened by secular values and science.[20]

Fortunately, private philanthropy permitted the much-needed breakthrough in higher education. Johns Hopkins was a wealthy owner of warehouses, the director of the Baltimore and Ohio railroad, and manager of several banks. He had amassed a fortune, but he was a bachelor and had no heirs. In 1867, he established a foundation to begin a hospital and a university. He wanted something new that would make a contribution to American society, and he asked the trustees to travel and define what a university should be. They asked Daniel Coit Gilman to lead this study (Figure 5). Gilman, a Quaker like Johns Hopkins, had been a librarian at Yale and a professor of physical geography there from 1850 to 1860. He was the secretary for the Sheffield Scientific School at Yale and later president of the University of California in 1872. In 1875 he set sail for Europe. He was influenced by the ideas of both Newman and Huxley, and while in Great Britain, he visited its

Figure 5. *Daniel Coit Gilman was already a noted educator when asked by Johns Hopkins to establish a university that would be new in its approach. The Gilman model was an amalgam of German Ph.D. scholarship and British liberal arts background education with a uniquely American interdisciplinary, multiprofessor department. The American Ph.D. became both a credential for college teaching and a research degree. (Courtesy of the Bancroft Library, University of California, Berkeley.)*

major scholars, including Spencer, Hooker, Kelvin, Tyndall, Jowett, and Huxley. He interviewed scholars in Paris, Berlin, Heidelberg, Strasbourg, Freiburg, Leipzig, Munich, and Vienna.[21] He also attempted to recruit top scholars or their best students for positions at the new Johns Hopkins University (Figure 6). An amalgam began to form in his mind: "Baltimore

Figure 6. *The Johns Hopkins University. Gilman built his university from the top down, beginning with its graduate school. He attracted his faculty from Europe and the United States and challenged them to "create and disseminate" new knowledge. The Hopkins model became the incentive for higher education in the United States. (Courtesy of the Ferdinand Hamburger Archives of The Johns Hopkins University.)*

afforded an opportunity to develop a private endowment free from ecclesiastical or political control, where from the beginning the old and the new, the humanities and the sciences, theory and practice, could be generously promoted."[22] Gilman was confident that the new university would attract students and faculty from all regions of America. President

Charles William Eliot of Harvard was not so sanguine and told Gilman that it would not succeed because all American colleges were regional and could not serve a national scope.[23] It was a debate that had earlier roots with Thomas Jefferson's failed efforts to establish a national American university. Jefferson had to settle, instead, for the University of Virginia.

Gilman Emphasizes Scholarship at Johns Hopkins

Gilman opened the university on February 22, 1876, serving as its first president. He chose his faculty without regard to their nationality or their religious beliefs; only their scholarship mattered. His first seven faculty members, English and American, taught mathematics, literature, chemistry, physics, zoology, astronomy, and classical languages. He recruited 20 fellows for his first graduate class. He also appointed nine associate professors: two from Germany, two from England, and five Americans. This new model, of departments staffed by more than one professor, was a major innovation. The junior faculty had opportunities to become full professors in the same departments with the initial full professors. As Gilman noted in his book, *The Launching of a University*, "We did not undertake to establish a German University, nor an English University, but an American University... ."[24] Gilman's zoology professor (and department chair) was H. Newell Martin (Figure 7), a student of Michael Foster (educated at Cambridge) and T.H. Huxley, and like them, an enthusiast for laboratory science. Martin made the study of introductory biology with a laboratory course a prerequisite for admission to the new medical school affiliated with the university. Of interest was Gilman's fidelity to scholarship as the criterion for appointment. At the time, his mathematics professor,

J.J. Sylvester, was the most distinguished mathematician in Great Britain, but his career was stymied (he was not allowed to earn a B.A.) because he was Jewish. Sylvester was brought to Gilman's attention by J.D. Hooker (then president of the Royal Society

Figure 7. *H. Newell Martin was a student of T.H. Huxley and devoted his career to research in physiology. He studied the effects of temperature, alcohol, and other variables on the mammalian heart and promoted physiological research as a model for all of the life sciences. (Courtesy of the National Library of Medicine.)*

and one of Darwin's closest friends).[25] Charles D'Urban Morris, who taught Greek and Latin classics, also introduced something new to the university. He suggested assigning graduate advisors for student projects and programs of study.

In his inaugural address, Gilman spelled out his philosophy: "Teachers and pupils must be allowed great freedom in their methods of work. Recitations, lectures, examinations, laboratories, libraries, field exercises, travel, are all legitimate means of culture."[26] He also stressed the role of the "creation and dissemination of knowledge" as the major role of his new university. "In forming all these plans we must beware lest we are led away from our foundations; lest we make our schools technical instead of liberal and impart a knowledge of methods rather than principles. If we make this mistake, we have an excellent *polytechnicum*, but not a *university*."[27] Also speaking at the inauguration in 1876 was T.H. Huxley. He spoke of his ideals of liberal education, echoing the principles laid down in his famous essay, and he emphasized the importance of new knowledge that would emerge from the Johns Hopkins model. The American press received Huxley's talk poorly. He was condemned for stressing biology (then a term associated with evolution and the mechanistic view of life) and for declining to start his lecture with a prayer. His presence at the inauguration was considered an insult to American traditions and his views on evolution a threat to American higher education.[28]

Biology Is a New Science at Johns Hopkins

The two key faculty recruited by Gilman were H. Newell Martin and William Keith Brooks.[29] Martin was a physiologist and enjoyed the experimental method that Huxley preached to a new generation of British biologists. Brooks was American, a student of Louis and Alexander Agassiz at Harvard. Louis Agassiz (1807–1873) (Figure 8) came to the United States after creating a distinguished reputation in Europe. He thrilled European readers by his daring. To study glaciers, he descended by rope and pulley through a hole drilled down the glacier and proved that there was running water underneath. From his analysis of the soil throughout Europe, he startled European readers who learned that a sheet of ice had once covered most of northern Europe during what was popularly called the Ice Age. Agassiz was a critic of evolution by natural selection and preferred a model of successive creations by God performed over reasonably long periods of time. He studied fossil fishes and other marine fossils, classifying them and comparing them to living forms. Despite his own religious leanings, he was a champion of scientific freedom and all of his students became evolutionists. Preservation, dissection, and exhibition of specimens were important to Agassiz and his school from the 1850s to the 1880s because they greatly increased the understanding of comparative anatomy and classification.[30]

Brooks admired Agassiz and the younger Agassiz, who followed the same tradition. He liked their broad liberal arts European tradition and he avidly took to philosophy, an inclination he had picked up at Hobart College. His idol was George Berkeley (1685–1753), an idealist whose ideas verged on the Platonic.[31] Berkeley's philosophy put limits on what we could determine with our five

Figure 8. *Louis Agassiz (left) was a superb naturalist who made major contributions to taxonomy, comparative anatomy, and the geographical distribution of animals. He worked out the detailed anatomies of fossil fishes and looked on their extinct species as victims of past catastrophes (with at least seven such episodes in his scheme) followed by new acts of divine creation. His son, Alexander (right), obtained degrees in engineering and zoology. He directed the museum that his father established at Harvard. Alexander Agassiz invested in a copper mine and helped make it into a major industry. He used his private wealth to fund the museum and other scientific research. (Reprinted, with permission, from [left] Jaffe B. 1944. Men of Science in America. Simon and Schuster, New York. [Right] Courtesy of the Harvard University Archives.)*

senses, and this made Brooks skeptical about materialist claims of his fellow scientists. Brooks was born with a cardiac malformation, had the cyanotic pallor and clubbed fingers characteristic of advanced heart disease, and thus led a sheltered life. He sometimes engaged in discussions with his colleagues and students while stretched prone on a couch, like a Roman dignitary without a toga. He was known for reading his mail while he sat on a toilet, leaving the nonpersonal mail for others to read. His mentoring habits were based on a sink-or-swim philosophy. He gave a graduate student a table and some equipment and told the student to find an interesting organism to work on.

Brooks (Figure 9) had broken with Agassiz's tradition. He believed that preserving and displaying was not where the new biology should go.[32] Instead of describing the anatomy of new species, he told his students to study the cell lineages from fertilized eggs onward to adult development. To Brooks, natural history was stale, but biology, particularly the biology being developed in German universities, was exciting. Microscopy was as important a feature for the new biologist as was dissection. He was not an experimentalist and it would be another 15–20 years before the German Entwicklungsmechanik (developmental mechanics), introduced in the 1890s by Wilhelm Roux and Hans Driesch, would find a strong foothold in American universities.[33] He believed that working with living things was real science, because it required a student's skills to figure out how development worked and what was going on inside cells. Every summer he would go with the students to a summer station usually in Chesapeake Bay. The station moved and could be easily set up or taken apart as a tented camp, unlike the fixed and

Figure 9. *William Keith Brooks studied with Louis Agassiz at the Penikese Island Station and received his Ph.D. working with Alexander Agassiz. Brooks taught at Hopkins with H. Newell Martin, and the two established dual activities for their students, with field work at the Chesapeake Bay Mobile Station and experimental work with Martin. Brooks took an interest in cell lineage studies and used crustaceans, mollusks, and tunicates for his work. He favored Darwin's theory of pangenesis but was open to the possibility that heredity could be successfully analyzed through experimentation. (Reprinted from Brooks W.K. 1909. Are Heredity and Variation Facts? The Last Public Address of William Keith Brooks. Reprinted from the Proceedings of the Seventh International Zoological Congress.)*

expensive housing that the Naples Station and similar European institutes had organized in the 1870s and 1880s. Some students liked the independence he fostered and others felt abandoned and had to rely on their fellow students. Some drifted away to Martin for a more structured introduction to experimental physiology.

Brooks liked to talk. He read widely and was tolerant of divergent views, so tolerant that at times he seemed "too philosophic" to be a good scientist. Nevertheless, he inspired students by making them think about the big picture and not just the narrow details of their work. Heredity, he argued, should not wallow in speculation and theory, as it had for most of the nineteenth century. Heredity should be accessi-

ble to experimentation. He inspired William Bateson, who visited after receiving his M.A. at Cambridge, and set him to work on the hemichordate *Balanoglossus* and work out its developmental biology and tie it to evolution (Figure 10).[34] His willingness to think of sports and discontinuities in evolution when discussing these ideas with Bateson gave Bateson the focus to shift from embryology to heredity when he returned to England. Bateson acknowledged that he might not have written his 1894 *Materials for the Study of Evolution* had it not been for Brook's encouragement.[35] Brooks worked on marine invertebrates and created a good deal of local attention by studying the sex life of oysters, of economic importance to the Baltimore fisheries industry and titillating to a public heavily censored by Victorian mores on what they could learn about the biology of sex.

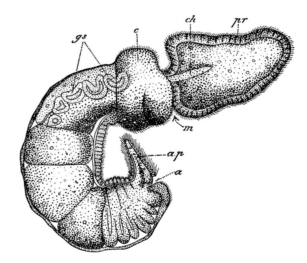

Figure 10. *The acorn worm Balanoglossus is a hemichordate. As an embryo, it has a notochord (a vertebrate feature), but as an adult, the notochord regresses and the organism resembles an invertebrate. Bateson studied Balanoglossus with Brooks, and this illustration is based on Bateson's study. (Reprinted from Willey A. 1894. Amphioxus and the Ancestry of the Vertebrates. Macmillan, New York.)*

Besides Bateson, four prominent American biologists owed their training to Brooks and the Hopkins biology department. E.B. Wilson, R. Harrison, E.G. Conklin, and T.H. Morgan were the first generation of Americans trained in the European style with the added bonus of the Hopkins interdisciplinary approach to life sciences.[36] Harrison became known for his work on tissue culture, and many believe that he should have shared a Nobel Prize with Alexis Carrel for that work. Conklin was a major contributor to cell lineage studies and kept his interests in developmental biology, cytology, evolution, and genetics throughout his career. Wilson and Morgan started out in embryology, but Wilson shifted to cytology and Morgan, in midcareer, shifted to genetics. All four were close friends and it was fortunate for the development of classical genetics in 1910–1925 that both Wilson and Morgan ended up doing their most productive work at Columbia University.

Wilson Is a Pioneer in the New Biology

Edmund Beecher Wilson (1856–1939) (Figure 11) came from a middle class family in Illinois and was raised mostly by his infertile aunt and uncle while his biological parents and the rest of their large family moved to Chicago.[37] He taught elementary school when he was 16 years old and then starting in 1874 spent a year at Antioch, a year at the University of Chicago, and three years at the Sheffield Scientific School, as the science program at Yale was called. He received a Ph.B. and did work on the anatomy of sea spiders (*Pycnogonids*) for his undergraduate thesis. While there, he was fascinated by E.L. Mark's microscopic studies of mitosis in the cells of snails. In 1878, he received a Bruce Fellowship to attend Johns Hopkins University. From Brooks and Martin, he learned to explore their basic question of "how the individual lies in the germ."[38] It excited him to think about their challenge. In a few days or weeks, a cell gives rise to an embryonic mollusk or a person. This was both a developmental question, requiring knowledge of embryology, and a genetic question, requiring a still unknown mechanism to account for the expression of hereditary factors. He chose a colonial polyp, *Renilla*, to study and plunged into the largely unknown history of its development. *Renilla* had an irregular cleavage and produced only two embryonic tissue layers (it lacked a mesoderm). When Wilson finished the paper and his dissertation, he took two years' leave to visit England and the Naples Station. While there he met Huxley, who read his paper and recommended that it be read to the Royal Society. The experience at Naples taught him the value of European experimental science, but it took two years before he could get into it. He taught at Williams College and at the Massachusetts Institute of Technology and finally got a job at Bryn Mawr College (the university for women inspired by the Hopkins model), where he launched into research on the earthworm *Lumbricus*. He noted that its mesoderm is a product of a group of cells (teloblast), whereas most organisms with three embryonic layers produce their mesoderm from pouches of an early embryonic organization (the archenteron). Wilson recognized that this might have evolutionary implications, with the annelids, flatworms, arthropods, and mollusks being teloblast in their mesodermal history and echinoderms and chordates having their mesoderm from a corresponding enteroblast. In both

Figure 11. *Edmund Beecher Wilson was an established scholar in 1890 (age 33)* (left) *just before moving from Bryn Mawr to Columbia.* (Right) *Wilson in later years.* (Reprinted, with permission, from Muller H.J. 1943. Edmund B. Wilson—An Appreciation. The American Naturalist **77:** 5–37.)

Lumbricus and a marine annelid, *Nereis*, Wilson did a cell lineage study, following, division by division, the formation of the blastomeres and the tissue cells of the developing embryo.

Wilson loved his stay in Naples, where he also participated in both playing his cello and going to many concerts to satisfy his deep love of classical music.[39] He made two more trips in 1891 and 1903, and on both occasions met his colleagues at Naples and in Germany to discuss their work and bring back the excitement of Entwicklungsmechanik to the United States. In his obituary for Wilson, Muller emphasized the importance of these trips abroad for Wilson's career, "It cannot be doubted that in the humbleness and in the internationalism of attitude exemplified in Wilson's European visits, in the willingness to learn from scientists in other countries, and in the increased opportunity thus afforded for integration of the best to be found everywhere, the United States owes much of its rise to a place second to none in the field of experimental and theoretical biology."[40]

A major issue that Wilson studied involved what

was called determinate and indeterminate development. In determinate cleavage, the first two cells (blastomeres) are fated differently (e.g., one may form the left side and the other the right side of the body). In indeterminate cleavage, each blastomere formed by the fertilized egg is totipotent (i.e., each can form an identical twin). Wilhelm Roux in 1888 had used a hot needle to obliterate a blastomere at the two-cell stage in a frog embryo.[41] He was surprised to see that the tadpole was actually a half-tadpole, as if the left-right symmetry had been established in the first cleavage division (hence the term determinate cleavage) (Figure 12). Three years later, in 1891, Hans Driesch (pronounced "dreesh") found that he could separate sea urchin two-celled embryos into separate blastomeres, and each formed normal identical twins.[42] This suggested that the first cleavage was indeterminate or totipotent. Totipotency was such an unexpected surprise for Driesch that it made him accept vitalism as an explanation. Some life essence had to pervade each cell to make a later cell form a full organism. Wilson was not persuaded that either model was correct. He found that cleavage could be determinate but still modified. When he agitated sea urchin eggs by violent shaking, he dislodged the nuclei but the embryo was still normal, suggesting a nuclear (rather than cytoplasmic) predominance in the determination of development. He noticed that development is epigenetic and that there is a progressive shift from totipotent to determinate development in the same organism as tissues formed. Wilson cut embryos, centrifuged them, fertilized portions of eggs, with and without nuclei, and tried many other techniques to explore these basic embryological questions. He remarked that "The scientific method is the mechanistic method."[43]

In 1891, Wilson was asked to head the Department of Zoology at Columbia and he shifted his research to the dividing cell. He spent several years studying the nonchromosomal components (asters, centrosomes, and spindle fibers), and he fol-

lowed the European developments on mitosis with great interest. In 1896, he put together, from his lecture course notes, a major work, *The Cell in Development and Inheritance.*[44] It was an immense success because it showed the directions in which life sciences were headed and stimulated a generation of students, especially those fortunate to take Wilson's courses, in the importance of experimental science and the interdisciplinary approach to broad problems in science. The beneficiaries of this approach were Morgan and his students as they formed the fly lab from 1909 on. Morgan's students, in particular, also benefited because by then Wilson, as we shall see later, provided the chromosomal findings that launched the chromosome theory of heredity and the conceptual basis of Morgan's new findings with fruit flies, at a time when Morgan was still skeptical that such a connection existed.

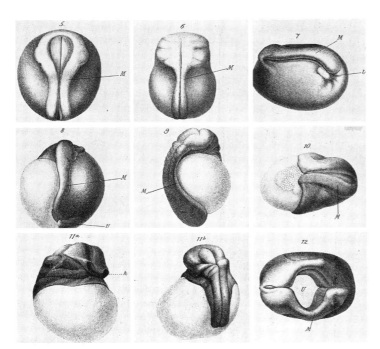

Figure 12. *Entwicklungsmechanik as a new approach to biology. Wilhelm Roux produced half-embryos by destroying one blastomere at the two-cell stage. Embryos 5–7 represent normal embryos during the early, middle, and late gastrula stages. Embryos 8–12 represent the half-embryos that Roux produced. The light areas in these deformed embryos consist of yolk and scar tissue. Roux was the first of several prominent German biologists to use experimental methods to study the development of embryos. Both Wilson and Morgan were enthusiasts for this new approach. (Reprinted from Willier B.H. and Oppenheimer J.M., eds. 1964. Foundations of Experimental Embryology. Prentice-Hall, Englewood Cliffs, New Jersey.)*

Morgan Influences the New Biology

Thomas Hunt Morgan (1866–1945) was born in Lexington, Kentucky, but grew up on a hemp farm that his father ran (Figure 13).[45] His father had been U.S. Consul to Italy (Messina) at the time of the reunification by Garibaldi, but his career ended when the Civil War broke out and he joined the Confederate army serving with his brother, General John Hunt Morgan. The two brothers, leading what is still celebrated as "Morgan's Raiders," gained infamy in Indiana (where they were known as horse thieves) for their pillaging of local farms. Young T.H. Morgan had little interest in family history, however, and despised the endless stream of war cronies that came to the farm. He attended the University of Kentucky and got his B.S. in 1886. From there he

Figure 13. *T.H. Morgan as a young scholar. Morgan began his career as an embryologist and was inspired to study evolution by Brooks and Martin. He later shifted to heredity after visiting Hugo de Vries in Holland in 1900. (Reprinted from Shine I. and Wrobel S. 1976. Thomas Hunt Morgan: Pioneer of Genetics. The University Press of Kentucky, Lexington.)*

spent a summer in Annisquam, Massachusetts, before entering Johns Hopkins.

Brooks had a strong interest in marine biology and that rubbed off on Morgan, who liked the ideas of studying embryology on live specimens and working out the fate of cells in early stages of development. He also enjoyed the experimental and physiological approaches of Martin. Conklin and Harrison were his fellow graduate students during this time. For his dissertation, Morgan worked on the embryology of *Pycnogonids*. He spent some time in Jamaica and the Bahamas and then took a position at Bryn Mawr. This was as valuable a place to him as it was to Wilson because it gave him an opportunity to do research. Also at Bryn Mawr was Jacques Loeb (Figure 14), a German biologist who had emigrated to the United States.[46] Loeb was an enthusiast for research and, like Morgan, a mechanist, who believed that all biological problems could be resolved through experimentation. They were also rivals and some competition for the same problems were inevitable because they constantly discussed what they were reading. Loeb is given credit for having been the first to publish a successful method of inducing parthenogenesis (it was widely publicized in the popular press because virgin birth had religious connotations that reporters were inevitably interested in pursuing).[47] According to Sturtevant, Morgan was working on the same problem and felt that Loeb pumped him for some of his ideas; Loeb, of course, got the credit for the discovery. Despite this rivalry and Morgan's feeling that Loeb was "secretive," the two got along well as colleagues while they were at Bryn Mawr, and certainly Loeb's influence on Morgan in steering him toward experiments and not just descriptive embryology was important in his formative years as a young professor.[48]

Figure 14. *Jacques Loeb. Jacques Loeb and Thomas Hunt Morgan were colleagues and competitors in the field of experimental embryology at Bryn Mawr College. Morgan shifted to genetics at Columbia College and Loeb, at The Rockefeller University, retained his studies of embryonic changes, particularly the chemical and physical induction of parthenogenesis (development and birth from unfertilized eggs). (Courtesy of Mrs. H.J. Muller to Elof Carlson.)*

Morgan also had some ambivalence toward Brooks. When Brooks talked about embryology and raised the basic questions of the day on differentiation, the preformation-epigenesis debates, the need to follow cell lineages to work out the fate of cells, and to explore the determinate and indeterminate status of the embryonic cells, Morgan felt excited by the ideas that could be explored. But there was another side to Brooks: He liked to bounce his scientific ideas off his wide range of philosophic readings. If he could find nuggets of biological thought in a Platonic dialogue, he would toss them into his lectures or his writings. His popular writings are often filled with tangential musings about the true meaning of biological ideas. He was almost Talmudic in his ability to see every nuance and contradiction in Darwinian natural selection, in Lamarckian acquired characteristics, in Galton's regression to the mean, or Galton's concept of midparents and other statistical approaches to heredity. This did not mean that Brooks was so wishy-washy that he had no idea what he was reading. In fact, he could be brilliantly accurate, as in his analysis of data (probably Alexander Graham Bell's) that he discusses in his monograph, *The Foundations of Zoology* (1899).[49] He notes that of some 2459 pupils at the American Asylum for the Deaf, 600 had married and produced some 800 children, 12% of whom were also deaf. He then fractionated the data into four groups and classified the deaf as "adventitious" or as having acquired deafness and as "congenital" (or as having apparently innate deafness). He states that "In 55 marriages, with 139 children, both parents are reported as adventitiously deaf, while in 52 marriages, with 151 children, both were congenitally deaf. In the latter group, 52 children, or 31.78 percent, are congenitally deaf, only 88 are stated to hear, and no facts are given about the hearing of 15 of them. In the first group only 4 of the 139 children, or 3.87 percent, are reported as congenitally deaf, 129 are reported as hearing, and 6 are not reported."[50] After going through his analysis and comparisons of the various groups from the data, he concludes that "While too few to give quantitative results, these statistics prove that it is the congenital and not the adventitious deafness which descendants have to fear."[51] That is careful, methodological, statistical, and informative. But such nuggets of information are embedded in pages of wordy digressions, such as "If we find in nature no reason why extended things should have weight, except that the fact is so, need we wonder if we fail to discover any ultimate or final reason why sensitive things should respond, for does not every scientific explanation rest upon something which is granted if unexplained?"[52] It was this frequently fuzzy quality to Brooks' thinking and writing that turned off

Morgan, who considered Brooks "too philosophic" for his tastes.

We explore Morgan's embryological studies later (Chapter 13) and see how they led him to become a geneticist. It is quite remarkable that both Morgan and Wilson entered genetics in their middle age, when both had already established international reputations in their prior fields. Fortunately they shared, through the Hopkins experience, a shove out of the nest to visit laboratories in Europe, which was part of the Hopkins tradition (the equivalent of present-day freshly minted Ph.D.s in the sciences going to another laboratory for a postdoctoral experience). They also shared the Gilman-mandated research work ethic to "create and disseminate" new knowledge and from Brooks, whatever their views on his musings, to appreciate that it is the broad, important, and seemingly unanswerable questions that can be dissected, analyzed, and tested by the creative imagination of experimental and descriptive science.

The Hopkins Model of Higher Education Achieves Rapid Success

Gilman recognized that teaching and research were both essential functions in his faculty. He argued that "the best teachers are usually those who are free, competent, and willing to make original researches in the library and in the laboratory." That outlook was imprinted in the graduate students receiving Ph.D.s at Hopkins. They saw themselves as teacher-scholars, with a commitment to do a substantial amount of research. For the universities copying the Hopkins model, this meant a reshaping of duties for the faculty. It was not uncommon in nonresearch colleges for faculty to teach as much as seven courses per semester. In the research universities, that "teaching load" was greatly reduced, and the very term "load" implies the future lopsided emphasis that would be given to research in the Hopkins-inspired university.

Johns Hopkins' new university quickly established its value. From 1876 to 1885, 69 Ph.D.s were granted. Among them were internationally known figures such as Woodrow Wilson and John Dewey. Also among the first generation of graduates were Edmund Beecher Wilson and Thomas Hunt Morgan, who would become major contributors to the founding of classical genetics. The Hopkins monopoly on Ph.D.s did not last long. Harvard, Cornell, Princeton, and Yale adopted the graduate model of scholarly research for its faculty and the awarding of Ph.D.s for their students who sought a career as scholars. The leading land-grant universities, such as Michigan and California, adopted the model. The new Stanford University, headed by Indiana's David Starr Jordan, also stressed the importance of graduate Ph.D. education as a major function of its faculty. Bryn Mawr copied the Hopkins model in full detail so that a similar program could be provided for female scholars (in those days, female scholars were usually excluded from the all-male enclave associated with colleges and universities). Post-Civil-War American wealth was finding a new outlet for its philanthropy, and millionaires did not blush when they saw their names adorning the fruits of what Andrew Carnegie called "the Gospel of Wealth."[53]

End Notes and References

[1] Ryan W.C. 1939. *Studies in Early Graduate Education*, p. 3. Carnegie Foundation for the Advancement of Teaching, New York.

[2] Emerson R.A. 1930. The land grant colleges and plant genetics. *Journal of Proceedings and Addresses of the Association of American Universities, 32nd Annual Conference*, pp. 119–132; see p. 121. Stanford University and University of California.

[3] Emerson R.A. 1930. op. cit. p. 121.

[4] Emerson R.A. 1930. op. cit. p. 121.

[5] Ryan W.C. 1939. op. cit. p. 4.

[6] Jordan D.S. 1922. *Days of a Man: 1851–1899*, volume 1. World Book Company, Yonkers-on-Hudson, New York.

[7] Botting D. 1973. *Humboldt and the Cosmos.* Harper and Row, New York.

See pp. 232–237 for Humboldt's impact on German university scholarship, especially in the sciences. Humboldt held the nature philosophy school of science in scorn. That Romantic movement was led by Goethe, Hegel, and Schelling.

[8] Gilman D.C. 1906. *The Launching of a University*, p. 6. Dodd, Mead, and Company, New York.

[9] Goldschmidt R. 1956. *Portraits from Memory*. University of California Press, Berkeley.

Goldschmidt gives a wonderful and witty account of life in the German universities when he was a student in the late nineteenth century.

[10] Dickens C. 1854. *Hard Times.* Bradbury & Evans, London.

Dickens opens with schoolmaster Gradgrind treating his pupils as empty vessels into which he pours "facts, facts, facts." Dickens loathed the utilitarian emphasis on things instead of broad liberal arts insights into human feelings and values.

[11] Newman J.H. (Cardinal). 1853 (1982 U.S. reissue). *The Idea of a University*, p. 77. University of Notre Dame, South Bend, Indiana.

[12] Newman J.H. (Cardinal). 1853. op. cit. p. 134.

[13] Newman J.H. (Cardinal). 1853. op. cit. p. 135.

[14] Desmond A. 1994. *Huxley: From Devil's Disciple to Evolution's High Priest.* Addison-Wesley, Reading, Massachusetts.

Desmond gives a thorough account of Huxley's life and work and the circumstances under which he had to work.

[15] Desmond A. 1994. op. cit., see pp. 361–363 for an account of Huxley's inaugural address as principal of the Working Men's College, South London. The address was made on January 4, 1868. The quote is from Huxley T.H. 1868. "A Liberal Education and Where to Find It." In Huxley's *Lectures and Lay Sermons,* p. 58. Everyman's Library, J.M. Dent and Sons, London.

[16] Huxley T.H. 1868. op cit. p. 60.

[17] Huxley T.H. 1868. op cit. p. 70.

[18] Huxley T.H. 1868. op cit. p. 72.

[19] Anonymous. 1909. Anton Dohrn. *Science* **30:** 833.

Dohrn was a student of Ernest Haeckel. After discussions with T.H. Huxley and other prominent biologists of his day, he organized an international Marine Biology Station at Naples. The Woods Hole Marine Biology Station, developed from a smaller station on Penikese Island, near Cape Cod, was funded in 1873 by tobacco money (from John Anderson) but it failed after Louis Agassiz's death late in 1873. The Woods Hole (originally called Woods Holl) Station in Cape Cod used the Naples Station as its inspiration. To sustain its operations it provided specimens for colleges and high schools through Turtox, a Chicago-based biological supply house.

[20] Ryan W.C. 1939. op. cit. p. 9.

[21] Gilman D.C. 1906. op. cit. p. 13.

[22] Gilman D.C. 1906. op. cit. p. 4.

[23] Gilman D.C. 1906. op. cit. p. 49.

[24] Gilman D.C. 1906. op. cit. p. 66.

[25] Ryan W.C. 1939. op. cit. p. 29.

26 Ryan W.C. 1939. op. cit. p. 29.

27 Gilman D.C. 1906. op. cit. p. 20.

28 Ryan W.C. 1939. op. cit. p. 33.

29 Martin went back to England after spending several years at Hopkins; Brooks spent his career at Hopkins. There is some debate about Brooks' influence. Some, like Bateson, considered it stimulating and formative to their careers, and others had mixed feelings, like Morgan who dedicated a book to him but also felt distanced because he was so speculative. Some regard Brooks as inadequate but believed his success was due to being "the only game in town." I believe the latter is a bit unfair, because Brooks did write monographs on oysters (he was an acknowledged scholar) and other marine life and knew how to steer students to good organisms for research.

30 A sampling of Agassiz's writings and a brief biography are presented in Davenport G. 1963 *The Intelligence of Louis Agassiz*. Beacon Press, Boston.

    Louis' son, Alexander, became a wealthy investor in mines and used a portion of his fortune to fund the Agassiz Museum at Harvard.

31 Berkeley argued that we only know an object by its sense data, and we have no way of knowing the existence or properties of matter except through the limited access of our senses. He did not deny an external reality; he limited what we could ever know about it. This appealed to Brooks, who frequently used this skeptical outlook for biological theories and phenomena.

32 Agassiz should not be seen as the last of the "preserved specimen tradition" that prevailed until the 1870s. His founding of the Anderson Station was a stimulus for Brooks to do the same in Chesapeake Bay. Although Agassiz wanted students to have a hands-on approach, there was nothing beyond descriptive and comparative biology in his approach. Brooks was open to following the entire life cycle and following the cells in their developmental histories. Brooks had the emerging view that biology went beyond taxonomy and species distribution.

33 Wilhelm Roux (1850–1924) was a student of Ernest Haeckel and shared a materialist view of life. In the 1880s, he began to seek causes of developmental processes and thus went beyond following cell lineages by actually disturbing the process of development to see its consequences. For an assessment of Roux's influence, see Frederick Churchill's Chabry, Roux, and the Experimental Method in Nineteenth Century Embryology. 1973, In *Foundations of Scientific Method* (ed. R.N. Giere and R.S. Westfall). Indiana University Press, Bloomington, Indiana. Roux believed that hereditary factors sorted out into the blastomeres but Driesch, who produced twins by separating blastomeres, argued that there is some sort of holistic property to life that allows a cell originally destined to form half a body to instead regulate itself and form an entire organism.

34 Bateson's views on Brooks and his work on *Balanoglossus* are described by his wife in Bateson W. 1928. *Essays and Addresses* (ed. B. Bateson) with her biographical memoir, pp. 28–73. Cambridge University Press, United Kingdom.

35 Bateson W. 1894. *Materials for the Study of Evolution*. Cambridge University Press, United Kingdom.

36 Also important was the proliferation of the Hopkins model to other universities and thus opportunities for Hopkins graduates to find academic positions. American higher education experienced rapid growth from the 1880s on. Both privately endowed and land-grant universities endorsed the new "teaching and research" model of faculty careers.

37 Muller H.J. 1943. Edmund B. Wilson—An Appreciation. *American Naturalist* **77:** 5–37; 142–172.

38 Muller H.J. 1943. op. cit. p. 4.

39 Wilson's musical abilities were outstanding and he was considered the most talented nonprofessional cellist in New York City. He sometimes played in quartets with professional musicians, but Wilson's later life was marred by a crippling arthritis that limited his movements.

40 Muller H.J. 1943. op. cit. p. 46.

41 Roux W. 1888. Beitrage zur Enticklungmechanik des Embryos. Ueber die kunstliche Hervorbringung halber Embryonen durch Zerstormung einer der beiden ersten Furchungskugeln sowie über die Nachtentwickelung (Postegeneration) der fehlenden Korperhalfte. *Virchow's Archiv für Pathologische Anatomie und Physiologie und für Klinische Medizin* **114:** 113–153 (translated into English in Willier B.H. and Oppenheimer J. 1974.

Foundations of Experimental Embryology, 2nd edition, Prentice-Hall, Englewood Cliffs, New Jersey.

42 Driesch H. 1894. *Analytische Theorie der Organischen Entwicklung.* Leipzig.

43 Muller H.J. 1943. op. cit. p. 20.

44 Wilson E.B. 1896. *The Cell in Development and Inheritance.* Macmillan, New York.

45 Allen G.E. 1978. *Thomas Hunt Morgan: The Man and his Science.* Princeton University Press, New Jersey. Also see Shine I. and Wrobel S. 1976. *Thomas Hunt Morgan, Pioneer of Genetics.* University of Kentucky Press, Lexington.

46 An account of Loeb's life and influence is in the introduction to Loeb J. 1912. *The Mechanistic Conception of Life* (reprinted 1964, with introductory essay by Donald Fleming). Harvard University Press, Cambridge, Massachusetts.

47 Loeb J. 1900. On the artificial production of normal larvae from the unfertilized eggs of the sea urchin. *American Journal of Physiology* **3:** 434–471.

    Loeb altered the ionic content of water to bring about parthenogenetic development.

48 Morgan worked at Bryn Mawr only one year while Loeb was there (1891), but they met each other regularly at Woods Hole during the summers.

49 Brooks W.K. 1899. *The Foundations of Zoology.* Macmillan, New York.

50 Brooks W.K. 1899. op. cit. p. 173.

51 Brooks W.K. 1899. op. cit. p. 174.

52 Brooks W.K. 1899. op. cit. p. 187.

53 Carnegie argued that a philanthropist should give money to institutions that benefit society. He also suggested using the bulk of an estate for such purposes. Too much money left to heirs, he believed, would corrupt them. He made a third demand on his fellow millionaires—that they should supervise the way their donated money was being used rather than leave the money upon their death. See Carnegie A. 1901. *The Gospel of Wealth and Other Timely Essays.* Reprinted in 1962 by Belknap Press, Harvard University Press, Cambridge, Massachusetts.

The Sex Chromosomes

Upon the assumption that there is a qualitative difference between the various chromosomes in the nucleus, it would necessarily follow that there are found two kinds of spermatozoa which, by fertilization of the egg, would produce individuals qualitatively different. Since the number of each of these varieties of spermatozoa is the same, it would happen that there would be an approximately equal number of these two kinds of offspring. We know that the only quality which separates the members of a species into these two groups is that of sex. I therefore came to the conclusion that the accessory chromosome is the element which determines that the germ cells of the embryo shall continue their development past the slightly modified egg cell into the highly specialized spermatozoon.

CLARENCE E. McCLUNG*

VIRTUALLY ALL COLLEGE STUDENTS who have taken a biology course know that human females have 44 autosomes (chromosomes other than sex chromosomes) and two X chromosomes. Human males have 44 autosomes and an X and a Y chromosome. Those who take a genetics course represent the human chromosome number as 46,XX for females and 46,XY for males. We also know from this that human males determine the sex of their children because half of their sperm bear X chromosomes and half bear Y chromosomes. Their union will be with a normal egg containing one X chromosome. Faculties teach this with confidence but rarely do they know the history of how the X and Y chromosomes came to be named and how these ideas evolved.

Bugs Have a Role in the Cytology of Sex Differences

The discovery of sex chromosomes came after a decade of observations, discussions, and debates on both sides of the Atlantic. Most of these were waged using data from studies of bugs, particularly two suborders of the hemipterans: the homopterans (e.g., grasshoppers and locusts) and the heteropterans (e.g., bedbugs, water bugs, and stinkbugs). Later, the

*1901. In Notes on the accessory chromosome. *Anatomischer Anzeiger* **20**: 220–226; see p. 225.

dipterans (especially flies) had an important role in clarifying the story. Six major players (out of many dozens) stand out in this analysis: Hermann Henking, Clarence E. McClung, Thomas H. Montgomery, Jr., Edmund B. Wilson, Fernandus Payne, and Nettie Stevens.[1] Of these, the most prominent was Edmund B. Wilson. The symbolic representation, X and Y, and the designation of these as sex chromosomes are certainly Wilson's. Yet it was those who knew Wilson well—Stevens, McClung, and Montgomery—who made major contributions that enabled Wilson to put the story together and get it across to other geneticists and students.

In 1891, Hermann Henking (1858–1942) studied the fire wasp, *Pyrrhocoris*, a hemipteran bug. Henking was born in Jerxheim, Germany, and attended the University of Göttingen (1878–1882), where he studied with E. Ehlers and received a Ph.D. for his anatomical studies of a mite. He stayed there until 1892 and then got a job with the German Fisheries Association, a position that took him away from basic research. Thus, Henking initiated a new field, but he dropped out very quickly after his first publications in this field. When he tried staining *Pyrrhocoris* with picro-acetic acid, he noticed that in the process of sperm formation, some cells contained 12 chromosomes and some cells contained 11. This seemed at first a contradiction to the then new concept of the constancy of chromosome number for a species. He referred to the extra chromosome (which was wadded into a chromatin body associated with the nuclear envelope and which took on a red color in what otherwise looked like a round yellow body) as the "X element" (i.e., the unknown element) because he did not know how to classify it (Figure 1).[2]

Henking noted that this nucleolus-like body seemed to shrink during meiotic reduction division and migrated to become a chromosome aligned as a single unpaired chromosome in metaphase. Note the similarity to two relatively modern concepts: the origin of the nucleolus and the process of lyonization or X inactivation. The first is the association of the nucleolus with a nucleolar-organizing region (NOR), as in the "p" or short arm of the X chromosome of the fruit fly (or the multichromosomal NORs involving chromosomes 13, 14, 15, 21, and 22 in humans). The second is the association of condensed chromosomal material with the nuclear envelope in mammalian cells associated with female cells bearing two X chromosomes, one of which is condensed as a Barr body or chromatin spot. Whether either or both of these phenomena were associated with *Pyrrhocoris* is not known. During anaphase the chromosomes seemed to divide or separate, but not the X element, which moved as a single unit to one of the two cells. Henking made no association of the X element with sex determination.

Figure 1. *Henking's X element. Henking abandoned cytology for fisheries management shortly after his discovery of the enigmatic X element, which had properties of both a chromosome and a nucleolus. Henking also noted that the X element was only seen in males, which had one more chromosome than the females. (Reprinted from Wilson E.B. 1911. The sex chromosomes. Archiv für Mikroskopische Anatomie 77: 249–271.)*

The Accessory Chromosome
Is Misinterpreted as Male Determining

Clarence Erwin McClung (1870–1946) had a major role in identifying (but not naming) what later became known as the X chromosome (Figure 2). He was born in Clayton, California and as a youth, he learned surveying and worked for an uncle who was a pharmacist. McClung entered the University of Kansas School of Pharmacy in 1890 and received a degree of Ph.G. in 1892. He then shifted to the liberal arts, getting a B.A. in 1896, and there picked up an interest in histology while studying biology. In 1896, he entered graduate school at the University of Kansas, earning an M.A. in 1898 and a Ph.D. in 1902. As part of his doctoral training, he spent one

Figure 2. *Clarence Erwin McClung obtained his Ph.D. at Kansas but spent a semester with Wilson at Columbia, who got him interested in chromosome studies. His discovery of the accessory chromosome in grasshoppers led him to propose a theory of sex determination, in which the male receives the accessory chromosome. The relationship of sex to chromosomes turned out to be more complex. (Courtesy of the National Library of Medicine.)*

semester with E.B. Wilson at Columbia University. He obtained his extensive knowledge of entomology from a semester at the University of Chicago, where he took courses with William Morton Wheeler.[3]

McClung taught as an instructor and assistant professor from 1898 to 1900 at the University of Kansas and rose to the rank of full professor in 1906. He also did a brief stint as acting dean of the medical school (1902–1906). In 1912, he moved to the University of Pennsylvania and retired in 1940; he died six years later. He was well admired by other scientists, and in 1919, he founded *Biological Abstracts*, which made it easier for his fellow scientists around the world to locate journal articles related to their areas of interest.

In 1899, McClung was working on the locust, *Xiphidium fasciatum*, and found an X element similar to the one found by Henking, but because it looked more like a chromosome than some nucleolar body associated with the nuclear membrane, he renamed it "the accessory chromosome" (Figure 3).[4] By 1902, McClung was quite sure that the accessory chromosome was a chromosome, and he discarded competitive terms with improper associations, such as those that Henking and others had discussed. Henking had actually assigned several names to his X element. At first, he thought that it was a nucleolus. He also saw it as a "double element X," which moved as a unit to form one spermatid, whereas the other spermatid did not receive the double element X. At the turn of the century, Thomas Harrison Montgomery, Jr., then at the University of Pennsylvania, was McClung's rival and tried to clarify the situation for McClung by pinning the accessory chromosome to spermatogenesis,

THE ACCESSORY CHROMOSOME—SEX DETERMI-
NANT?

C. E. McCLUNG.

PART I. OBSERVATIONS AND COMPARISONS.

The peculiar chromatic element discussed under this name in
several recent papers is one that gives promise of throwing considerable light upon the nature of the chromosomes. So long
as all chromosomes of the nucleus were observed to pass
through a cycle of changes apparently identical in each case,
there was little chance to gain an insight into their interrelations.
With the discovery of the accessory chromosome and the recognition of its true chromosomic character, however, there has
been offered an opportunity to draw comparisons and so to formulate conclusions which, in time, are certain to materially increase
our knowledge of these most important nuclear structures.

In recognition of this fact, and with the hope of hastening such
a desirable end, I have devoted some time to the study of the
accessory chromosome and have also encouraged students in my
laboratory to direct their attention to it. Much material has
been collected, and is still being accumulated, in order that as
broad a view of the subject as possible could be obtained.

Since, however, the difficulties involved in securing and preparing material from widely different forms would unduly delay the
attainment of any comparative results, I have confined my studies largely to the Orthoptera. This has been done in the belief
that more substantial good can be derived from a thorough knowledge of a limited group than from a superficial acquaintance
with a wider field. Once the basic principles underlying the cellular phenomena of one group are discovered, their recognition
in other forms will be rendered much easier.

As a result of the studies so far pursued, it has been found
that individual forms rarely present all the details of a problem
equally well. Different species excel in the clearness with
which certain points are brought out. A feature obscure in one
species will appear distinct enough in another while for the elu-

Figure 3. *The title page of McClung's paper, which raised
the possibility that a specific chromosome might be associated with a hereditary trait—sex determination. Note the question mark on the title, reflecting the uncertainty of its role.
(Reprinted from McClung C.E. 1902. The accessory chromosome—Sex determinant? Biological Bulletin 3: 43–84.)*

Figure 4. *Although American born and raised, Thomas
Harrison Montgomery, Jr. went to Berlin for his Ph.D. and
picked up an interest in cytology there. He chose hemipteran bugs, and their smaller chromosome number and varied
sizes led him to describe homologous pairing of chromosomes in meiosis. He favored Henking's interpretation of the
X element rather than McClung's, because his own work
indicated that the alleged chromosome remained condensed
and otherwise differed from the other chromosomes in the
bugs that he studied. (From the Collections of the University
of Pennsylvania Archives.)*

where it acted as a chromosome "that pursues a
somewhat different course from the others" (Figure
4).[5] Unfortunately, Montgomery (whose training
was from the German school) sided with Henking's
interpretation, and he felt that it was sufficiently
different to deserve being called a "chromatin nucleolus." Montgomery's reasoning was that it changed
shape differently from the other chromosomes and
seemed to move at its own pace; hence, it could not
be just a chromosome. Montgomery used a different
hemipteran for his study, a Pentatoma, *Euschistus
harpalus.*

The Small Chromosome
Is Seen as the Accessory Chromosome

Adding to the confusion over terminology was the work of F.C. Paulmier, a student of E.B. Wilson, who noticed in 1899 that this unusual chromosome went to only one of the spermatids during the second meiotic division in the bug *Anaxas tristis*, and thus, he argued that it could not be a true nucleolus.[6] He used two terms in his article to represent what he saw: The accessory chromosome was a "degenerating chromatin," because of its change in shape and size, and it was the "small chromosome," because it was notably shorter than the other chromosomes. In 1900, McClung's student, Walter Sutton, used a more favorable organism, the lubber grasshopper, *Brachystola*, and demonstrated that McClung's accessory chromosome was indeed a true chromosome that retained its individuality.[7] In *Brachystola* the chromosomes stained alike and moved alike, and thus, the accessory chromosome had no differentiating qualities other than being unpaired at the metaphase of reduction division. The differing interpretations of Henking, McClung, Montgomery, Paulmier, and Sutton reflect the choice of organism each had used. It is a matter of luck that Sutton had the best material. Recall that in the study of hybridization, Mendel had the good luck to work with *Pisum* (garden peas) and Nägeli had the bad luck of working with *Hieracium* (hawkweeds). In the first stages of working out a new finding or new biological concept, there may be inconsistencies or contradictions caused by the diverse ways in which life has adapted to past circumstances.

Montgomery and McClung Debate the
Chromosomal Status of the Accessory Chromosome

Thomas Harrison Montgomery, Jr. (1873–1912) was born in New York City.[8] His family came from New Jersey and dated their ancestry there to the early 1700s. The family was prominent in the clergy, anthropology, law, and business. His father was an insurance company executive and a scholar, who had obtained a degree in history from the University of Pennsylvania. He moved the family to Chester, Pennsylvania, where the young Montgomery attended school. He received his B.A. from the University of Pennsylvania (1889–1891), but studied in Berlin for his Ph.D. in 1894 with F.E. Schultze, H.W.G. Waldeyer (who named chromosomes), and O. Hertwig (who helped work out meiosis). Montgomery's dissertation was on the anatomy and development of a nemertean worm. The German cytological school's influence on him was profound. When Montgomery returned to the United States, he worked at the Wistar Institute at the University of Pennsylvania (1895–1898) and taught at the University of Pennsylvania (1897–1903). He moved to the University of Texas, where he stayed from 1903 to 1908, and returned to the University of Pennsylvania in 1908, dying four years later in his

39th year. Of the 81 papers published in his short life, 25 were on cytology. Like McClung, Montgomery worked on bugs, especially hemipterans and heteropterans. A lengthy debate between McClung and Montgomery took place as their papers attempted to point out each other's alleged errors.[9]

McClung did extensive studies of Hemiptera and Orthoptera. Like his German counterparts earlier, he usually used spermatogenesis for study rather than oogenesis, not because of a sexist bias, but because the cells in testes are easier to obtain and process than those in ovaries. The egg is filled with immense amounts of fat, making it impermeable to many of the dyes (usually water soluble) that are readily taken up by the developing sperm. It is not uncommon for a mature egg to be 300 times the size of a mature sperm, making the nucleus difficult to locate. Eggs also have a more complex relationship to meiosis with the production of one or more discarded polar bodies, involving very asymmetric partitioning of the cells. At the time McClung and Montgomery were studying chromosomes in spermatogenesis, this phenomenon was not yet worked out with respect to chromosomal events.[10]

McClung, in 1901, stated his case: "Being convinced from the behavior in the spermatogonia and the first spermatocytes of the primary importance of the accessory chromosome and attracted by the unusual method of its participation in the spermatocyte mitoses, I sought an explanation that would be commensurate with the importance of these facts. Upon the assumption that there is a qualitative difference between the various chromosomes in the nucleus, it would necessarily follow that there are formed two kinds of spermatozoa, which by fertilization of the egg, would produce individuals qualitatively different. Since the number of each of these varieties of spermatozoa is the same, it would happen

that there would be an approximately equal number of these two kinds of offspring. We know that the only quality which separates the members of a species into these groups is that of sex. I therefore came to the conclusion that the accessory chromosome is the element which determined that the germ cells of the embryo shall continue their development past the slightly modified egg cell into the highly specialized spermatozoa."[11] A year later, McClung reinforced his belief that the accessory chromosome was male determining: "My conception of the function exercised by the accessory chromosome is that it is the bearer of those qualities which pertain to the male organism, primary among which is the faculty of producing sex cells that have the form of spermatozoa."[12]

If we use the designation X for the accessory chromosome and O for the nonexistent "homolog," this would lead to two types of sperm, X or O bearing, in equal amounts, and one type of egg, O, for which fertilization would lead to two types of zygotes, XO and OO. McClung assumed that the OO type lacked something essential for male determination and as a result, it formed females. Today, we would say that McClung assumed that the female was the default sex. McClung's model is similar to the Y-determining role involved in male determination in mammals, but it is clearly not the way sex is determined in fruit flies.[13] It is not known whether the bugs used in these studies had the Y-determining mechanism or some other mechanism (such as a quantitative model of X chromosomes) for their sex determination. McClung's second error was in assuming that the sex determination was not actually achieved at fertilization by entry of the X- or O-type sperm. Instead, he was sympathetic to a widely held European model that the egg selected the sperm and that eggs were preconditioned to becoming male (X accepting) or female (O accepting). He actually wavered on this idea, because he had no physiological basis for predicting which egg would accept which sperm. However, McClung was confident that once the zygote was established, the sexuality of the future offspring was irreversibly fated because identical twins were always of the same sex.

Montgomery and McClung reveal their personalities in their articles. Until 1919, articles were written in the first person and authors freely attacked each other's ideas, sometimes in tactless or painful prose. After 1919, journals adopted a policy of using the third person and editing out personal attacks. Both McClung and Montgomery had sniping styles, but Montgomery tended to be defensive and easily wounded, despite his own sarcasm doled out to oth-

ers. After McClung took Montgomery to task for several errors that he pointed out in his 1902 paper, Montgomery replied, concluding his defense with this plaintive assessment: "McClung has recently (1902) made an embittered attack upon my studies on spermatogenesis, due in large part to a misunderstanding of my position. He states at various places that my views were conflicting, and that he is unable to harmonize them. Certain serious mistakes I made in my first paper (1898) I took pains to correct in two others (1899, 1901), and in these my position is stated very definitely and without contradictions. Were he as frank in admitting mistakes, there would be great unanimity."[14]

Wilson's Idiochromosomes Reveal
Two Chromosomes Associated with Sex

E.B. Wilson played a key part not only in resolving the accessory chromosome role, but in shaping the chromosome theory of heredity.[15] Both issues were connected through his students and through these debates, which clarified the functional significance of chromosomes, not only those that paired as seemingly identical homologs, but among those that did not. Edmund Beecher Wilson (1856–1939) was born in Geneva, Illinois. His father was a lawyer who became Chief Justice of the Appellate Court of Chicago. He was the fourth of five children, and his parents decided that it would be easier on them, and a favor to his mother's infertile sister, if he was raised by his uncle and aunt. This led Wilson to the happy situation of having four parents who adored and encouraged him. Wilson liked nature studies and music and he became a gifted cellist. His true love, however, was science. He moved from Antioch to Chicago and finally to Yale, where he completed a Ph.B. in 1878.

He was not impressed by E.L. Mark's work at Yale on the developmental biology of the snail. "Here's a man who has written two hundred pages about the development of the snail and has only gotten as far as the two cell stage."[16] But he was impressed by his first observation of the mitotic process that was described in Mark's monograph. He went on for his doctoral work at Johns Hopkins University and became a student of William Keith Brooks and H. Newell Martin. The "see-for-yourself" approach of the Hopkins Ph.D. program greatly appealed to him. From there, he went to England and continental Europe to take a tour of the major laboratories in cell biology and developmental biology. He taught at Williams College, the Massachusetts Institute of Technology, and Bryn Mawr before coming to Columbia University, where he also chaired the Department of Zoology.

Wilson greatly admired the European tradition of academic scholarship, and he introduced those mental habits and technical skills to his own research and his students. He had an encyclopedic knowledge of the literature and he was fluent in several languages. In 1896, he published his vast learning in a monograph, *The Cell in Development and Inheritance* (Figure 5).[17] This became the standard text that dominated cytology and the new biology for a generation of students in graduate and undergraduate classes. He returned from his European visit convinced that microscopy studies would provide insights into heredity and development. At the time (before 1900), Mendelism was still buried in forgotten literature, but the cytological work describing meiosis, the production of gametes, the constancy of chromosome numbers, and the equal contribution of sperm and eggs made the chromosomes in the nucleus Wilson's overwhelming favorite for generating a theory of heredity.

Wilson started out in developmental biology and studied cell lineages, tracing how cells in the early embryo ended up in specific organs of the later embryo. He took an interest in the components of the cells, especially the apparatus that moved chromosomes during cell division. But most of all he looked for ways to confirm his strong commitment to the hereditary model of the cell through its chromosomes. In his 1896 monograph, he noted "that the nucleus contains the physical basis of inheritance, and that the chromatin, its essential constituent, is the idioplasm postulated in Nägeli's theory."[18] Today, we would rephrase that sentence and refer to DNA, not chromatin, and to genes, not idioplasm, without challenging our most accepted principle of heredity. When the chromosome studies of McClung began to appear, Wilson shifted to the cytology of insects and encouraged his students to pursue this new line of work which clarified the meiotic process. He was rewarded with both Sutton's and Cannon's clear contributions of cytology to heredity and shared with them, through repeated discussions, the correlation of

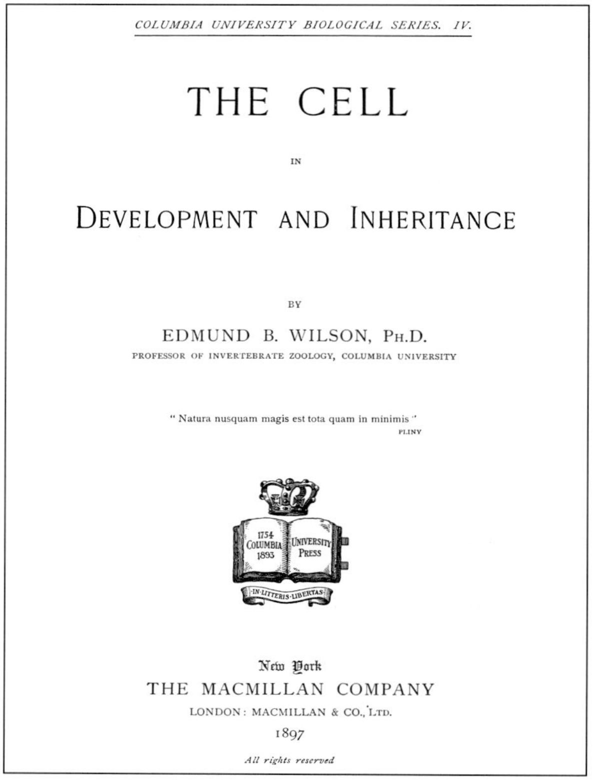

COLUMBIA UNIVERSITY BIOLOGICAL SERIES. IV.

THE CELL

IN

DEVELOPMENT AND INHERITANCE

BY

EDMUND B. WILSON, PH.D.
PROFESSOR OF INVERTEBRATE ZOOLOGY, COLUMBIA UNIVERSITY

" Natura nusquam magis est tota quam in minimis "
PLINY

New York
THE MACMILLAN COMPANY
LONDON: MACMILLAN & CO., LTD.
1897

All rights reserved

Figure 5. *Title page of Wilson's monograph,* The Cell in Development and Inheritance. *The 1897 issue is a reprint of the 1896 edition along with a page of corrigenda. The book had a profound effect on both fields in American graduate education and research. (Reprinted from Wilson E.B. 1897.* The Cell in Development and Inheritance. *Macmillan, London.)*

chromosome distribution and Mendelian outcomes.

From 1905 to 1910, Wilson published many papers on the accessory chromosome problem. As McClung had interpreted it, there was a sex-determining (male-determining) chromosome in animals. The hemipterans that Wilson studied were more complex, and they revealed a far more interesting story. Wilson established that in *Anasa tristis*, using

Paulmier's own slides, he could clearly distinguish that the first meiotic division separated the accessory chromosome into one cell and its counterpart cell did not receive it. He argued that other bugs, such as *Lygaeus turcicus*, did not show this pattern. Instead, there was a distribution of two unequal-sized chromosomes: "One of the two conjugating chromosomes is much smaller than the other."[19] This made the so-called accessory chromosome one part of a pair of chromosomes involved in the sex-determining process. He called these unequal partners "idiochromosomes" or idiosomes. To clarify the confusion with competing terms, Wilson recommended that the term chromosome be applied to any "chromatin mass" that is present in the equatorial plate. In those days, the plane of alignment of chromosomes at metaphase was called a plate. A chromosome could then be bivalent (two chromosomes present in a paired state) or univalent (one chromosome present), depending on the stage of cell division in which it is present. The term chromatid was reserved for the basic chromosome thread, one being present in a mature sperm or egg chromosome and four being present in a bivalent. He discarded Montgomery's "chromatid nucleolus" and assigned the nucleolus to the "resting stage" (i.e., interphase) nucleus. In Wilson's clarification for *Lygaeus*, "Each spermatid-nucleus thus receives seven chromosomes, one half from the spermatogonial number and no accessory chromosome, in the usual sense of the word, is present; but the spermatids nevertheless consist of two groups, equal in number, one of which contains the smaller, the other the larger of the idiosomes."[20]

This was true of several bugs, including *Euchistus*, *Brochymena*, *Podisus*, and *Trichopepla*. In one species, *Nezara*, the two idiosomes were of the same size, but functionally different in their sex-determining roles. Wilson later proposed an evolutionary model in which a pair of chromosomes (starting with the *Nezara* type) evolved through a loss of

chromatin in one of the sex-determining chromosomes, leading to unequal-sized chromosomes. Finally, the loss of the smaller chromosome altogether would lead to a sex-defining mechanism with one idiochromosome, superficially similar to the type McClung had observed (Figure 6).

Studies on Chromosomes.   403

that as far as the Hemiptera are concerned neither the suggestion I have made, nor the hypothesis of McClung has at present any direct support in observed fact.[1]

The practical interest of the idiochromosomes lies in the very definite basis that they give for an examination of the question by the study of fertilization, for their disparity in size gives us the hope of determining their history by direct observation. There is good reason to believe that such a study will yield interesting results.

SUMMARY OF OBSERVATIONS.

1. In Lygæus turcicus, Cœnus delius, Euschistus fissilis, Euschistus sp., Brochymena, Nezara, Trichopepla and Podisus spinosus all of the spermatids receive the same number of chromosomes (half the spermatogonial number), and no accessory chromosome is present; but the spermatozoa nevertheless consist of two groups, equal in number, which differ in respect to one of the chromosomes, which may conveniently be called the "idiochromosome."

2. In all of the forms named, excepting Nezara, half the spermatozoa receive a larger, and half a smaller, idiochromosome. In Nezara the idiochromosomes are of equal size, but agree in behavior with the unequal forms.

3. In all of the forms the idiochromosomes remain separate and univalent in the first maturation-division, while the other chromosomes are bivalent; this division accordingly shows one more than half the spermatogonial number of chromosomes. They divide separately in the first mitosis, but at the close of this division their products conjugate to form a dyad, which in all the forms save Nezara is asymmetrical. The number of separate

―――――
[1] The discovery, referred to in a preceding foot-note, that the spermatogonial number in Anasa is 21 instead of 22, again goes far to set aside the difficulties here urged. Since this paper was sent to press I have also learned that Dr. N. M. Stevens (by whose kind permission I am able to refer to her results) has independently discovered in a beetle, Tenebrio, a pair of unequal chromosomes that are somewhat similar to the idiochromosomes in Hemiptera and undergo a corresponding distribution to the spermatozoa. She was able to determine, further, the significant fact that the small chromosome is present in the somatic cells of the male only, while in those of the female it is represented by a larger chromosome. These very interesting discoveries, now in course of publication, afford, I think, a strong support to the suggestion made above; and when considered in connection with the comparison I have drawn between the idiochromosomes and the accessory show that McClung's hypothesis may, in the end, prove to be well founded.

Figure 6. *In 1905, Wilson and Stevens independently discovered nonhomologous chromosomes in males that Wilson called "idiochromosomes" and Stevens called "heterochromsomes." Wilson cites Stevens' independent discovery in a footnote addendum. (Reprinted from Wilson E.B. 1905. Studies on chromosomes. I. The behavior of the idiochromosomes in Hemiptera.* Journal of Experimental Zoology *2: 371–405.)*

Nettie Stevens Uses Diptera
to Describe Two Heterochromosomes

At the same time (1905) that Wilson was working out his hemipteran series of dissimilar idiochromosomes, Nettie M. Stevens was working on Diptera (flies) and other insects, including *Tenebrio*, at Bryn Mawr.[21] Nettie Maria Stevens (1861–1912) (Figure 7) was born in Cavendish, Vermont, where her father was a carpenter. During a summer course in Martha's Vineyard where she was studying to become a teacher, she took an interest in science and went to Stanford University for her B.A. (1899) and M.A. (1900). She became one of T.H. Morgan's students at Bryn Mawr (the best he had had in 15 years of teaching up to that time) and completed her Ph.D. in 1903. She received a fellowship to travel to Europe, studied at the Naples Station, and spent some time with Theodor Boveri at the University of Würzberg. She admired Boveri's work.

In 1905, after she returned, Stevens' first recognition of what she called heterochromosomes

Figure 7. *Nettie Stevens in 1904, the year before she discovered what were later named the X and Y chromosomes. (Courtesy of the Carnegie Institution of Washington.)*

was in the mealworm, *Tenebrio*. That same year she studied the housefly, *Musca domestica*. It had five pairs of chromosomes, which all paired as homologs, and one pair, which she designated as h1 and h2, that did not actually pair fully, but which did separate during meiosis I (Figure 8). She later reported a similar finding for *Calliphora vomitoria*, *Sarcophaga sarracinae*, and *Drosophila ampelophila*.[22] The latter, the fruit fly, was to be renamed *Drosophila melanogaster* and in Morgan's laboratory would become one of the most studied organisms on earth. She had more difficulties with *D. melanogaster* than with any of the other Diptera she studied and used thousands of specimens to isolate a few good cells with details of meiosis. She noted that in the somatic cells, the chromosome number was always a constant eight. In the reproductive cells, the reduction of chromosome number provided, in oogenesis, a constant four chromosomes in all cells. But in the spermatogenesis of the males, she noted what seemed like a smaller chromosome of a pair that she called "heterochromosomes" and gave them the same h1 and h2 designation that she had for *M. domestica*.

Stevens drew some conclusions: "Here, as in similar cases previously described, it is perfectly clear that an egg fertilized by a spermatozoon containing the smaller heterochromosome produces a male, while one fertilized by a spermatozoon containing the larger heterochromosome develops into a female."[23] She also studied the sex ratio of fruit flies bred on grapes and bananas and found no significant differences, the two sexes each fluctuating slightly about the expected 50%, a phenomenon that she also found to hold true for the sex ratio among houseflies. There was no profound effect of

THE GERM CELLS OF DIPTERA   PLATE III
        N. M. STEVENS

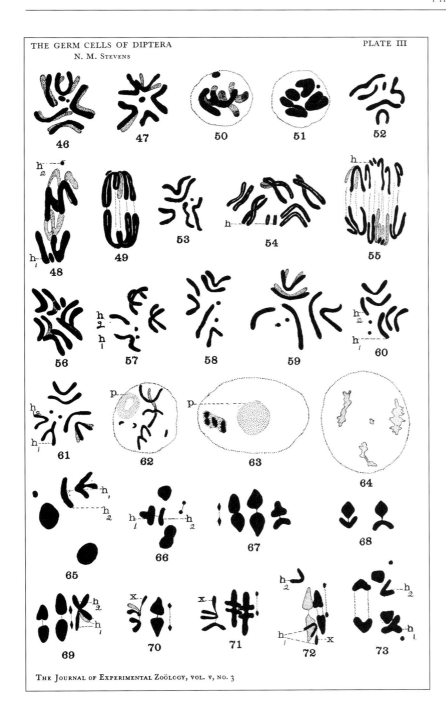

Figure 8. *Nettie Stevens had great difficulty obtaining good chromosomes of Drosophila ampelophila (later, D. melanogaster). From her successful specimens, she believed that two chromosomes (she called them heterochromosomes), which she designated as h1 and h2, corresponded to what Wilson later called the X and the Y in males (items 57 and 58 are D. melanogaster males). (Reprinted from Stevens N.M. 1908. A study of the germ cells of certain Diptera with reference to the heterochromosomes and the phenomena of synapsis.* Journal of Experimental Zoology **5**: 359–374.)

food on sex determination and the sex-determining mechanism seemed tied to the heterochromosomes. Stevens added a final note in her paper that was misinterpreted by Morgan and his students in the early days of their study of *D. ampelophila*. She stated, "A preliminary statement in regard to the chromosomes of *Drosophila* was made at the International Congress of Zoology in Boston, August 21, 1907. The question as to whether an odd chromosome or an unequal pair of heterochromosomes was present in the cells was then unsettled."[24]

Morgan thought he would play it conservatively and stayed with the "odd chromosome" interpretation. Until 1916, *D. melanogaster* was **XO** for males and **XX** for females. Note that McClung's idea of an accessory chromosome was now dead. Most species that were examined did not have an extra solitary chromosome involved in male determination. Instead of **XO** (= male) and **OO** (= female) for McClung's symbolism, the typical situation was **XO** (= male) and **XX** (= female) when one sex had a single sex chromosome.[25]

Multiple Sex Chromosomes
Are Found by Wilson and Payne

Complicating the story was the work of Fernandus Payne, who greatly extended Wilson's story by discovering multiple sex chromosomes in a variety of bugs. Payne worked with both Wilson and Morgan at Columbia University. For his doctoral work, he studied reduviid bugs. These were unusual because they had more than two sex chromosomes. In 1910, Payne studied *Acholla multispinosa*.[26] In the somatic tissue of males, the chromosome number was 26 for males and 30 for females. What we would today call autosomes numbered 20 in both sexes. In the female, ten sex chromosomes were present: four medium and six small. In the male, in addition to the 20 autosomes, there were six sex chromosomes: one large, two medium, and three small. Payne argued that the eggs uniformly had ten autosomes and five sex chromosomes (three small and two medium). In addition, the males produced two classes of sperm; those that were female determining had three small

and two medium along with the ten autosomes, like the eggs, but the male-determining sperm had only one large sex chromosome in addition to their ten autosomes. Payne realized that in living things, diversity, not uniformity, was a common occurrence, and he doubted the efforts of looking for one determining mechanism of sexuality (Figure 9): "Those who believe that the odd chromosome is merely a delusion in the minds of a few investigators still cling to the universality of van Beneden's law. However, the law is no longer of universal application. Not only the odd chromosome but a number of other irregularities have been recently described, the present case of *Acholla* giving the greatest variation in number."[27] Despite Payne's criticism, van Beneden's law was essentially correct—the somatic or diploid chromosome number is usually even and the gametic or haploid chromosome number is half that of the somatic number and can be odd or even.

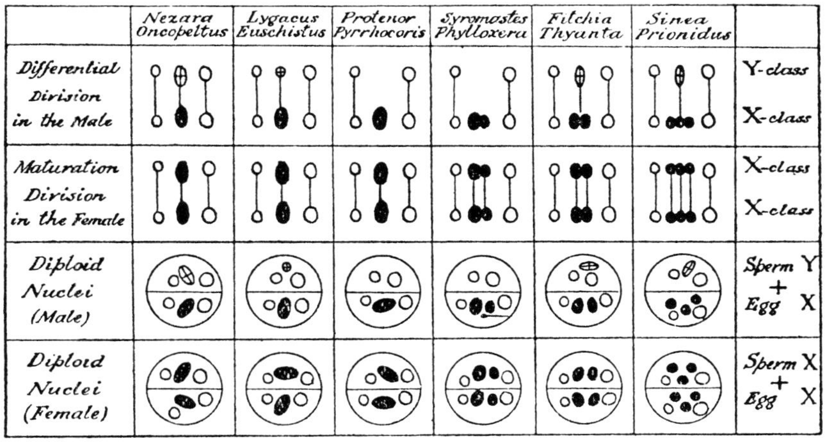

Figure 9. *Multiple sex chromosomes. Payne extended Wilson's observation that some bugs (the three columns on the right) had multiple sex chromosomes. Their existence made both Payne and Morgan skeptical of universal models of sex determination. (Reprinted from Wilson E.B. 1909. The chromosomes in relation to the determination of sex. Science Progress 4: 90–104.)*

Wilson's Solution:
Sex Chromosomes Are Represented by X and Y

Wilson's clarification came to him in a series of brilliant papers that he published between 1905 and 1911. In 1905, he had independently found chromosomes of unequal size (idiochromosomes) present in the males of some insects he studied, just as Stevens had found what she called heterochromosomes present in the males of her insects. Wilson was not sure of the full significance of the function of these chromosomes, but he was certain that they established evidence for the continuity of the chromosomes because the two forms, larger and smaller, could be followed into the germ cells and among the subsequent progeny. They also separated during the first meiotic division, as did the solitary accessory chromosome. Wilson thus stressed the meiotic

implications in his first paper. Although he was sympathetic to a sex-determining role for them, he argued "that as far as the Hemiptera are concerned, neither the suggestion I have made, nor the hypothesis of McClung has at present any support in observed fact."[28] Wilson added a footnote citing Stevens' work on the beetle *Tenebrio* that had just come to his attention (Figure 10). "She was able to determine, further, the significant fact that the small chromosome is present in the somatic cells of the male only, while in the female it is represented by a larger chromosome."[29]

Wilson, using Stevens' phrasing, but assigning the sex chromosome distinction to somatic cells instead of the spermatogonial, was almost there. He

STEVENS.   PLATE VI.

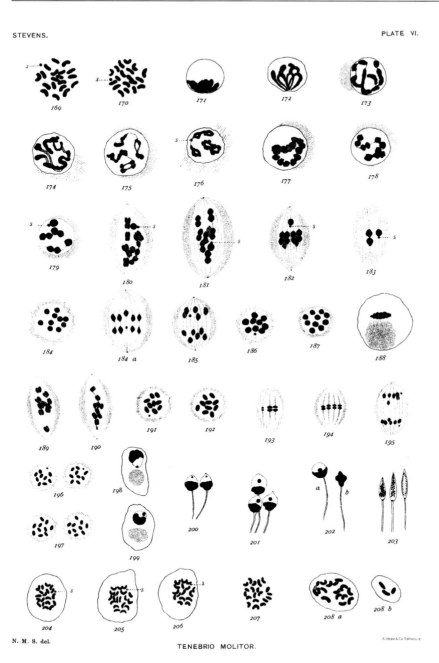

N. M. S. del.

TENEBRIO MOLITOR.

A. Hoen & Co. Baltimore.

Figure 10. *Nettie Stevens discovered two different sex chromosomes in Tenebrio. Stevens' 1905 illustrations of chromosomes in* Tenebrio molitor *implied the existence of sex chromosomes. Note that 184 shows the female having ten chromosomes of about the same size. In 184a, there are nine such chromosomes and a small dot. Stevens inferred that two of the ten chromosomes in the female, and one of the nine chromosomes of uniform size in the male, were what Wilson later called sex chromosomes. (Reprinted from Stevens N.M. 1906. Studies in spermatogenesis. Carnegie Institution of Washington.)*

did not immediately see in his mind *two* larger chromosomes in the female and, until he and Stevens both saw this, the proof of sex determination was still tentative. If he did see this relationship, his phrasing was misleading. I believe that Wilson in that same paper does make that inference when he states that in *Anasa*, "Each spermatid-nucleus thus receives seven chromosomes, one half the spermatogonial number, and an accessory chromosome, in the usual sense of the word, is present; but the spermatids nevertheless consist of two groups, equal in number, one of which contains the smaller and the other the larger of the idiosomes."[30] The next year, both extended their observations and stressed that association. Stevens did so with her dipteran studies and Wilson with further studies on his bugs.

The February 1906 paper gave Wilson an opportunity to rethink the way he could describe Stevens' heterochromosomes, his own idiochromosomes, and McClung's accessory chromosome. "They may also be designated (whenever it is desirable to avoid circumlocution) as sex-chromosomes or 'gonochromosomes.'"[31] Fortunately, the former, and not the latter, of these two terms became the preferred phrase for discussion among geneticists and cytologists. Wilson recognized that his designation did not imply that the larger sex chromosome was female determining and the smaller chromosome male determining, because the male had both chromosomes in the somatic cells and in the diploid germ cells as they entered spermatogenesis. He felt uncomfortable with interpreting sex determination as a simple case of dominance and recessiveness (patency and latency), remarking that "...we are still ignorant of the action and reaction of the chromosomes on the cytoplasm and on one another, and have but a vague speculative notion of the relations that determine patency and latency in development."[32]

Wilson Describes Homogametic and Heterogametic Sexes

By April 1910, Wilson was confident in his interpretation: "In many species of insects there are two classes of spermatozoa, equal in number, which in the early stages of their development, differ visibly in respect to the nuclear constitution; while there is but one class of egg, which is of the nuclear type identical to one of the classes of spermatozoa. That is to say, if the two kinds of spermatozoa be designated as the 'X-class' and the 'Y-class,' respectively; the eggs are all of the X-class. The male may, accordingly be designated as the *heterogametic* sex, the female as the *homogametic*."[33] Wilson also now used his symbolism to represent the sexual karyotypes: "The female diploid groups contain accordingly XX, the male XY (being otherwise identical); and upon reduction each mature egg contains one X; while half the spermatozoa contain X and half Y."[34] Wilson came up with a plausible suggestion. "Could we regard the sexual differentiation as due primarily to factors on a quantitative rather than a qualitative nature, most of these difficulties would disappear; and such a conception would be in harmony both with the cytological facts and with the experimental evidence regarding sex-heredity" (Figure 11).[35] This

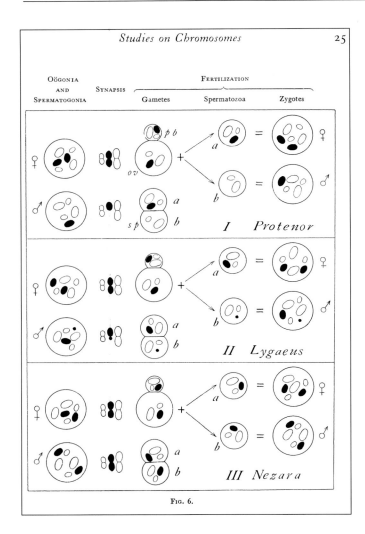

Figure 11. *Wilson identified three ways in which sex chromosomes (his idiochromosomes) may appear. In Protenor, the male is XO and the female is XX; in Lygaeus, the male is XY and the female is XX; in Nezara, the male XY idiochromosomes are equal in size and the XX female cannot be cytologically distinguished from the male. The Protenor form (XO) suggests a quantitative basis for sex determination. (Reprinted from Wilson E.B. 1906. Studies on chromosomes. III. The sexual differences of the chromosome groups in Hemiptera, with some considerations on the determination and inheritance of sex.* Journal of Experimental Zoology **3:** *1–40.)*

quantitative model, indeed, was absorbed by Morgan and his students in their interpretation of sex determination in *Drosophila*. (For a summary of the contending interpretations of sex chromosomes, see Table 1.)

Just as there is some confusion regarding who should have priority for working out Mendelism—

Mendel, de Vries, Correns, or Tschermak von Seysenegg—with each having some flaw of omission or interpretation, so too does the Wilson–Stevens discovery present the same problem. Stevens had stated her confusion on the "odd chromosome" or "heterochromosomes" in her fruit fly work; Wilson is flawed in initially seeing the female represented by

Table 1. *Contending interpretations of the sex chromosomes.*

Researcher	Genus	Comment
Henking	*Pyrrhocoris*	X element as a nucleolus or chromosome.
McClung	*Xiphidium*	Accessory chromosome as a male-determining chromosome; female = OO male = XO.
Paulmier	*Anaxas*	Short chromosome from a degenerating chromatin.
Montgomery	*Euschistus*	Chromatin nucleolus.
Sutton	*Brachystola*	Accessory is a true unpaired chromosome.
Wilson	*Anaxas*	The short chromosome is an accessory chromosome.
Stevens	Diptera and *Tenebrio*	Heterochromosomes (h1 and h2) in male but females have two h1. Ambiguity in fruit flies (odd versus heterochromosomes is unsettled).
Wilson	*Lygaeus*	Two chromosomes of unequal size are male; the larger represented twice is female; Wilson calls these idiosomes.
Wilson	Many bugs (Hemiptera)	A range of idiosomes: some with equal size in both sexes, some with unequal size, and some with idiosomes as two of same size in females and one of same size as in females but solitary in male.
Payne	Reduviids	Multiple heterochromosomes.
Wilson	Many bugs and flies	XX = female; XY or XO = male.

Henking had no idea what the X element was and speculated on its possible role as a nucleolus. McClung renamed the X element as an accessory chromosome and assigned it a male-determining role. Montgomery sided with Henking and saw it as a nucleolus. Paulmier noted that it was a shorter chromosome present in the males (thus, male determining, as McClung claimed). Sutton saw the accessory as a true chromosome and not a nucleolus. Wilson recognized the existence of idiosomes (which Stevens independently called heterochromosomes), and both claimed that the male had one of each and the female had two of the larger of the two idiochromosomes or heterochromosomes. Wilson finally called them sex chromosomes and assigned XX to females and XY to males. Payne worked out numerous cases of multiple sex chromosomes in bugs. Most researchers of macroscopic animal organisms use the XY and XX system, but whether the heterogametic sex is male or female varies widely. At the time of this cytological study (1890–1910), all of the heterogametic bugs and Diptera were male. Genetic but not cytological studies of moths and poultry in Great Britain in the early 1900s suggested that this was not universal.

"a larger chromosome" instead of two larger chromosomes. Making the transition from an accessory chromosome to a sex chromosome took some additional time and work with more organisms for both Stevens and Wilson to arrive at identical results: the **XX** female and the **XY** male as we see them today. The working out of a clean story minus its flaws requires a process of comparison and repetition. It is

less likely that the complete story emerges in a single observation or experiment.

The survey of the literature from the 1890s through 1910 reveals how the sex chromosome story evolves piecemeal from the many contributions of scientists, sometimes colleagues, sometimes rivals, and each with errors generated by speculations based on incomplete knowledge (Table 1). Yet each successive finding narrows the interpretive range and in the span of 20 years, a coherent story emerged that has essentially remained unchanged since 1910 (Figure 12). Wilson's 1910 description could appear almost unchanged in an introductory biology course today. It is not a story of the victory of one class of scientists over another. It is not the story of a dying out of competitors. It is the "winning of the facts" that triumphs.

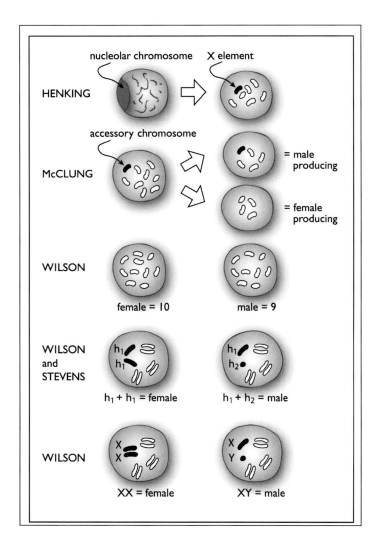

Figure 12. *Clarifying the idea of sex and chromosomes. The working out of the sex chromosomes spanned about 17 years (1893–1910). Henking identified an X element that behaved as a chromosome at one stage of meiosis, but seemed to be a nucleolar-like object near the nuclear envelope at an earlier stage. McClung, using more favorable material, saw a unique chromosome in males that he did not see in females. He called this the accessory chromosome and believed that it was male determining. In his disputes with McClung, Montgomery argued that the accessory chromosome was Henking's X element. Wilson noted that in some beetles, the male chromosome count was one less than the female, but because the chromosomes were about the same size, he could not make an interpretation associating one particular chromosome with sex. By 1907, both Wilson and Stevens had identified organisms with what Stevens called heterochromosomes (h1 and h2), which were morphologically distinct. In these species, the same chromosome number was present in both males and females, unlike Wilson's earlier series. Wilson clarified the story by calling the heterochromosomes sex chromosomes and giving them the modern designation of X and Y. He also argued that neither X nor Y was specific for sex determination. Instead, he believed that a quantitative relation existed, with one sex XX (the female in the species studied by Stevens and himself) and the other sex with a single X (either XO or XY). (Figure drawn by Claudia Carlson.)*

End Notes and References

[1] See Voeller B.R., ed. 1968. *The Chromosome Theory of Heredity: Classic Papers in Development and Heredity.* Appleton-Century-Crofts, New York for some of the hard-to-find early papers on cell division, fertilization, and the nature of the chromosomes.

[2] For a biographical account of Henking's life, see Berger J.D. 1970. Hermann Henking. In *Dictionary of Scientific Biography* (ed. C. Gillespie), volume 6, pp. 267–268. Scribner's, New York. The work on *Pyrrhocoris* is in Henking H. 1891. Untersuchunger ueber die ersten Entwicklungsvogange in der Eiern der Insekten. II. Ueber Spermatogenese und der Beziehung zur Entwicklung bei Pyrrhocoris apterus L. *Zeitschrift für Wissenschaftliche Zoolige* **51**: 685–736

[3] Allen G. 1970. Charles Erwin McClung. *Dictionary of Scientific Biography* (ed. C. Gillespie), volume 6, pp. 267–268. Scribner's, New York.

[4] McClung C.E. 1899. A peculiar nuclear element in the male reproductive cells of insects. *Zoological Bulletin* **2**: 187; and McClung C.E. 1902. The accessory chromosome—Sex determinant? *Biological Bulletin* **3**: 43–84.

[5] Montgomery Jr., T.H. 1898. The spermatogenesis in *Pentatoma* up to the formation of the spermatid. *Zoologische Jahrbuchern* (abstract) **12**: 1–99.

[6] Paulmier F.C. 1899. The spermatogenesis of *Anasa tristis*. *Journal of Morphology* (supplement) **15**: 223–272.

[7] Sutton W.S. 1900. The spermatogonial divisions of *Brachystola magna. Kansas University Quarterly* **9**: 73–100.

[8] Pollister A. 1970. Thomas H. Montgomery. In *Dictionary of Scientific Biography* (ed. C. Gillespie), volume 9, pp. 496–497. Scribner's, New York. An obituary for Montgomery by Conklin E.G., 1913, can be found in *Science* **38**: 207–214.

[9] Montgomery's major papers in this debate are 1898. The spermatogenesis in *Pentatoma* up to the formation of the spermatid. op cit. 1899. *Journal of Morphology* **15**: 204–265; 1899. Comparative cytological studies with especial regard to the morphology of the nucleolus, pp. 265–582. University of Pennsylvania. Contributions from the Zoological Laboratory for 1899; 1901. A study of the chromosomes of the germ cells of the metazoa. *Transactions of the American Philosophical Society* **20**: 154–236; and 1903. The heterotypic maturation mitosis in amphibia and its general significance. *Biological Bulletin* **4**: 259–269.

[10] The polar body is a discarded haploid nucleus in a small wad of cytoplasm. The egg reserves most of the cytoplasm at the reduction division (meiosis I), the polar body being the grossly smaller cell in an unequal cytokinesis in telophase. To achieve this, the nucleus migrates to the periphery of the cell and cytokinesis then eliminates one nucleus as a polar body. The second division (an equation division in today's usage) produces a second polar body from the haploid egg. Depending on the species, the first polar body may divide (producing a third polar body) or it may fail to divide, which is what happens in the human egg.

[11] McClung C.E. 1901. Notes on the accessory chromosome. *Anatomischer Anzeiger* **20**: 220–226; see p. 225.

[12] McClung C.E. 1902. The accessory chromosome—Sex determinant? *Biological Bulletin* **3**: 43–84. See pp. 72–74 for the quote and the article for its analysis of the extant literature. See McClung C.E. 1902. op. cit. for his concerns, p. 73, citing male elements as catabolic and female elements as anabolic and p. 75 for "Here they are approached by the wandering male elements from which they may choose—if we may use such a term for what is probably a chemical attraction—either the spermatozoa containing the accessory chromosome or those from which it is absent. In the female element, therefore, as in the female organism, resides the power to select that which is for the best interest of the species."

Even if there is a sexist basis for that European influence on American cytologists, that bias eventually collapsed after Wilson and Stevens showed conclusively the more familiar (to us) model of the male as the determiner of the offspring's sex when the male is the heterogametic sex (i.e., either **XY** or **XO**).

[13] Note that when **XO** = male and **OO** = female, the X becomes a male-determining chromosome. Because McClung did not or could not study oogenesis, he did not have a clear comparison of male and female karyotypes. In addition, many difficulties exist when studying insect somatic cells in the two sexes. Most of the cells in the adult are nondividing (they are in G_0 of the cell cycle) and metaphase chromosomes are difficult to obtain. In the G_0, the chromosomes would be in the unwound or chromatin state of interphase.

[14] Montgomery T.H. 1903. The heterotypic maturation mitosis in amphibia and its general significance. *Biological Bulletin* **4**: 259–269.

[15] Wilson's career is discussed later in more detail with respect to members of the fly lab (see Chapter 13 for Wilson's recognition that human color blindness must be associated with the human X chromosome). Both Morgan and Muller revered Wilson and each wrote a lengthy biographical memoir for him. See Morgan T.H. 1941. E.B. Wilson 1856–1939. In *Biographical Memoirs of the National Academy of Sciences* **21**: 315–342. Also see Muller H.J. 1943. Edmund B. Wilson—An Appreciation. *American Naturalist* **77**: 5–32; 142–172.

[16] Morgan T.H. 1941. op. cit. p. 318.

[17] Wilson E.B. 1896. *The Cell in Development and Inheritance*. Macmillan, London.

   A second edition was published in 1900 and a third edition in 1925.

[18] Wilson E.B. 1905. Studies on chromosomes. I. The behavior of the idiochromosomes in Hemiptera. *Journal of Experimental Zoology* **2**: 371–405; see p. 374.

[19] Wilson E.B. 1905. op. cit. pp. 375–376.

[20] Wilson E.B. 1905. op cit. p. 398.

[21] For a biography of Stevens' life, see Brush S.G. 1974. Nettie M. Stevens and the discovery of sex determination by chromosomes. *Isis* **69**: 163–172.

[22] Stevens N.M. 1905. *Studies in spermatogenesis with especial reference to the accessory chromosome*. Carnegie Institution of Washington Publication No. 36. Also, see her review: Stevens N.M. 1908. A study of the germ cells of certain diptera with reference to the hete-rochromosomes and the phenomena of synapsis. *Journal of Experimental Zoology* **5**: 360–374.

[23] Stevens N.M. 1908. op. cit. p. 370.

[24] Stevens N.M. 1908. op. cit. p. 373.

[25] Note that the various relationships of the sex chromosomes to symbolism at this time were based on actual observations of some sort of unusual chromosome appearances that departed from the expected uniformity of pairs of homologous chromosomes. The finding of hereditary traits associated with these chromosomes came later, with Morgan's finding of white eyes in males but not in females in his 3 red to 1 white F_2 results. That made the debate even more prolonged because the British found such traits in the females and not in the males. Curiously there was no such effort in Europe or Great Britain to enter the cytological study (after Henking) until Morgan's extension from 1910 on.

[26] Payne F. 1910. The chromosomes of *Acholla multispinosa*. *Biological Bulletin* **18**: 174–179.

[27] Payne F. 1910. op. cit. pp. 174–179. van Beneden's law proposed a constancy of chromosome number for a species, with the male and female gametes having half that number. The somatic chromosome number is even but the gametic number may be odd or even. As is often true with biological laws, it is generally true but diversity always reveals exceptions.

[28] Wilson E.B. 1905. op. cit. p. 403.

[29] Wilson E.B. 1905. op. cit. p. 403.

[30] Wilson E.B. 1905. op. cit. p. 403.

[31] Wilson E.B. 1906. Studies on chromosomes. III. The sexual difference of the chromosome groups in Hemiptera, with some considerations on the determination and heredity of sex. *Journal of Experimental Zoology* **3**: 1–40; see p. 28.

[32] Wilson E.B. 1906. op. cit. p. 38.

[33] Wilson E.B. 1910. The chromosomes in relation to the determination of sex. *Science* **4**: 570–592; see p. 572.

[34] Wilson E.B. 1910. op. cit. p. 579.

[35] Wilson E.B. 1910. op. cit. p. 587.

The Rediscovery of Mendelism

The ever-recurring number relationship of 3:1 could naturally not escape me, any more than the number relation of 1:1 on back crossing of green-seeded peas with hybrid pollen of the F1 generation.

E. Tschermak von Seysenegg*

On the rediscovery and confirmation of Mendel's law by DeVries, Correns, and Tschermak two years ago, it became clear to many naturalists, as it certainly did to me, that we had found a principle which is destined to play a part in the study of evolution comparable only with the achievement of Darwin—that after the weary halt of forty years we have at last begun to march.

William Bateson†

THE INTERPLAY OF CYTOLOGY AND BREEDING ANALYSIS between 1890 and 1910 is discontinuous. At first, the two fields were independent of each other, with advances in mitosis and meiosis moving at a fast clip in Continental universities. Breeding analysis had an ongoing and inconsistent record in the nineteenth century, with multiple intentions: some breeding for practical reasons (commercially useful new varieties), some to look for the relationship of hybrids to true species, and some involving

Mendel's intent to "atomize" heredity into fundamental hereditary-transmissible units. However, Mendel's intention was put aside because it was not universal, as Nägeli (and then Mendel himself) pointed out. By the start of the twentieth century, the two tributaries were beginning to merge. The rediscovery of Mendel's laws was a key component. Equally important, as we saw in the discussion of sex chromosomes, was the recognition of chromosomes as homologs and chromosomes having at least an indirect connection to heredity through sex determination. In the years 1900–1905, the sex chromosomes, the rediscovery of Mendelism, and the chromosome theory of heredity became subjects of intellectual sparring matches for theorists and experimentalists who presided over the

---

*To Roberts H.F., January 7, 1925, on his results of 1899. In Roberts H.F. 1929. *Plant Hybridization before Mendel*, p. 343. Princeton University Press, New Jersey.
†1902. In *A Defense of Mendel's Principles of Heredity*, p. 104. Cambridge University Press, United Kingdom.

birth of what would be named genetics by Bateson in correspondence (1905) and publication (1906). As is true in all histories, there is an irony in Bateson's superb contribution. He gave the name in Great Britain but excluded the chromosomal component to what would quickly intertwine with the new Mendelism in American universities. To understand the chromosome theory of heredity, we now interrupt the cytological story to see the reemergence of Mendel's work in 1900 and its influence on cytology.

We speak of the rediscovery of Mendel's laws in 1900 and assign that event to three European botanists: Hugo de Vries, Carl Correns, and Erich Tschermak von Seysenegg.[1] From the time that their three papers emerged, however, the participants and historians have never agreed on whether the event was an independent discovery or a confirmation for one, two, or all three of the participants. Despite that nuanced search for motivation and alleged warping of ethics, one fact is not in dispute. After some 35 years, Mendel's laws had been dusted off from near oblivion, and the confirmations created enormous excitement for both European and North American scientists. Breeding analysis quickly shifted from an art to a science.

de Vries' Rediscovery Utilizes Numerous Plant Species to Support His Theory of Intracellular Pangenesis

Of the three, there is no doubt about the stature and motivation of Hugo de Vries (1848–1935) (Figure 1).[2] He was born in Holland to a family of prominent scholars and statesmen. His father was Minister of Justice to William III, and his uncle was a prominent philologist. After his education in Haarlem, he attended school at the Hague and the University of Leiden, where he studied plant physiology. He studied also in Germany at Heidelberg and Würzberg, returning to teach in Amsterdam in 1871. His work on osmosis in plant cells (exploring the related problems of turgor, osmosis, and plasmolysis in cells) led to new insights in physical chemistry by van't Hoff and others. By 1890, he switched fields and dropped physiology for the relation of heredity to evolution. He quickly recognized Darwin's flawed model of pangenesis with its Lamarckian dependence. Instead, he proposed a model of "intracellular pangenesis," which united both Weismann's theory of the germ plasm and the model of hereditary units proposed by Spencer, Darwin, and—unknown to him at the time—Mendel.

After he published his *Intracellular Pangenesis*, de Vries began breeding a variety of plants.

Somewhere between 1890 and 1900, he read Mendel's papers, and this is where the controversy continues. The earliest possible reference in the lit-

Figure 1. *Hugo de Vries modified Darwin's theory of pangenesis by confining the hereditary units to the nuclei and adopting Weismann's theory of the germ plasm. He used his model of intracellular pangenesis to design the experiments in plant breeding that led to his independent confirmation of Mendel's laws. (Courtesy of Cold Spring Harbor Laboratory Archives.)*

erature to that reading would be 1896, the year that he showed his students the results of his crosses with poppies (*Papaver somniferum*). de Vries always maintained that he found Mendel's results before he read the paper, but it is not clear from different accounts when that reading first took place. In the most-cited account, he claims that he read the paper in 1900 and felt obliged to publish his results.[3]

de Vries' first paper appeared in the highly regarded French journal *Comptes Rendus de l'Academie des Sciences*[4] (Figure 2). It is a short paper and was read by the secretary, because de Vries could not attend that session. The title, "Sur la loi de disjonction des hybrides" (On the law of hybrid separation), reflects the intense interest in hybridization and evolution, a theme explored by Darwin. It also reflects the practical interest in hybridization as a source of varieties and breeds and similar problems of horticultural and agricultural interest. de Vries provided a table with 11 different plant species, all of which gave a 3:1 ratio in the F_2 generation from parental crosses of species or varieties. He recognized that the traits crossed yielded one dominant and one recessive in attribute ("l'un dominant, l'autre latent"). He also used the French phrase, "caractère récessif."[5] He took the F_2 dominant class and showed that it produced a ratio of 2 heterozygotes to 1 homozygote for the dominant trait. In that paper, the terms homozygous and heterozygous do not appear; they were supplied later by Bateson. Instead, he used "hybride." He used a symbolism different from that used in Mendel's original paper. He generalized the traits (and inferred units) as D and R (for dominant and recessive), and gave the generalized algebraic polynomial formula $(D + R)(D + R) = D^2 + 2DR + R^2$, without any doubleness for what we would designate as the DD or RR classes. He correctly inferred that the polynomial expansion gives a 1 dominant pure:2 dominant hybrid:1 recessive pure ratio in the F_2 generation. He did not discuss the use of two or more characters and what Mendel described in his second law (and to which we still refer) as independent assortment. But he did point out that he confirmed his theory of intracellular pangenesis: "L'ensemble de ces expériences met donc en évidence la loi de disjonction des hybrides et vieut confirmer les principes que j'ai énoncés sur les caractères specifiques considerés comme des unités distinctes" (All of these experiments support the law of disjunction of hybrids and serve to confirm the principles that I proposed for specific characters being associated with distinct units.)[6]

A larger paper appeared in a German journal, *Berichte der Deutschen Botanischen Gesellschaft*,

---

● **BOTANIQUE**

Sur la loi de disjonction des hybrides.

Note de M. HUGO DE VRIES, présentée par M. Gaston Bonnier.

———

« D'après les principes que j'ai énoncés ailleurs (*Intracellulare Pangenesis*, 1889), les caractères spécifiques des organismes sont composés d'unités bien distinctes. On peut étudier expérimentalement ces unités soit dans des phénomènes de variabilité et de mutabilité, soit par la production des hybrides. Dans le dernier cas, on choisit de préférence les hybrides dont les parents ne se distinguent entre eux que par un seul caractère (les monohybrides), ou par un petit nombre de caractères bien délimités, et pour lesquels on ne considère qu'une ou deux de ces unités en laissant les autres de côté.

Figure 2. de Vries' first paper, the first confirmation of Mendelism, was an abstract (in French) of a larger paper (in German) that describes his results with many species of plants. This abstract does not mention Mendel's priority. (Reprinted from de Vries H. 1900. Sur la loi de disjonction des hybrides. Comptes Rendus de l'Academie des Sciences **130**: 845–847.)

● II. Hugo de Vries:

DAS SPALTUNGSGESETZ DER BASTARDE.

Vorläufige Mittheilung.¹)

Eingegangen am 14. März. 1900.

———

Nach der Pangenesis ist der ganze Charakter einer Pflanze aus bestimmten Einheiten aufgebaut. Diese sogenannten Elemente der Art oder Elementar-charaktere denkt man sich an materielle Träger gebunden. Jedem Einzelcha-rakter entspricht eine besondere Form stofflicher Träger.²) Uebergänge zwischen diesen Elementen giebt es ebenso wenig wie zwischen den Molecülen der Chemie.

Figure 3. *The second paper by de Vries on plant breeding acknowledged Mendel's priority. It was convincing in launching Mendelism because of the numerous species de Vries used to demonstrate its near universality. (Reprinted from de Vries H. 1900. Das spaltungsgesetz der bastarde.* Berichte der Deutschen Botanischen Gesellschaft 18: 83–90.)

dated March 14, 1900.[7] A similar title was used: "Das spaltungsgesetz der bastarde" (The disjunctional law of hybrids) (Figure 3). Apparently, the March 14th date is also the one he attributed to the sending of the *Comptes Rendus* paper. He intended the *Comptes Rendus* paper to be an abstract. In this more extensive paper, he used 19 different genera, which yielded 3:1 ratios, and gave the dates of the crosses (1892–1899). There is no doubt that de Vries had done extensive work before publishing and had confirmed the segregation of hybrids among many species of plants. In his letter to H.F. Roberts, the first historian of genetics, dated December 18, 1924, he claimed that his first success with the finding occurred in a cross of *Oenothera lamarckiana* x *Oenothera brevis* in 1893.[8] After he had carried out the crosses in some 20 different species, he began thinking of publishing and when he got to reading references for his paper, he looked at L.H. Bailey's *Plant Breeding* (1895) and there found the reference to Mendel "...and accordingly looked it up and studied it. Thereupon I published in March 1900 the results of my own investigations."[9] This is a bit ambiguous to interpret because it is not clear if he had all the theoretical implications (one doubts nei-

ther the data nor the ratios he obtained). Although Sturtevant took a somewhat conspiratorial interpretation (but hedged his accusation) and explored a possible shady motivation (de Vries wanted priority and thought he could get away with it in his *Comptes Rendus* paper, which does not mention Mendel), I am more inclined to accept that de Vries probably had his results and saw them as a confirmation of his pangene theory with his own *Intracellular Pangenesis*. We do know that in his 1899 publication, "Sur la fécondation hybride de l'albumen," also in *Comptes Rendus*, de Vries did state Mendel's ratios for his maize crosses using a sweet and waxy cross: "Environ un quart des graines étaient sucrées, les trois autres quarts étaient amylacées. Les premières étaient revenues au caractère de la grand-mère, les dernières montraient celui du père et du grand-père." (Approximately one fourth of the kernels were sugar corn, the remaining three-fourths were waxy. The former had retrieved the grandmother's trait, whereas the latter showed the father's and grandfather's traits.)[10]

In the *Berichte* paper, de Vries claimed that the Mendel papers were seldom referred to and "...I myself for the first time came to know about it after I had closed the majority of the experiments and had derived therefrom the principles contributed in the text." (Diese wichtige Abhandlung wird so selten citirt, dass ich sie selbst erst kennen lernte, nachdem ich die Mehrzahl meiner Versuche abgeschlossen und die Text mitgetheilten Satze daraus abgeleitet hatte.)[11] In my reconstruction, he had found the laws in the early 1890s, probably read the Mendel paper about 1896 (shortly after Bailey's reference work came out), realized that there was no rush to publish what was already in the literature, built a solid case for the confirmation of his theory of intracellular pangenesis, and published the *Comptes Rendus* paper, a short abstract, without reference to Mendel's paper, because he was more interested in promoting his own (and to him, more important) theory of heredity in *Intracellular Pangenesis*. Sturtevant may be right that de Vries had added the reference to Mendelism in his page proofs (or at least made it appear in the text more prominently) in the *Berichte* paper, once he heard about (or received a preprint of) Correns' paper, which came out as a confirmation of Mendel. Instead of attributing malicious self-serving to de Vries, I prefer to see his behavior of interpreting Mendel's paper as a footnote to his own more extensive work. Mendel's was a forgotten work and one could attribute to Mendel's second paper a retraction of his first work as universally applied to all hybrid crosses. The second paper was a lengthy description of *Hieracium* (hawkweed) crosses. Many of them, including the seeds to carry them out, had been provided by Nägeli in their correspondence and many others were hawkweeds that Mendel collected from his local surroundings. Because *Hieracium* did not "Mendelize" (their mode of reproduction, unbeknownst to Mendel or Nägeli, was largely parthenogenetic), this maternal mode of inheritance discouraged Mendel from pursuing plant hybrid studies. Even if de Vries' motivation is suspect, from our more privileged vantage point, one cannot deny the power of de Vries' model of *Intracellular Pangenesis* nor the power of his 20 or more confirmations or extensions of Mendel's first law. de Vries blew away the initial criticism of Mendel by Carl Nägeli that Mendel's work was idiosyncratic and that *Hieracium* was more likely to be the norm, a suggestion that sadly plagued Mendel when his next organism, bees (*Apis*), turned out to be equally non-Mendelian. In his lengthy analysis, de Vries does briefly discuss the 9AB:3Ab:3aB:1ab independent assortment cases that Mendel had found for peas and he uses *Datura* and *Trifolium* to confirm them. He also reports what Bateson later called epistasis, modified 9:3:3:1 ratios for flower color in *Antirrhinum, Silene,* and *Brunella*. He also mentions the test cross results for the monohybrid, with its 1 dominant:1 recessive outcome, noting the former to all be hybrids and the latter to all be pure ("rein").

Correns' Rediscovery Is Motivated by His Interest in Endosperm Formation

Carl Correns (1864–1933) was born and raised in Munich, where he had his undergraduate education (Figure 4). He liked botany and studied with Nägeli and other plant physiologists, taking an interest in the formation of plant cell walls. He shifted to studying the life cycle of mosses, and at Tübingen (where he became a professor), he shifted again to the study of hybrids. It was there that he discovered Mendel's

Figure 4. *Carl Correns confirmed Mendelism while studying hereditary traits in the endosperm. His later career involved the difficult problems of sexuality in plants and the apparent non-Mendelian phenomenon of cytoplasmic (maternal) inheritance as seen in variegated leaves. (Courtesy of the American Philosophical Society, Curt Stern Papers.)*

laws. Although he was a student of Nägeli's, he did not know of Mendel's work on hybrids in peas. He knew from Nägeli that Mendel had corresponded with him about the *Hieracium* crosses. Correns wrote (or changed) his title after the de Vries *Comptes Rendus* paper appeared. It was submitted April 24, 1901 also to the *Berichte der Deutschen*

Botanischen Gesellschaft. Correns called his work "G. Mendel's regel über das verhalten der nachkommenschaft der rassenbastarde" (G. Mendel's law on the distribution of traits of hybrids) (Figure 5).[12] He claimed that he was, like de Vries, under the illusion that he had discovered something new when he had completed his series of crosses on peas and maize. He did not use Bailey for his reference; he had found it in Focke's monograph on plant hybridization (*Pflanzenmischlingen*).[13] Most of Correns' paper is written as a review of Mendel's findings with peas (*Pisum*) and beans (*Phaseolus*). He identifies de Vries' "law of disjunction of hybrids" as being identical to Mendel's law of segregation. He does not say when he had worked on the problem, but in all likelihood, it was done between 1895 and 1899. He had studied a related problem, xenia, which involved the endosperm traits and which sometimes showed the presence of paternal (pollen) traits in what was originally believed to be maternal tissue.[14] At that time, the double fertilization in flowering plants was either not known or not well worked out. It is now well established that endosperm tissue in seed plants is polyploid, the most familiar form (triploid) arising from two embryo sac nuclei merging with one pollen tube nucleus. Correns made a brief reference to Mendel and Darwin in his 1899 paper on xenia. He told Roberts that although he had seen the reference

---

● 19. C. CORRENS:

G. MENDEL'S REGEL ÜBER DAS VERHALTEN DER NACHKOMMENSCHAFT DER RASSENBASTARDE.

Eingegangen am 24. April 1900.

———

Die neueste Veröffentlichung Hugo de Vries': „Sur la loi de disjonction des hybrides",[1] in deren Besitz ich gestern durch die Liebenswürdigkeit des Verfassers gelangt bin, veranlasst mich zu der folgenden Mittheilung.

Figure 5. *Correns made sure that Mendel's name was included in his title to pay homage to Mendel's priority. The paper is analytic in its treatment of Mendelism, but Correns had not extended it to numerous organisms as had de Vries. (Reprinted from Correns C. 1900. G. Mendel's regel über das verhalten der nachkommenschaft der rassenbastarde. Berichte der Deutschen Botanischen Gesellschaft 18: 158–168.)*

to Mendel in 1899, he had not read the paper until his own results on hybrids in peas and maize were essentially completed. Correns mentioned the dihybrid and trihybrid results of Mendel but limited his

confirmation of the 9:3:3:1 ratio to crosses in *Zea* (maize). He also reported blending inheritance (e.g., pink) for the F_1 from crosses of white and red flowers.[15]

Tschermak von Seysenegg's Rediscovery Is Motivated by Darwin's Work on Peas

Erich Tschermak von Seysenegg was the last of the three to submit a rediscovery paper. His paper also appeared in 1900. Tschermak von Seysenegg (1871–1962) (Figure 6) was an Austrian botanist working in an agricultural institute in Ghent (Belgium), where he had begun his hybrid studies. In 1899, as he wrote his paper, he looked through Focke's reference and came across the mention of Mendel's paper on peas. Like de Vries and Correns, he chose *Berichte der Deutschen Botanischen Gesellschaft* for publication. His article is dated June 2, 1900, and he used the title "Ueber künstliche kreuzung bei Pisum sativum" (On experimental crosses of *Pisum sativum*) (Figure 7).[16] He claimed that he began his studies in 1898 as a follow-up to reading Darwin's attempts to study pea hybrids. He too obtained 3:1 ratios using his pea crosses and, like de Vries and Correns, in almost identical words, "thought I had found something new" (Auch ich dachte noch im zweiten versuchsjahre etwas ganz neues gefunden zu haben).[17] In his first paper, he did not study the 9:3:3:1 ratio in his crosses. He did not realize that others had scooped him until de Vries sent him a copy of the *Comptes Rendus* abstract.

Curt Stern demoted Tschermak von Seysenegg from the trio of rediscoverers. He felt Tschermak von Seysenegg had obtained his results, but not the interpretation, up to the time that he read Mendel's paper.[18] In a sense, one can say that of all three. I doubt both Sturtevant's and Stern's harsh judgments. All three of

the rediscoverers claimed that they had been working for at least two years on plant hybrid crosses prior to reading Mendel and that all of them had ratios of 3:1. Tschermak von Seysenegg did not go beyond the F_2,

Figure 6. *Using peas, Erich Tschermak von Seysenegg confirmed Mendelism after reading Darwin's analysis of traits in peas. Although his paper was not as elaborate in scope as de Vries' nor as insightful as Correns', it led Tschermak von Seysenegg to a lifelong career in applied Mendelism. He studied cereal grains, pumpkins, and legumes. He also pioneered methods for the systematic introduction of genes into commercial varieties. (Reprinted from Wilks W. 1906. Report of the Third International Conference 1906 on Genetics. Royal Horticultural Society/Spottiswoode and Company, London.)*

● 26. E. Tschermak:

UEBER KÜNSTLICHE KREUZUNG BEI PISUM SATIVUM[1]).

Eingegangen am 2. Juni 1900.

———

Angeregt durch die Versuche Darwin's über die Wirkungen der Kreuz- und Selbstbefruchtung im Pflanzenreiche, begann ich im Jahre 1898 an *Pisum sativum* Kreuzungsversuche anzustellen, weil mich besonders die Ausnahmefälle von dem allgemein ausgesprochenen Satze über den Nutzeffect der Kreuzung verschiedener Individuen und verschiedener Varietäten gegenüber der Selbstbefruchtung interessirten, eine Gruppe, in welche auch *Pisum sativum* gehört. Während bei den meisten Species, mit welchen Darwin operirte (57 gegen 26 bezw. 12), die Sämlinge aus einer Kreuzung zwischen Individuen derselben Species beinahe immer die durch Selbstbefruchtung erzeugten Concurrenten an Höhe, Gewicht, Wuchs, häufig auch an Fruchtbarkeit übertrafen, verhielt sich bei der Erbse die Höhe der aus der Kreuzung stammenden Pflanzen zu jener der Erzeugnisse von Selbstbefruchtung wie 100 : 115. Darwin erblickte den Grund dieses Verhaltens in der durch viele Generationen sich wiederholenden Selbstbefruchtung der Erbse in den nördlichen Ländern. In Anbetracht des geringen Beobachtungsmateriales bei Darwin (es wurden nur vier Erbsenpaare gemessen und verglichen) erschien es mir, zumal Darwin die Blüthen nie castrirte, angezeigt, diese Versuche in grösserem Massstabe und mit grösserer Genauigkeit zu wiederholen.

Figure 7. *Erich Tschermak von Seysenegg's rediscovery paper. Curt Stern argued that his paper was the least informative and the most likely to have been written after he had seen Mendel's paper. I believe that this might be too harsh a judgment. (Reprinted from Tschermak E. 1900. Ueber künstliche kreuzung bei* Pisum sativum. *Berichte der Deutschen Botanischen Gesellschaft* 18: 232–299.)

---

whereas Correns went to the F_6 to show the purity of extracted recessives and the constant segregation of recessives from the hybrid dominant plants. Tschermak von Seysenegg reported a 9:3:3:1 ratio for *Pisum* in a second paper published in 1900, but he does not allude to that work in his first paper. Neither Correns nor Tschermak von Seysenegg had worked on the problem as long and with as much detailed experimentation as had de Vries. All three certainly confirmed Mendel's first law if they had prior knowledge of his interpretation—and they deserve that minimal credit. The confirmation from three independent sources was the convincing reason that led so many plant breeders to turn to Mendelism with approval. What was important was not how they came to write their papers and conclusions, but that they did so; they

all paid homage to Mendel for his initial discovery, and whatever their reasons for doing so, before or after reading Mendel's paper, they brought his work back from the dead, thereby celebrating his name and his laws and launching a field of breeding analysis.

The rediscovery year of 1900 may be considered a capstone for the nineteenth century or the initiation of the twentieth century. For the new field of genetics that would soon emerge from this rediscovery, it was both. The nineteenth century had solved the problem of hybridization. The new twentieth century saw an immediate explosion of work in hybridization, with several objectives among those taking up this new approach. There was an interest in extending Mendelism to animals and other plants. There was intense interest in using Mendelism to produce commercially valuable traits. A third major interest was championed, particularly by Gregory Bateson in Great Britain. He sought in Mendelism a new way to study evolution. These were the expectations of those who did the breeding. To their surprise, a very differ-

ent connection was about to made in 1902 and 1903 by cytologists who did no breeding at all. They were all connected to the theoretical work and influence of cytologist E.B. Wilson. They would usher in what they called the chromosome theory of heredity.

End Notes and References

[1] The major rediscovery papers, in their original languages and pagination, are available in Krizenecky J. and Nemec B., eds., 1965. *Fundamenta Genetica*. Publishing House of the Czechoslovak Academy of Sciences, Prague.

The editors provide 28 publications in the peri-rediscovery period, as well as Mendel's first paper.

[2] van der Pas P.W. 1970. Hugo de Vries. In *Dictionary of Scientific Biography* (ed. C. Gillispie), volume 14, pp. 95–105. Scribner's, New York. See also the obituary by Cleland R. 1935. Hugo de Vries, 1848–1935. *Journal of Heredity* **26:** 289–297; and Allen G. 1969. Hugo de Vries and the reception of the mutation theory. *Journal of History of Biology* **2:** 55–87.

[3] For accounts of the rediscovery and interpretations of the ethics or timing of the rediscoveries in relation to a first reading of Mendel's paper, see Roberts H.F. 1929. *Plant Hybridization Before Mendel* (1965 reprint, Hafner, New York). Princeton University Press, New Jersey.

Roberts wrote to all three of the rediscoverers for their comments on when they first encountered the Mendel reference and read Mendel's papers. See Sturtevant A.H. 1965. *A History of Genetics*. Harper and Row, New York. Chapter 5, The Rediscovery, pp. 25–32. Also see Stubbe H. 1975. The rediscovery of Mendel's laws of heredity. In *History of Genetics from Prehistoric Times to the Rediscovery of Mendel's Law*, Chapter 9, pp. 261–290. The Massachusetts Institute of Technology Press, Cambridge.

[4] de Vries H. 1900. Sur la loi de disjonction des hybrides. *Comptes Rendus de l'Academie des Sciences, Paris* **130:** 845–847.

[5] de Vries H. 1900. op. cit. p. 845.

[6] de Vries H. 1900. op. cit. p. 847.

[7] de Vries H. 1900. Das spaltungsgesetz der bastarde. *Berichte der Deutschen Botanischen Gesellschaft* **18:** 83–90.

In this paper, de Vries uses the following genera: *Papaver* (poppy), *Antirrhinum* (snapdragon), *Polemonium, Chelidonium, Oenothera* (evening primrose), *Lychnia, Datura* (jimsonweed), *Zea* (maize), *Lychnis, Veronica, Agrostemma, Hyoscyamus, Aster, Chrysanthemum, Coreopsis, Solanum* (potato), *Viola, Clarckia,* and *Silene.*

[8] Roberts H.F. 1929. op. cit. p. 323.

[9] Roberts H.F. 1929. op. cit. p. 323.

[10] Bailey L.H. 1895. *Plant Breeding.* Macmillan, New York. Bailey got the Mendel reference from Vergl G. and Focke A. in Focke W.O. 1881. *Die Pflanzenmischlinge, Ein Beitrag zur Biologie der Gewächs,* p. 110. Borntraeger, Berlin. de Vries' prediscovery paper is in Krizenecky J. and Nemec B. 1965. op. cit. pp. 194–195. de Vries H. 1899. Sur la fécondation hybride de l'albumen. *Comptes Rendus de l'Academie des Sciences, Paris,* p. 975. This paper, like the 1900 *Comptes Rendus* article, was read by the secretary, Gaston Bonnier.

[11] de Vries H. 1900. *Berichte der Deutschen Botanischen Gesellschaft,* op. cit. p. 85.

[12] Correns C. 1900. G. Mendel's regel über das verhalter der nachskommenschaft der rassenbastarde. *Berichte der Deutschen Botanischen Gesellschaft* **18:** 158–168.

[13] Focke W.O. 1881. *Die Pfanzen-mischlinge ein Beitrag zur Biologie der Bewächse.* Gebrüder Borntraefer, Berlin.

[14] Both de Vries and Correns had independently worked on this xenia problem. The name xenia was assigned by Focke. The solution to this was worked out genetically by both these rediscovers and it was worked out cyto-

logically (what is known among students of first year biology courses as "the double fertilization" in embryo sacs of flowering plants) by Navashin S.G. 1899. Neue beobachtungen uber befruchtung bei *Fritellaria tenella* und *Lillium martagon*. *Botanische Zentralblatt* **77**: 62; and Guignard L. 1899. Sur les anthérozoides et la double copulation sexuelles chez les vegetaux angiosperms. *Revue Generale Botanique* **11**: 129–135.

[15] Roberts H.F. 1929. op. cit. p. 335.

In his January 23, 1925 letter to Roberts, Correns said that he was working on xenia and did crosses from 1894 on and off without success until August 1899 when the interpretation of his ratios came "in a flash" while in bed one morning. He didn't remember exactly when he came across the Mendel paper after that insight, but it was "a few weeks later." Ironically, Correns knew of Mendel as a hawkweed breeder from his teacher, Nägeli, with whom Mendel corresponded, but Correns had not bothered to read Mendel's paper on *Hieracium* until September 1899.

[16] Tschermak von Seysenegg E. 1900. Ueber künstliche kreuzung bei *Pisum Sativum*. *Berichte der Deutschen Botanischen Gesellschaft* **18**: 232–239.

[17] Tschermak von Seysenegg E. 1900. op. cit. p. 239.

[18] Stern C. and Sherwood E. 1966. *The Origin of Genetics: A Mendel Source Book*. W.H. Freeman, San Francisco.

The Chromosome Theory of Heredity

Thus, only one possibility remains, namely that not a definite number, but a definite combination of chromosomes is essential for normal development, and this means nothing else than that the individual chromosomes must possess different qualities. At the moment, we are unable to give a more definite setting to this irrefutable conclusion.

THEODOR BOVERI*

ALTHOUGH HENKING'S "X ELEMENT" or McClung's "accessory chromosome" pointed the way to a chromosome theory of heredity, neither they nor Wilson in 1901 were ready to make a commitment relating the chromosomes to sex determination. They were even less inclined to relate the newly found Mendel's laws to the chromosomes. That association would soon be made, and it would bring together the first two tributaries—cytology and breeding analysis—in the yet-to-be-named field of genetics. Four important contributions put the story together, and the players were Theodor Boveri, William Austin Cannon, Walter Sutton, and Edmund Beecher Wilson. Both Cannon and Sutton were Wilson's students and Wilson was a friend to Boveri and a great admirer of his work (Wilson's classic, *The Cell in Development and Inheritance*, appearing in 1896, was dedicated to "my friend, Theodor Boveri").

Boveri's Work Contributes to a Hereditary Role for the Chromosomes

Boveri took an interest in cell biology and began a series of papers in 1886 that explored the mitotic apparatus, nucleo-cytoplasmic relations, and the relationship of chromosomes to development and heredity (Figure 1). He was attracted to Entwicklungsmechanik (developmental mechanics), the experimental approach to biology that opened up a flood of new knowledge in European universities and attracted the notice of American scholars emerging from what some of its envious critics or beneficiaries called the Johns Hopkins "Ph.D. factory." In 1889, he broke eggs of sea urchins into fragments by agita-

*1902. *Über mehrpolige mitosen als mittel zur analyse des zellkerns.* Verhandlungen der Physikalisch-medizinischen Gesellschaft zu Würzburg Neu Folge **XXXV**: 67–90.

Figure 1. *Theodor Boveri's experiments provided indirect evidence for the role of the nucleus and the chromosomes in heredity. His friendship with Wilson shifted Wilson from embryology to cytology and cytogenetics. (Reprinted from Stubbe H. 1963. Kurze geschichte der genetic bis zur wiederentdeckung der vererbungsregeln Gregor Mendels. G. Fischer, Jena.)*

tion and fertilized both the nucleated and non-nucleated components. The enucleated cytoplasmic remnant, to his surprise, had a normal development, producing what is known as a pluteus larva. The only difference he could detect from a normal fertilization resulting from the union of male and female pronuclei and the other necessarily androgenous fertilization was one of size. The pluteus larva from the fertilization of the non-nucleated egg fragment was a dwarf in size. This suggested to Boveri that "...it is not a given number of chromosomes as such that is required for normal development, in as much as these fragments, although they contained only half the normal amount of chromatin and half the number of elements, namely the chromosomes of one sperm nucleus, still give rise to normal plutei."[1]

In 1895, Boveri also ruled out a major role of the cytoplasm in determining hereditary characteristics of the offspring. He used the non-nucleated cytoplasm of *Sphaerechinus granularis* and fertilized it

with the sperm of *Echinus microtuberculatus*. As he expected, he got dwarf larvae, but they showed the physical characteristics of the *Echinus* father and none of the physical attributes of the *Sphaerechinus* mother. In Wilson's monograph of 1896, this work stood out, and Wilson leaned heavily toward the idea that "...the maternal cytoplasm has no determining effect on the offspring, but supplies only the material in which the sperm nucleus operates. Inheritance is, therefore, affected by the nucleus alone."[2] Although the relationship of the nucleus and the chromosomes in some general way might be associated with heredity through these experiments, there was still no evidence that the individual chromosomes differed from one another (Figure 2). Some sort of quantitative model could be invoked to explain these results without making a more direct role for the chromosomes as the bearers of hereditary determinants. The chromosomes could just be tagging along, with the nuclear sap in some way having the actual hereditary role.

In 1902, Boveri came out with a brilliant experiment in which he used multiple fertilizations of eggs to produce what we would today call triploid zygotes, formed by two sperm entering one egg. These produced an unusual spindle formation, with three or sometimes four asters or centers of origin of the spindle fibers instead of two. Sometimes when the cell with the four asters underwent cleavage, it produced four blastomeres; these aborted when separated. In contrast, a normal diploid zygote, after undergoing cleavage divisions to produce four blastomeres, produced four normal pluteus larvae when the blastomeres were separated. "...The next question was whether this unequal distribution of the chromatin is of any influence upon the properties of the four cells...."[3] He pointed out that this was indeed the case: "...While the four blastomeres of a normally divided egg are absolutely equivalent to each other, it is seen that the properties of the blastomeres of a doubly fertilized one are different from each other in diverse ways, and to varying extent."[4] His conclusions now pointed toward the emerging chromosome theory that "...All that remains is that not a definite number but a *definite combination of chromosomes* is necessary for normal development, and this means nothing other than that the *individual chromosomes* must possess different qualities."[5]

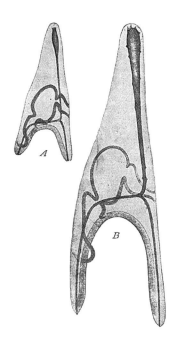

Figure 2. *Abnormal larvae implied a hereditary role for the chromosomes. Boveri's dwarf larvae were produced by fertilization of enucleated eggs. A normal pluteus larva is shown on the right. The smaller larva (left) is androgynous and expresses the characteristics of its nuclear father (from an* Echinus microtuberculatus *sperm) but not of its cytoplasmic mother (from a* Sphaerechinus granularis *enucleated egg). (Reprinted from Wilson E.B. 1896. The Cell in Development and Inheritance. Macmillan, New York.)*

Bateson Avoids a Chromosomal Association with Heredity

In this paper, Boveri did not associate Mendel's laws with the chromosomes, although in his mind there was no doubt that each chromosome carried different determiners of heredity. His was a cytological basis for the chromosome theory of heredity, and that basis rested on the individual differentiation of the chromosomes. William Bateson had the equivalent of a postdoctoral stay at the Chesapeake Bay Marine Station associated with Brooks and the Hopkins school, yet he felt less enthusiastic about chromosomes than he did about developmental patterning, homeotic mutants,

and the metamerism (repeated segments) present in the embryos during their period of organogenesis. It was a different developmental interpretation from what had attracted Wilson to Boveri's work. In 1902, Bateson was immersed in Mendelism and saw it as a future tie to evolution and developmental biology. In his book, *Mendel's Principles of Heredity—A Defence*, Bateson recognized that Mendel's quantitative findings had to have some broader biological significance. "It is impossible to be presented with the fact that in Mendelian cases the cross-bred produces on average equal numbers of gametes of each kind, that is to say, a symmetrical result, without suspecting that this fact must correspond with some symmetrical figure in the distribution of the gametes in the cell divisions by which they are produced."[6] For the next 20 years, Bateson pursued a will-o'-the-wisp model of differential cell divisions to produce the puzzling new non-Mendelian ratios that he encountered.

Cannon Suggests a Flawed Chromosome Theory of Heredity

It was in the United States that the association of chromosomes and Mendel's laws came together. Wilson kept edging closer to it as he read the debates between McClung and Montgomery on the accessory chromosome. Although he spurned the idea that it was a sex determinant as McClung had tentatively put forward, Montgomery provided clear evidence for the individuality of the chromosomes and the observation of their pairing as homologs (Figure 3). Cannon was the first to publish, bringing together the work of McClung, Montgomery, and Mendel (Figure 4). His title was as direct as could be: "A cytological basis for the Mendelian laws." Cannon used cotton (*Gossypium*) for this paper. A year later, he would extend his theory with chromosome studies of peas. He grew his plants at the New York Botanical Garden and discussed his work with his mentor, Wilson. "The

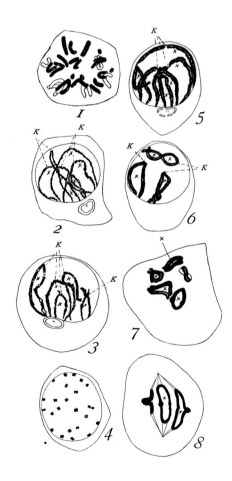

Figure 3. *Although he was doubtful about calling Henking's X element a chromosome, Montgomery recognized that chromosomes paired during meiosis and he was the first to shift the reduction division to the first meiotic division. (Reprinted from Montgomery T.H. 1903. The heterotypic maturation mitosis in amphibia and its general significance. Biological Bulletin* **4**: *259–269.)*

Figure 4. *William Austin Cannon was the first to recognize that Mendel's principle of segregation was explainable through meiosis. He did not interpret independent assortment because he erred, as did his contemporaries, in believing that the paternal and maternal chromosomes separated in the same association as they entered at fertilization. (Courtesy of the Stanford University Archives.)*

maternal chromosomes, and no cells receive chromosomes from both the father and the mother."[7] Cannon's prose is unwieldy and plods along in fits and starts throughout his paper. He limits himself to the hybrid **Aa** and the parental types **AA** and **aa** for his characters. His chromosome theory addresses Mendel's law of segregation for a hybrid and the familiar 3:1 ratio. He does not address the 9:3:3:1 ratio associated with two factors, each assorting independently. He could not because he had made a serious error and accepted a belief shared by many cytologists that the paternal chromosomes aligned in one plane and the maternal chromosomes aligned in the other plane when a metaphase alignment took place in the first meiotic division. The two sets would then drift apart and retain their paternal or maternal integrity. In 1903, he corrected that error and accepted Sutton's model for a chromosome theory of heredity. He also had an important insight that would resonate with T.H. Morgan in the coming decade: "If then the chromosomes maintain their individuality certain characters might be associated together in the same chromosome, and might be coupled together in the hybrid organism, not being capable of separating from each other... ."[8]

chromosomes derived from the father and the mother unite in synapsis and separate in the metaphase of one of the maturation divisions, and also a single longitudinal, division occurs, so that the end is attained that the chromatin is distributed in such a way that two of the cells receive the paternal and two cells pure

Sutton Focuses on Mendel's
Second Law and Its Tie to Meiosis

In 1902, Walter Sutton (1877–1916) worked with the lubber grasshopper, *Brachystola magna*. Its chromosomes were large, clearly identifiable, and not too numerous (11 in the haploid gametes). He confirmed that chromosomes paired through synapsis as homologs and that each new generation after fertilization retained the chromosome number of the species as well as the physical sizes and shapes that allowed him to identify each chromosome pair. He inferred Mendelism before he read Mendel's rediscovery papers: "The general conceptions here advanced were evolved purely from cytological data, before the author had knowledge of the Mendelian principles, and now are presented as the contribution of a cytologist who can make no pretensions to complete familiarity with the results of experimental studies on heredity."[9] After acknowledging McClung's and Montgomery's work in clarifying the meiotic process and the reality of the individuality of the chromosomes and their existence as a paternal and maternal set in the germ cells, he challenged Cannon's error of maintaining the paternal and maternal sets intact: "On the contrary, many points were discovered which strongly indicate that the position of the bivalent chromosomes in the equatorial plate of the reducing division is purely a matter of chance—that is, that any chromosome pair may lie with maternal or paternal chromatid indifferently toward either pole irrespective of the positions of other pairs—and hence that a large number of different combinations of maternal and paternal chromosomes are possible in the mature germ-products of an individual" (Figure 5).[10]

Sutton (Figure 6) was inspired by a paper of W.J. Moenkhaus, who studied fish cytology. Moenkhaus noted that in some of his hybrids, the paternal and maternal members in a pair of homologs looked different. Although acknowledging that in *Brachystola* "Absolute proof is impossible in a pure bred form...," he urged a closer inspection of Moenkhaus' fish hybrids: "...if the same can be done in the maturation divisions the question of distribution of chromosomes in reduction division becomes a very simple matter of observation."[11] Sutton shows how staggering the combinations might be from $F_1 \times F_1$ hybrids, if each independently assorting unit is carried by a different chromosome pair. A hybrid diploid organism with a chromosome number of 10 would have 1,024 different zygotic combinations; for 20 chromosomes, the number jumps to 1,048,576, and for 30 chromosomes, it approaches an astronomical size: 68,719,476,736. "It is this possibility of so great a number of combinations of maternal and paternal chromosomes in the

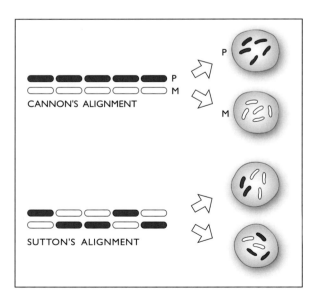

Figure 5. *Sutton versus Cannon. The Cannon model for obtaining Mendel's segregation ratio was based on his recognition that chromosomes pair as homologs and undergo a reduction division. Cannon incorrectly assumed that the paternal set of chromosomes is discarded in the formation of ovules. Cannon's model does not explain independent assortment. Sutton's model, which came out shortly after Cannon's paper appeared, did explain both of Mendel's findings. Sutton assumed an independent alignment of the maternal and paternal pairs of chromosomes, and he calculated that most species had a dozen or more pairs of chromosomes and, hence, a large number of possible recombinants through this meiotic process. Neither author nor Wilson, who supervised their work, conceived of a linear sequence of hereditary factors and a mechanism to account for their recombination within a pair of homologs—that came later with Morgan's insight from breeding analysis, which he tied to Janssen's figures of chiasmata during meiosis. (Figure drawn by Claudia Carlson.)*

Figure 6. *The clarity of Walter Sutton's insight into the chromosome theory of heredity is impressive. He inferred Mendel's laws from the cytology of meiosis before reading or knowing of Mendel's paper. (Courtesy of The Alan Mason Chesney Medical Archives of The Johns Hopkins Medical Institutions.)*

gametes which serves to bring the chromosome theory into final relation with the known facts of heredity; for Mendel himself followed out the actual combinations of two or three distinctive characters and found them to be inherited independently of one another and to present a great variety of combinations in the second generation."[12]

Like Sutton, Boveri had approached his work from cytological findings, but he had used sea urchin blastomeres with irregular sets of chromosomes.

Sutton decided to write to Boveri and test out his theory. He was rewarded with support: "It appears from a personal letter that Boveri had noted the correspondence between chromosome behavior as deductible from his experiments and the results on plant hybrids."[13] Like Cannon, Sutton recognized the implications of the individuality of the chromosomes and the likelihood that each chromosome contained many hereditary factors: "...We have not before inquired whether an entire chromosome or only part of one is to be regarded as the basis of a single allelomorph. The answer must unquestionably be in favor of the latter possibility, for otherwise the number of distinct characters possessed by an individual could not exceed the number of chromosomes in the germ products, which is undoubtedly contrary to fact."[14] He believed that his chromosome theory would explain coupling of characters through this association on a common chromosome. His only error occurred when he tried to explain the non-Mendelian results of de Vries' *Oenothera* (evening primrose) studies and the strange results that Carl Nägeli had reported for hawkweeds (to Mendel's chagrin) when Mendel corresponded with him. Sutton tried to explain away the apparent constancy of hybrids (from F_1 to F_2) in these two forms as a fusion of homologs rather than a separation of them. It took many years to work out these strange cases. The *Oenothera* story is complex, but the story was resolved decades later through the linked chromosome rearrangements that had forced their separation in meiosis in clusters.[15] The *Hieracium* (hawkweed) work was resolved when it was shown that the pollen tube stimulates the ovaries in the embryo sac to undergo parthenogenetic development, but the pollen nuclei did not fertilize the embryo sac nuclei.[16] Sutton suggested as an alternate possibility the loss of a pronucleus just before nuclear fusion, leading to parthenogenetic development.

Wilson Promotes the
Work of Sutton, Cannon, and Boveri

It helped both Cannon and Sutton that in 1902, Wilson published a short article calling attention to their theories. His title also states the relation: "Mendel's principles of heredity and the maturation of the germ-cells" (Figure 7).[17] He attributed to Sutton the recognition that reduction of the hybrid homologous chromosomes "gives a physical basis for the association of dominant and recessive characters in the cross-bred, and their subsequent isolation in separate germ-cells, exactly such as the Mendelian principle requires."[18] He cited Cannon as having made such a conclusion as a "botanist, on account of the fact that most recent botanical workers in this field have reached the result that transverse or reducing divisions do not occur in the maturation of the germ cells in higher plants."[19] His optimism fills his last remark: "Should the study of the maturation-divisions indeed reveal the basis of the Mendelian principle we shall have another and most striking example of the intimate connection between the study of cytology and experimental study of evolution."[20] Very clearly, Wilson was anticipating the birth of classical genetics—a fusion of cytology, evolution, breeding analysis, and reproductive developmental biology. He also cited, with pleasure, the work of his friend, Boveri.

Why did Wilson put forth the work of his students and that of his friend Boveri? Some scholars have suggested that it was a way to claim credit for the theory, because he was a mentor to Cannon and Sutton. I believe it was an effort by Wilson to use his considerable prestige (his book *The Cell* was an instant classic) to endorse an idea he did not want to see languish or dismissed because it lacked experimental evidence. There is no doubt that Wilson, like Boveri,

> ### MENDEL'S PRINCIPLES OF HEREDITY AND THE MATURATION OF THE GERM-CELLS.
>
> IN view of the great interest that has been aroused of late by the revival and extension of Mendel's principles of inheritance it is remarkable that, as far as I am aware, no one has yet pointed out the clue to these principles, if it be not an explanation of them, that is given by the normal cytological phenomena of maturation; though Guyer and Juel have suggested a possible correlation between the variability or sterility of hybrids and abnormalities in the maturation-divisions, while Montgomery has recognized the essential fact in the normal cytological phenomena, though without bringing it into relation with the phenomena of heredity.   Since two investigators, both students in this University, have been led in different ways to recognize this clue or explanation, I have, at their suggestion and with their approval, prepared this brief note in order to place their independent conclusions in proper relation to each other and call attention to the general interest of the subject.

Figure 7. *1902 title page of Wilson's support for the chromosome theory. Wilson introduced the chromosome theory as developed by his associates Cannon, Sutton, and Boveri, and he used his reputation to stress the importance of their work. (Reprinted from Wilson E.B. 1902. Mendel's principles of heredity and the maturation of the germ-cells. Science 16: 991–993.)*

saw an association between chromosomes and heredity. He had made such indirect associations since his 1896 edition of *The Cell*. The work of Cannon established such a one-to-one relationship for a pair of homologs. The work of Sutton established it for both of Mendel's laws. The work of Boveri demonstrated that different chromosomes carried different genetic factors. For that reason, the Sutton-Cannon-Boveri-Wilson chromosome theory of heredity is aptly named. Today, we just assign full credit to Sutton and look upon the other contributions as being flawed or ancillary. Without their work, however, Sutton might not have made that association.

As is often the case with science, each optimistic hope for the survival and spread of a new theory is likely to encounter opposition. It would take many years before the attacks and doubts would dissipate. The first salvo came from an American, O.F. Cook, in 1903: "The notion that heredity, variation, and evolution are the functions of special organs or mechanisms of cells has no ascertained basis of fact."[21] Much more resistance would be encountered from the Europeans, continental and British, who, with the exception of Boveri, took a dim view of this forced marriage of disciplines. In addition to resistance was indifference. The new Mendelism had immediate applications to horticulture, agriculture, and animal breeding. Whatever connection it had to chromosomes was seen as a theoretical interest. It would take another decade before the chromosome theory became a central focus of the newly emerging field of genetics.

End Notes and References

[1] Cited in Boveri T. 1902. On multiple mitoses as a means for the analysis of the cell nucleus. *Verhandlungen der Physiklische-Medizinischen Gesellschaft zu Würzberg* **35**: 67–90 (translated into English in Voeller B., ed. 1968. *The Chromosome Theory of Inheritance*, pp. 87–94. Appleton-Century-Crofts, New York).

[2] Wilson E.B. 1896. *The Cell in Development and Inheritance*. Macmillan, New York

[3] Boveri T. 1902. op. cit., Voeller B., ed., p. 87.

[4] Boveri T. 1902. op. cit., Voeller B., ed., p. 91.

[5] Boveri T. 1902. op. cit., Voeller B., ed., p. 93.

[6] Bateson W. 1902. *Mendel's Principles of Heredity: A Defence*, p. 30. Cambridge University Press, United Kingdom.

[7] Cannon W.A. 1902. A cytological basis for the Mendelian laws. *Bulletin of the Torrey Botanical Club* **29**: 657–666; see p. 660.

[8] Cannon W.A. 1903. Studies in plant hybrids: The spermatogenesis of hybrid peas. *Bulletin of the Torrey Botanical Club* **30**: 519–543; see p. 533.

[9] Sutton W.S. 1903. The chromosomes in heredity. *Biological Bulletin* **4**: 231–251; see p. 231.

Sutton decided to enter medical school after his initial moment of fame. He became a physician in Kansas City, but he died young, from a ruptured appendix.

[10] Sutton W.S. 1903. op. cit. pp. 234–235.

[11] Sutton W.S. 1903. op. cit. p. 234.

[12] Sutton W.S. 1903. op. cit. p. 235.

[13] Sutton W.S. 1903. op. cit. p. 236.

[14] Sutton W.S. 1903. op. cit. p. 240.

[15] See Chapter 10 for an account of the *Oenothera* work.

[16] See Chapter 5 for an account of the *Hieracium* work (and Reference 18 of that chapter).

[17] Wilson E.B. 1902. Mendel's principles of heredity and the maturation of the germ cells. *Science* **16**: 991–993.

[18] Wilson E.B. 1902. op. cit. p. 992.

[19] Wilson E.B. 1902. op. cit. p. 992.

[20] Wilson E.B. 1902. op. cit. p. 993.

[21] Cook O.F. 1903. Evolution, cytology, and Mendel's laws. *Popular Science Monthly* **63**: 222.

The Predominance of
Plant Breeding to 1910

Indeed there are botanists who apparently are not convinced that there is any relation between chromosomes and the genetic factors concerned in the development of even such characters as color of seeds and flowers or the numerous other qualities which are the stock in trade of geneticists.

R.A. EMERSON[*]

The most attractive theory of reduction phenomena assumes that specific characters are largely defined by the amount and nature of the chromatin in the nucleus and that a species, to keep true, must so provide that the chromatin content is relatively stable from generation to generation.

BRADLEY M. DAVIS[†]

THE REDISCOVERY OF MENDELISM in 1900 had an immediate effect on the plant breeders of Europe and North America, and they explored several issues. Although de Vries' rediscovery paper had shown that Mendelism was extensive, breeders wanted to know the universality of Mendel's two laws. Only the first law was extensively confirmed, and none of the three rediscoverers spent much time on dihybrid or more complex crosses. They also wanted to know if a simple latency-patency (dominant-recessive) relationship was the norm. Both Mendel and de Vries encountered examples in which the F_1 generation was intermediate rather than dominant in appearance. They were curious about the traits used, which were necessarily discontinuous to test for Mendelism. Were most traits continuous, as Darwinians had claimed throughout the last decades of the nineteenth century? A second group of breeders was interested in the relationship of Mendel's laws to evolution. Was there any relevance at all, or were these traits that demonstrated Mendelism classifiable as sports, monstrosities, and pathologies of no significance for evolu-

---

[*]1924. In A genetic view of sex expression in the flowering plants. *Science* **59**: 176–182; see p. 176.
[†]1903. The origin of the sporophyte. *American Naturalist* **37**: 411–429; see p. 14.

tionary change but just the dross of unfit individuals about to be sifted by natural selection for an oblivious end? A third group was interested in the Mendelian rediscovery because they were primed to see if practical uses could be derived from domesticated varieties. Could yield and the quality of commercial crops be increased? Among those most interested, especially in the U.S. experimental agricultural field stations and their professors in land-grant universities, were growers of maize, cotton, tobacco, beans, cucurbits, tomatoes, potatoes, and other marketable foods.

The key players in the first generation were William Bateson and his students in Great Britain, Carl Correns in Germany, Erich Tschermak von Seysenegg in Austria, R.P. Gregory in Great Britain,

Wilhelm Johannsen in Denmark, Herman Nilsson-Ehle in Sweden, Hugo de Vries in Holland, and R.H. Lock in Ceylon.[1] In the United States were Rollins Adams Emerson, Edward Murray East, George Harrison Shull, Alfred F. Blakeslee, and Bradley Moore Davis. One peripatetic investigator, working in Canada, the United States, and Great Britain, was Reginald Ruggles Gates. They not only confirmed and extended Mendelism, they ushered in new discoveries, answering the original tasks of the postdiscovery period of breeders, and their work led to the emergence of genetics as a new discipline. From the locations of these investigators, it is clear that breeding analysis remained strong in Europe and had become international in scope through the publications of the rediscoverers of Mendel's laws.

Bateson's School Extends Mendelism to New Findings

William Bateson (1861–1926) (Figure 1) grew up in an academic family, his father being the Master of St. John's College at Cambridge.[2] Bateson's education was broad. He had a love of the classics as well as a passion for natural history, but he was weak in mathematics and this limited his scientific background. After receiving his B.A. and M.A. from Cambridge, where he studied embryology and its relation to evolution, he took one of the rare reverse trips across the Atlantic to study biology at Johns Hopkins University. He enjoyed his association with Martin and Brooks and from Brooks learned the importance of heredity as a subject fit for experimentation and scientific effort. Brooks was, like most scientists, a Lamarckist, but he had a gut feeling that the experimental method would yield answers on how inheritance worked, a subject of monumental importance to evolution. While at Hopkins, Bateson pursued the life cycle of *Balanoglossus*, a hemichordate, and tried to show the evolutionary significance of its temporary notochord (a

Figure 1. *William Bateson shifted from embryology to heredity after discussions with Brooks. He favored discontinuities in evolution through duplication of parts, and when Mendel's work appeared, Bateson began studying Mendelism in plants and animals. His vigorous promotion of Mendelism was equally resisted by his colleagues in Great Britain. (Reprinted from Wilks W. 1906. Report of the Third International Conference 1906 on Genetics. Royal Horticultural Society/Spottiswoode and Company, London.)*

characteristic vertebrate trait), which regressed and left its adult form much like that of an invertebrate. Bateson loved his stay in America and was always eager to visit and lecture there, but he was committed to an academic life in England where he held a Fellow's status (a lifetime sinecure) at Cambridge, which he hoped would launch his career.

Bateson followed Brooks' advice and took on the problem of variation when he returned by preparing an encyclopedic survey of discontinuities in plants and animals, *Materials for the Study of Variation* (1894) (Figure 2).[3] He did not feel comfortable with a gradual emergence of species over tens of thousands of generations. He was impressed by the numerous instances of duplication of organs, occurrences of homeotic mutations (many of them known as sports or sudden mutations), and especially the

tendency of organisms to evolve through a duplication of parts, such as the annular segments in annelids, which were superficially mimicked in our own vertebral columns and the body segments of arthropods. This process of duplication of parts in embryos was known among comparative anatomists as metamerism. Bateson was versatile and as comfortable doing research with plants as with animals, and in later life, he headed the internationally renowned botanical facility, the John Innes Institute. Bateson's training was in embryology, a field of intense interest to evolutionists of that era because Ernest Haeckel, among others, followed up Darwin's suggestion to seek in the embryo a history of life on earth. Haeckel's "biogenetic law" was a catchphrase for students: "Ontogeny recapitulates phylogeny." Although many zoologists soon ran into contradic-

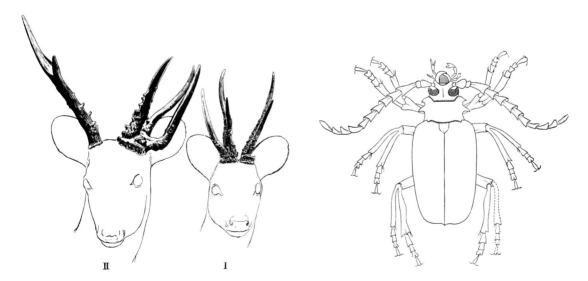

Figure 2. *Evolution through duplication of parts. In 1894, Bateson argued that discontinuities in evolution led to major changes in species. He illustrated this with hundreds of examples of what today would be called homeotic or malformation mutations. In his day, these were regarded by Darwinians as monstrosities of no evolutionary significance. (Reprinted from Bateson W. 1894. Materials for the Study of Variation Treated with Especial Regard to Discontinuity in the Origin of Species. Macmillan, New York.)*

tions, and others rejected the idea outright, it remained too speculative to generate a field of experimental evolution. Bateson realized this, and this was what made him initially attracted to de Vries' work in *Oenothera*. Breeding analysis was the key to evolutionary studies for Bateson, and plants were easier to study than animals because they were less costly and easier to raise.

Bateson Encounters Modified Mendelian Ratios

During the late 1890s, Bateson was working on plant and animal crosses. He was convinced that discontinuous variations should be studied and bred to determine their mode of inheritance. His colleagues, evolutionists like himself, but wedded to the idea that most evolutionary traits were continuous and yielded some sort of normal or skewed curves, were also keen breeders of animals and plants, with a stronger tendency than Bateson to rely on fieldwork surveys rather than to stress cross-breeding. In 1899, Bateson was close to establishing Mendel's laws himself, but he had not yet found a consistent dominance of traits or a consistent numerical ratio for a separation of the crossed traits through the hybrid offspring that emerged in his crosses. Early in 1900, he read de Vries' account of the rediscovery and quickly had Mendel's first paper translated for the English scientific community to read.[4]

Bateson was optimistic that Mendelism would provide insights into evolutionary theory, and he was convinced it would usher in a field of research that would dominate twentieth-century science. He was correct on both counts, but this optimism was not immediately embraced. Bateson encountered a wall of silence or rejection by most of his contemporaries. This came about, in part, for the same reason that the French did not jump on the Mendelian bandwagon. The French were not Darwinians; they were Lamarckists and evolution was theirs by priority. They believed Darwin's natural selection to be an extension of Lamarck's insights into the evolution of life. Fixed traits that did not change with the environment were trivialized as being of no evolutionary importance. It was not so much nationalism that guided them—it was, instead, being wedded to a belief in the transformation of species by the environment's intercession with the hereditary process. Similarly, Bateson's critics did not perceive Bateson as a traitor to Darwinism. Rather, they saw discontinuous traits as irrelevant to the evolutionary process. They would be selected out immediately, and almost all surveys of sports (spontaneous mutations) and other discontinuous traits picked up by breeders, whatever their commercial worth to hobbyists or crop breeders, were evolutionary dead ends for the competition in life demanded by natural selection.

Bateson had been sparring with the biometric group headed by Karl Pearson and W.F.R. Weldon (Figure 3).[5] They vied for Galton's support as did Bateson; curiously, Galton felt comfortable with both camps. Galton decided to smooth things between them by putting both factions on a committee on evolution for the Royal Society. Their meetings ended up as donnybrooks and left lacerating mental wounds on both sides. When Weldon died of a massive heart attack while riding his bicycle, his widow

Figure 3. *Karl Pearson (left) and W.F.R. Weldon (right) were the founders of the biometric school of evolution. They sought mathematical descriptions of continuous traits rather than experimental tests of discontinuous traits. They rejected the value of Mendelism for both heredity and evolution. ([Left] Courtesy of the National Library of Medicine. [Right] Courtesy of the Cold Spring Harbor Laboratory Archives.)*

nothing short of sensational. In addition to confirming Mendelism in animals (he used poultry), Bateson and his students found two new phenomena. They found modified 9**AB**:3**Ab**:3a**B**:1**ab** ratios, including 15:1, 9:7, 9:4:3, 12:3:1, and 10:6. These often involved color factors for flower petals. In the 15:1 case, only the **aabb** combination was white; any other combination containing one or more dominant factors had an indistinguishable color (blue or red, most often). In the 9:7 ratio, only those that had both factors **A** and **B** were colored. Those with an **aa**, **bb**, or **aabb** combi-

blamed his premature death on Bateson! In hindsight, it is difficult to see what the fuss was about, but at the time, the biometric school believed that everything was statistical and the shape of curves reflected evolution. Mendelism was seen as the latest of hybridist attempts to explain species by sudden origins. To the biometricians, it was not monstrosities that fueled evolution. Variations had to be subtle—barely perceptible—yet significant enough to make the difference between leaving offspring or fading out of the evolutionary struggle.

Bateson tried to get his articles and views published, but he met with resistance. He privately subsidized his small book, *Mendel's Principles of Heredity: A Defence*[6] (1902) (Figure 4) and he sent copies to all of the leading students of heredity to make sure that Mendel would not suffer another 35 years of neglect. He sent his research papers to the Royal Society's Evolution Committee, and they were obliged to print them as reports. What he and his students found was

MENDEL'S

PRINCIPLES OF HEREDITY

A DEFENCE

BY

W. BATESON, M.A., F.R.S.

WITH A TRANSLATION OF MENDEL'S ORIGINAL PAPERS ON HYBRIDISATION.

CAMBRIDGE:
AT THE UNIVERSITY PRESS.
1902

Figure 4. *Title page from the book* Mendel's Principles of Heredity: A Defence, *which Bateson made sure would be sent to his critics so that Mendel would no longer be ignored. (Reprinted from Bateson W.B. 1902.* Mendel's Principles of Heredity. *Cambridge University Press, United Kingdom.)*

nation were albino. Bateson interpreted these relationships as epistatic. Some traits required two or more factors to express a trait.[7]

The second finding was more strange, but it did not appear to follow Mendelian independent assortment. Two different factors, **C** and **D**, might go together as a unit to half the offspring in the dihybrid, **CcDd**, with the other half of the offspring receiving cd. He called this phenomenon "coupling." In other plants, the reverse seemed to be true, even for the same two factors. One half of the offspring would get **Cd** and one-half would get **cD**. The C and D seemed to repel each other like magnets with the same poles. He called this phenomenon "repulsion." Note that the same genetic symbolism, **CcDd**, applied for the coupling or repulsion state of the dihybrid.[8]

Bateson Provides a Terminology for Genetics

To assist the new breed of students of heredity, Bateson coined a flood of terms. In 1906, he called this new field "genetics" (Figure 5).[9] He described the hereditary units as "unit factors" or "character units," terms that would confuse geneticists with a naïve belief that an entire character is a product of a single factor, despite the modified 9:3:3:1 ratios he published, which showed that this was not true (Figure 6). He referred to the contrasting traits (i.e., A and a) as "allelomorphs" (more than 15 years later, Shull trimmed the term to "allele"). He modified some notation used by Galton for describing pedigrees and introduced the P_1, F_1, and F_2 symbols for the parental and filial (offspring) generations. He identified the genetic constitution as being "homozygous" (i.e., **AA** or **aa**) or "heterozygous" (i.e., **Aa**). This was an advance over the notation of Mendel and de Vries, who used the symbolism 1A:2Aa:1a instead of Bateson's format using 1AA:2Aa:1aa. Mendel and the rediscoverers were swayed by the purity of what we (and Bateson) call the "homozygote." For them, **A** and **a** were pure. For breeders since Bateson's symbolism, it has been the doubleness of the factors that is significant, and purity is represented through its doubleness as **AA** or **aa**. In Mendel's day, it was not yet proven that fertilization involved the union of one sperm (or pollen nucleus) and one egg (or embryo sac nucleus).[10]

Bateson was not a fan of the chromosome theory that emerged soon after he began sending his reports to the Evolution Committee. He felt it was grossly mechanistic, with too many factors and too few chromosomes; at best, the Sutton-Cannon-Boveri-Wilson

> Like other new crafts, we have been compelled to adopt a terminology, which, if somewhat deterrent to the novice, is so necessary a tool to the craftsman that it must be endured. But though these attributes of scientific activity are in evidence, the science itself is still nameless, and we can only describe our pursuit by cumbrous and often misleading periphrasis. To meet this difficulty I suggest for the consideration of this Congress the term *Genetics*, which sufficiently indicates that our labours are devoted to the elucidation of the phenomena of heredity and variation : in other words, to the physiology of Descent, with implied bearing on the theoretical problems of the evolutionist and the systematist, and application to the practical problems of breeders, whether of animals or plants. After more or less undirected wanderings we have thus a definite aim in view.

Figure 5. *The first public appearance of the word "genetics" was in Bateson's presidential address to the Royal Horticultural Society to introduce the term in 1906 and to rename the conference. (Reprinted from Wilks W. 1906. Report of the Third International Conference 1906 on Genetics. Royal Horticultural Society/Spottiswoode and Company, London.)*

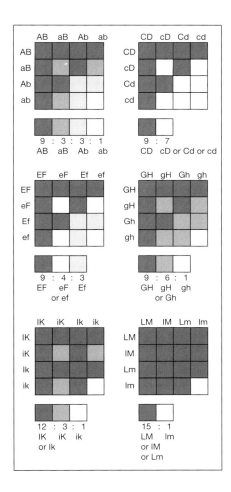

Figure 6. *Bateson discovers epistasis with a Mendelian basis. Epistasis is the name Bateson applied to modified 9:3:3:1 ratios for dihybrid crosses. It assumed an interaction between the two factors. Bateson and his school in Great Britain and Davenport, Emerson, and Shull in the United States encountered these ratios in a variety of plants and animals. The Punnett squares show how each of these modified ratios works. The standard noninteractive model is shown as 9AB:3aB:3Ab:1ab. The 9:7 ratio occurs when two dominant factors are needed for color (CD) and other combinations homozygous for cc or dd (or both) result in albinism. The 9:4:3 ratio arises when EF is associated with one color, ee with albinism (either eeF- or eeff), and E-ff as a color distinct from EF. In the 9:6:1 ratio, the GH individuals have one color, the G or H individuals have a second color, and the gghh are albinos. The 12:3:1 ratio arises when the dominant factor I is present (IK or Ik) for one color, iK is present for a second color, and iikk results in an albino. The last ratio of 15:1 assumes that the same color arises whether the individual is LM, Lm, or lM; only the homozygote llmm is an albino. More complex ratios were also obtained for trihybrid epistatic interactions of color factors. Determining these ratios in animals was a daunting process because they generally produce fewer progeny than do flowering plants. Ironically, the early Mendelians on both sides of the Atlantic Ocean had to use statistics (Chi-square methods, especially) developed by Pearson and Weldon at a time when they were in fierce opposition to Pearson's and Weldon's interpretations of Mendelism. Those statistical calculations were used to confirm or exclude the inferred ratios. (Figure drawn by Claudia Carlson.)*

theory was speculation. He wanted evidence, and the only experimental system he trusted was breeding analysis. Bateson did speculate about the nature of the Mendelian factors. Most of them seemed to be the loss of an attribute (color, height, or organ systems), and he assumed that they were precisely that—the factor was lost, and the lowercase letter represented the absence of a factor. When the factor was present, it was dominant; when it was lost, it was recessive. The heterozygote rescued the individ-

ual because at least the factor was still present in the cells. Bateson called this the "presence and absence hypothesis."[11] He invoked a different model for coupling and repulsion. In most instances, coupling and repulsion were not permanent, and the factors seemed to come together or separate with a constant but non-Mendelian ratio. Where no such flip-flopping occurred, he called the factors "spurious allelomorphs," and where the switching took place, he called the events "partial coupling or partial repulsion." Thus, if **CD** and **cd** were in a coupled state in a cross, a similar cross might mostly have **CD** and **cd** with a few **Cd** and **cD** progeny.

Bateson had to work out a mechanism for this new phenomenon.[12] His old embryological views came into play. He assumed that a differential multiplication of the factors or the cells bearing the factors in the reproductive tissue led to partial coupling and repulsion. Since most of this work was done in plants (which usually lack sex chromosomes), his crosses were largely F_1 x F_1 and this gave idiosyn-

cratic ratios, which he tried to resolve into fixed reduplication ratios for the gametes.[13] Bateson was the first geneticist to create the feeling that genetics was a tough field with arcane symbolism and a dizzying logic to it, much like intricate mathematical puzzles. But more important, Bateson was the key figure in getting Mendelism turned into an exact science and by brute force convinced most of his colleagues on both sides of the Atlantic that genetics was here to stay and that it dealt with more than trivial characters.

Correns Encounters and Fails to Solve Difficult Problems

Carl Correns shifted from peas and maize to a variety of plants to explore some evolutionary problems. Some plants are dioecious (the male and female organs are in males and females, separately) and others are monoecious (the organism has both male and female organs present in the same individual; it is hermaphroditic). Were there Mendelian factors involved in sex determination? Correns found the problem complex.[14] In some species, some branches or flowers were monoecious and the rest of the plant had female flowers (no anthers; only stigma, style, and ovary); there was no simple Mendelian resolution. Correns also became interested in variegated leaves of plants and flower petals. What caused the streaking or spotting pattern? In the plants he studied with white or yellow streaks in otherwise green leaves, he found that some of his species gave a maternal pattern of inheritance. The streaking or loss of chlorophyll followed the distribution of the embryo sac and not the pollen nuclei that were introduced. Correns pursued this as a developmental problem, known to embryologists who had encoun-

tered it earlier as "maternal inheritance" and to cell biologists as the study of "nucleo-cytoplasmic relations."[15] Just as Mendel had had a tough time with *Hieracium*, which discouraged him from further plant breeding, Correns had many publishable findings but no broad theories or discoveries for his meticulous pedigrees of plants in the search for a comprehensive theory of sex determination or a comprehensive theory of the relationship of the cytoplasm to the nucleus.

Sometimes the choice of organisms or the choice of a problem can defeat the most dedicated scientist's attempts to solve a problem. Correns chose an equally daunting problem for his third post-Mendelian attempt to find something new. He applied his breeding analysis to the problem of self-sterility.[16] Some plants require pollen from another plant to bring about fertilization. Other plants have no difficulties using their own pollen and ovules to bring about self-fertilization in the monoecious plant. This, too, defied a simple Mendelian interpretation.

de Vries Chases a False Evolutionary Model and Goes Down With It

Hugo de Vries also decided to follow a different direction. Mendelism was nice, but with Mendel's priority, he wanted something more significant to be associated with his name. Since the early 1890s, he had been breeding the evening primrose, *Oenothera lamarckiana* (Figure 7), an American import that was now growing wild in parts of Europe. de Vries noticed

a few atypical plants in one such field, and although they were clearly primroses, they differed in several factors. When he tried to cross them with *O. lamarckiana*, he found that some could not do so—they only bred with their own kind or through self-fertilization. Such an event, in which numerous traits were different and a sexual barrier existed between the two

Figure 7. *de Vries promoted his mutation theory of evolution based on the sudden origin of new species such as* Oenothera gigas *and* O. lata *from the stock species,* O. lamarckiana. *(Reprinted from de Vries H. 1909. The Mutation Theory. Open Court, Chicago.)*

varieties, suggested that they were new species, and de Vries thought that these were saltations or discontinuous events in evolution. He noted that such eruptions of new species of *Oenothera* came in bunches. He called these "mutating periods." He looked for, but had not yet found, other plants with this mode of speciation, but one was enough to explore for the time being. de Vries called his theory of evolution by saltations "mutation theory," and he published it at about the same time that he was confirming Mendel's laws.

de Vries pulled out of the extension of Mendelism effort and focused on his mutation theory.[17] The theory was impressive because he claimed that it established a basis for experimental evolution. Many scientists did not like the Darwinian process of natural selection by gradual modification because they had to accept the model on historical grounds, rather than experimental grounds. Here was an answer: You could show evolution in the laboratory or greenhouse with new species that you could document.

Gates Casts Doubts on de Vries' Interpretations of *Oenothera* Evolution

One of the first to test the new species of *Oenothera* was Reginald Ruggles Gates (1882–1962) (Figure 8).[18] He showed from his genetic crosses of *Oenothera gigas* that the Mendelian factors gave unusual ratios, suggesting a tetraploid (4N) chromosome number instead of the inferred diploid (2N) number of most species. He confirmed this in 1909, and the diploid number of 14 was a tetraploid 28 in *O. gigas*.[19] If that were true, no new genetic factors entered into the evolutionary story by doubling the chromosome number. It was a quantitative change, not a qualitative one. In Gates' own words, "This cannot be a common method of species formation, however, and bears no necessary relation to the general process of evolution in the group. It appears to be rather of the nature of an incident among evolutionary phenomena."[20] Gates attributed many of the other peculiar results that de Vries had obtained by 1910 to some unusual and yet unknown defect of cytology, rather than to new mutations. He considered *Oenothera* to be unrepresentative of Darwinian evolution—not all giant forms were tetraploids. R.P.

Figure 8. *Reginald Ruggles Gates was introduced to* Oenothera *work at Woods Hole in 1905 and dedicated much of his life to this organism. He used cytology to identify ring-shaped chromosomes in O. rubriveris, tetraploid (4N = 28) chromosomes in O. gigas, and aneuploidy (2N = 15 instead of the normal 2N = 14) in O. lata. In his later years, he switched to eugenics, human genetics, and racial anthropology. (Courtesy of the National Portrait Gallery, London.)*

Gregory, using *Primula sinensis*, found a giant form with a diploid chromosome number. The trait was caused by a simple dominant allele that had arisen by chance mutation.[21] Gregory read Correns' papers and plunged into the sex determination issue by studying *Valeriana dioica* (marsh valerian). After six years of unsatisfactory results, he found that he could not interpret his data (no consistent ratios were obtained and many of the crosses were infertile with poor yields). He announced in 1909 that "From what has been said above it will be clear that *Valeriana dioica* is not a favourable subject for experimental work which requires the rapid and accurate determination of the characters of offspring... . It has therefore been decided to abandon the experimental work begun in 1903."[22] Despite the lack of support from Gates and from Gregory, both de Vries and Correns continued to study these difficult problems.

Johannsen's Study of Pure Lines Stimulates Evolutionary Theories

In Denmark, Wilhelm Johannsen was working on a different problem (Figure 9). He was interested in a breeding analysis of a continuous trait to test Darwin's model of natural selection. He chose the size of beans, using *Phaseolus vulgaris*.[23] It was well known that beans varied in size and weight, and if they were counted and sorted by size, a Gaussian or normal curve resulted (frequently called a bell curve because of its characteristic shape). Johannsen selected large bean size and inbred the largest beans he found. He noticed that the beans from such inbred largest beans produced a noticeably different curve of bean sizes from that of the outbred beans in the field. The mean shifted to a larger size, around which was a new bell curve; the very smallest beans had disappeared (Figure 10). Johannsen repeated his selection of largest beans and his inbreeding of these beans for each of 12 generations. He found that the mean shifted at a more modest rate with each new generation, and by the eighth to tenth generations, the mean was repeated. When he chose any bean in such a curve (largest to smallest), the curve was con-

Figure 9. *The distinction between genotype and phenotype was made by Wilhelm Johannsen who used beans to study a continuous trait (size). Johannsen's work became a bridge uniting an underlying Mendelian mechanism for the production of continuous traits and thus the entry of genetics into evolution in Darwin's own natural selection for continuous traits. (Reprinted from Wilks W. 1906. Report of the Third International Conference 1906 on Genetics. Royal Horticultural Society/Spottiswoode and Company, London.)*

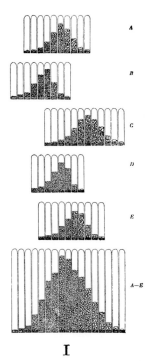

Figure 10. *The bean distributions of different pure lines obtained by Johannsen through intense selection and inbreeding. (Reprinted from Morgan T.H. 1919. A Critique of the Theory of Evolution. Princeton University Press, New Jersey.)*

what he called a phenotype. Some might be the middle-sized bean of an outbred line. Some might be one of the smaller beans selected for a pure line of large size. Some might be one of the largest beans of a pure line selected for small size. Johannsen said that they shared a common appearance or phenotype, but that each had a distinct genotype. Johannsen also argued that the variation found in a pure line was not associated with genetic differences. These environmental differences were associated with a host of factors including amount of sunlight, exposure to wind, nutrition of the soil, amount of water around the roots, and other variables to be found in an experimental garden.

Johannsen's work was brilliant because it demonstrated a relationship between heredity and the environment that was consistent with a Darwinian model of natural selection. Small differences could occur from continuous traits involving a large number of factors cooperating in a complex physiological process such as size or weight. Johannsen made one other contribution. He did not like the competing terms unit character, merkmale, or unit factor. He suggested instead, in 1909, to take the root (gene) from de Vries' pangenes, which in turn had been trimmed from Darwin's theory of pangenesis, and this simple noun, gene, would represent the hereditary unit used in breeding analysis. It would be free of defining attributes until chemistry and physiology had a better idea of the structure, chemistry, and function of the gene.[24]

stant. He called such a population of inbred beans a "pure line," and he inferred that the genetic constitution was rendered homozygous by breeding. He called that inferred genetic constitution the genotype. Medium-sized beans, however, represented

Nilsson-Ehle Resolves
Quantitative Traits through Mendelism

The other major contribution to genetics in this decade from a European geneticist was associated with the laboratory of the Swedish botanist,

Herman Nilsson-Ehle (Figure 11). Nilsson-Ehle discovered modified 9:3:3:1 ratios that suggested two different genes in which dominance was lack-

Figure 11. *Herman Nilsson-Ehle studied a quantitative trait, coat color, in cereal grains. He showed modified Mendelian ratios producing a distribution consistent with Johannsen's beans and the Mendelism of discontinuous traits. (Courtesy of the Genetics Society of America.)*

ing.[25] He identified wheat or oats with a reddish cereal grain coat color as being **AABB** and the blond or straw color of one variety as being **aabb** in genotype. The F_1 progeny were dihybrid, **AaBb**, and they showed an intermediate or pink color. Selfing or breeding such plants derived from pink grains yielded an unusual ratio of 1 red:4 dark pink:6 pink:4 light pink:1 blond. Nilsson-Ehle described this 1:4:6:4:1 ratio as quantitative inheritance, with neither white nor red being dominant and only the number of color factors present expressing their cumulative pigmentation to seed coat color (Figure 12). Nilsson-Ehle also discovered instances of three or more factors, with correspondingly more complex distributions resembling normal curves as the number of factors increased.

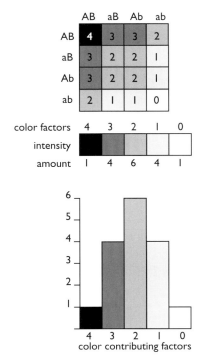

Figure 12. *Quantitative inheritance was independently worked out by Nilsson-Ehle in Sweden using cereal grains and by East in the United States using maize. Nilsson-Ehle's analysis was far more extensive and involved many different species of cereal plants (oats, wheat, and rye). Nilsson-Ehle demonstrated that in a two-factor determination, a simple addition of the color factors (later called genes) sufficed to produce a modified digenic ratio of 1 red:4 dark pink:6 pink:4 light pink:1 blond for the varieties of grain colors obtained. In the Punnett square, the amount of color factors is indicated in each square. Thus, AABB has four and aabb has none of these factors. The results are additive, with no dominance for the heterozygote. The resulting distribution can be organized as a bar graph; when the number of factors involved is large (four or five genes), the resulting bar graph resembles a Gaussian or bell curve. Davenport later made use of Nilsson-Ehle's model to interpret human skin color with an alleged ratio of 1 black:4 dark brown:6 brown:4 light brown:1 white in skin color. Social use of skin color in the United States either gave names (mulattoes and octoroons) associated with slave-holding days or lumped together any category other than aabb (white) as black or Negro. (Figure drawn by Claudia Carlson.)*

By 1910, the European Mendelians were on the march. They showed new phenomena such as epistatic ratios, the Mendelian basis for quantitative traits, the Mendelian basis for continuous traits, and a more complex relationship of genes to characters. They also weakened the case that de Vries had found something new to replace natural selection by gradual change, and they were forcing the biometrical school to take a new look at Mendelism.

American Plant Breeding Stresses Farm Crops

American motivation for plant breeding was long-standing. The land-grant colleges and the experimental stations were already studying effects of inbreeding and various ways to select for commercially useful traits. Strains of maize with more protein would be beneficial in preventing malnutrition. Selection of the best seed to plant was a constant problem, with farmers and scholars at the land-grant colleges discussing their own methods of getting more rows of kernels, more cobs per plant, and other valued traits. But there was no theory of heredity that was practical and breeding was an uncertain art with occasional gifted "selectionists" like Luther Burbank, who had a good eye for rare varieties and lots of patience to look through immense amounts of fruit and vegetables before finding what he thought was a worthwhile product. Without guaranteed seed quality, it was difficult to produce commercial seeds for farmers, and many farmers did what their ancestors did: They put aside the best seed for the next growing season. If seed could be sold commercially, it would relieve the farmer of storing and protecting the seed.

Emerson Extends Mendelism

In 1902, a young investigator named Rollins Adams Emerson (1873–1947) (Figure 13) worked in Nebraska. He was born in upstate New York, but most of his elementary and high school education took place in Nebraska, where his parents had moved.[26] He took an interest in plant breeding, and in 1902, he published some results on crosses of different varieties of beans (*Phaseolus vulgaris*). He did not look for any numerical relations, and he tried to see which traits showed in the hybrid parents and which were recessive or latent. But after he read and discussed Mendel's work, he realized that this new approach could be more fruitful: "...But at the present time, when problems of heredity are receiving so much attention from plant and animal breeders, as well as from biologists, any evidence, however meager, is of value. Mendel's experiments in hybridization opened a new line of inquiry."[27] The traits that Emerson studied included "stringiness of pods," "toughness of pods," color of pods, color of flowers, and color of seeds. After reading Mendel, he also tested what he called "pole" or "bush" habits, which included the axis of the plant, the twining of the stem, and the position of the pods. He found that these produced simple Mendelian ratios with, for example, axial dominant to terminal pods. Most of his traits showed a simple dominance,

Figure 13. *Rollins A. Emerson (right) with Thomas H. Morgan at the 1932 International Congress in Ithaca, New York. Emerson was the founder of the corn genetics school at Cornell University that produced Nobelists Barbara McClintock and George Beadle. Emerson was the first to devise methods to demonstrate a linkage of traits in maize. (Courtesy of the California Institute of Technology Archives.)*

but in a few cases, the F_1 hybrid was intermediate. "Where intermediates are found in the first generation, three types occur in the second generation, namely, intermediates and the two extremes."[28] In pod color, the green was dominant to yellow, but when he crossed two different yellow strains, he noticed something new (although Bateson was finding a similar phenomenon at about the same time): The F_1 progeny of the two yellow strains were purple in pod color. These purple strains never bred true (they were obligate hybrids).[29] Emerson also noted that some traits were associated. Thus, white flowers were correlated with white seed. Later, geneticists would call this pleiotropism, the many expressions on different organs of an individual gene. Some crosses yielded surprises: When he crossed "navy white" beans to "challenge black" beans, the F_1 progeny were a mottled dark brown-gray. The intermediate hybrid had characteristics different from those of either parent. In the F_2 generation, the ratios were 1 white:2 brown-gray mottled:1 black.[30] By 1909, Emerson had confirmed Bateson's results and found both modified 9:3:3:1 epistatic ratios and more complex ratios. He found some mottled strains that bred true and other mottled strains that were hybrids. Mottling turned out to have many factors associated with this trait. "Two sorts of behavior of a character can mean nothing else to me than that they are two characters each with its own distinct behavior. I have been driven to adopt the two-factor hypothesis, and to me, it seems to account for all results thus far secured."[31]

East Presents Many Experimental Approaches to Plant Genetics

Edward Murray East (1879–1938) (Figure 14), like Emerson, began his plant work with species other than maize, but he turned to maize as a more favorable organism.[32] Although Emerson was older than East, he had only a master's degree from Nebraska. He went to study with East, who was newly appointed to Harvard's Bussey Institute. East had started out in genetics studying potatoes. He was interested in a Darwinian problem that Darwin called bud sports. Darwin considered these to be poor materials for the source of new and adaptive functions in plants. But some were quite successful commercially, like seed-

Figure 14. *E.M. East was a first-generation maize geneti-cist who had earlier worked on tobacco and other plants. He made substantial contributions to the development of hybrid corn, and he used his Harvard credentials to speak out on social issues, including an uncritical enthusiasm for the American eugenics movement. (Reprinted, with permis-sion, from Crabb A.R. 1948. The Hybrid-Corn Makers: Prophets of Plenty. Rutgers University Press, New Brunswick, New Jersey.)*

less navel oranges, although these too were nonadap-tive and their sterility required that they be propagat-ed asexually through grafts of new branches on other trees. This was not the case for potatoes. All of his bud sports turned out to be pathological; they had loss of color, deep-set eyes, and spherical shape. He did not know the origin of the bud sports, but he did not rule out that some Mendelian segregation was taking place in somatic tissue. His guess was wrong. At the time, the idea of spontaneous mutations to a dominant form was not recognized.[33]

East also tried to induce directed mutational changes in tomatoes. He used yellow tomatoes and injected ovaries with macerated remnants of red seeds, fruit, and pollen in a fine suspension. He injected castrated plants and later pollinated them with yellow pollen. Like Galton, who experimented to test Darwin's theory of pangenesis, he obtained nega-tive results. There was no environmental influence on the hereditary factors when the specimens were injected into ovaries. All the specimens were yellow, although yellow was known to be recessive to red.[34]

East would later turn to corn genetics and make the two most valuable contributions of his career. He developed the double-hybrid cross for producing hybrid corn (for which he and Shull collaborated) and he discovered, independently of Nilsson-Ehle, the 1:4:6:4:1 ratio for quantitative traits.[35] As the number of factors involved increased, the rarity of the full multiple recessive became more apparent. This led East to suggest that some atavisms might be examples of such rare associations of four or more recessive factors associated with a quantitative trait. He also recognized that as the factors increased, the number of categories among the F_2 phenotypes began to produce what looked like a normal curve.

Plant Cytogenetics Is Introduced by Davis

Bradley Moore Davis (1871–1957) (Figure 15) start-ed his career in cryptogamic botany and studied evo-lutionary problems, such as the development of the sporophyte (diploid plant) from the gametophyte (haploid plant) in mosses and liverworts.[36] He took an interest in experimental evolution after reading de Vries' papers on *Oenothera*, and he began looking at the chromosomes of *Oenothera*. What he found was a surprise: They formed strange looking rings or con-tinuous clusters joined tip to tip. It was Davis who

Figure 15. *Cytology became cytogenetics when Bradley Moore Davis turned his attention to the cytology of Oenothera. He discovered irregularities (such as polyploidy) independently of Gates, and he also was mentor to Gates, getting him interested in Oenothera cytology as a way to study its evolution. (Reprinted, with permission, from Cleland R.E. 1958. Bradley Moore Davis. Genetics 43: 1–2.)*

suggested to Gates that he study *Oenothera* and they independently found that *O. gigas* was tetraploid.[37] Davis, like Gates, argued that de Vries' mutation theory rested on *Oenothera* because no other plant or animals had shown a similar type of saltation mechanism for generating new species. If cytological aberrations were the basis for these novel forms, then *Oenothera* was not typical in its evolutionary history.

de Vries remained faithful to his mutation theory of evolution, despite the findings of Davis and Gates. His original supporters, like Bateson, realized that *Oenothera* evolution was not typical and soon sought other ways to launch an experimental evolution. It embittered de Vries in his later years to see his name cited as a co-rediscoverer of Mendel's laws and as the founder of a failed theory of evolution.

We can conclude this survey and appreciate that the first wave of plant geneticists in North America found much the same phenomena as their counterparts had in Great Britain and continental Europe. The American plant breeders tended to use commercially valuable flowering plants and they worked primarily at land-grant colleges and experimental field stations. Their motivations for studying their problems were almost all drawn from Darwin's legacy—the effects of inbreeding on vigor, the nature of continuous traits, the relation of genes to characters, and the applications of Mendelism to evolutionary change. Breeding analysis in this decade yielded modified Mendelian ratios, complex characters, coupling and repulsion (partial or complete), quantitative inheritance, and the genetics of continuous traits through pure line studies. It was a heady collection of new phenomena and it proved the worth of Mendelism. Although there was a cytological movement of greater importance in the United States than in Europe, the chromosome theory was scarcely used during the first decade of plant breeders. Both Gates (then mostly in North America) and Davis were among the few who were applying microscopy to genetic problems.

End Notes and References

[1] Many more persons and countries were involved. I list those who had a major impact on the way breeding analysis (and to a lesser extent, cytology) was used to shape classical genetics among botanists.

[2] Bateson's life and career are discussed by his wife, Beatrice Bateson, in Bateson W. 1928. *Essays and Addresses*, pp. 28–73. Cambridge University Press, United Kingdom.

[3] Bateson W. 1894. *Materials for the Study of Variation.* Cambridge University Press, United Kingdom.

[4] The paper first appeared in the Royal Horticultural Society's proceedings for 1901. C.T. Druery was the translator. It was republished in Bateson W. 1902. *Mendel's Principles of Heredity: A Defence.* Cambridge University Press, United Kingdom.

[5] For accounts of this debate, see the very thorough study by Provine W. 1971. *The Origins of Theoretical Population Genetics.* Chicago University Press. A more abbreviated account is in Carlson E. 1966. The fight to legitimize genetics. *The Gene: A Critical History,* Chapter 2. Saunders, Philadelphia. See also Gillham N.W. 2001. *A Life of Sir Francis Galton.* From African explorer to the birth of eugenics, Chapter 21. Oxford University Press, United Kingdom.

[6] Bateson W. 1902. op. cit.

[7] The early findings of Bateson were published for the Evolution Committee. They are entitled *Reports to the Evolution Committee.* Bateson W. and Saunders E.R. 1902. I. Experimental studies in the physiology of heredity. **1:** 1–160; Bateson W., Saunders E., and Punnett R. 1904. II. Experimental studies in the physiology of heredity. **2:** 1–154; Bateson W., Saunders E., and Punnett R. 1906. III. Experimental studies in the physiology of sex. **3:** 1–53; and Bateson W., Saunders E., and Punnett R. 1908. IV. Experimental studies in the physiology of heredity. **4:** 1–60.

[8] Coupling and repulsion were first reported in Bateson W. 1906. The progress of genetics since the rediscovery of Mendel's paper. In *Progressus rei Botanicae* (ed. J.P. Lotsy), pp. 368–418. G. Fischer, Jena.

[9] Bateson W. 1906. Inaugural address. The progress of genetic research. *Third International Congress of Genetics,* pp. 90–97. London.

   Bateson renamed the conference the Third International Conference on Genetics because he had invited both plant and animal geneticists to the International Conference on Plant Hybridization sponsored by the Royal Horticultural Society.

[10] See Chapter 4 for that historical development.

[11] Bateson W. 1906. op. cit. See also Carlson E. 1966. op. cit., Chapter 8: The presence and absence hypothesis.

[12] Bateson W. 1904. et al. op. cit., Evolution Society Report III.

[13] Bateson used both gametic ratios (e.g., 7AB:1Ab:1aB: 7ab) and their predicted zygotic ratios from a $F_1 \times F_1$ or self-fertilization (e.g., 177**AB**:15**Ab**:15**aB**:49**ab**). He had to round off to obtain his gametic ratios because actual crossing-over was along a continuous chromosome thread, whereas partial coupling would assume some sort of differential duplication of units.

[14] Correns C. 1904. Experimentelle untersuchungen über die gynodioecie. *Berichte Deutsche Botanische Gesellschaft* **22:** 506.

[15] Correns C. 1909. Vererbungsversuche mit blass (gelb) grünen und bluntbattrigen Sippen bei *Mirabilis jalapa, Urtica pilulifera,* und *Lunaria annua. Zeitschrift fuer Inductive Abstammungs und Vererbungslehre* **1:** 291–329.

   For a historical account of nucleo-cytoplasmic relationships and their role in classical genetics (or, according to Sapp, their being barred unfairly from such a role), see Sapp J. 1987. *Beyond the Gene: Cytoplasmic Inheritance and the Struggle for Authority in Genetics.* Oxford University Press, United Kingdom.

[16] A similar frustration plagued physicist and molecular geneticist Max Delbrück. He solved the life cycle of the bacteriophage T_2 and demonstrated its worth as a useful genetic organism in less than five years. But when he turned to *Phycomyces,* a water mold, to study the molecular genetics of phototropism (how does a photon of light do this?), he spent 20 years without finding an answer to how that mold responds to light quanta.

[17] de Vries H. 1901–1903. *Die Mutationstheorie.* Viet and Company, Leipzig.

[18] Gates was an unusual figure in genetics. He was Canadian and did his early work at McGill and at the Marine Biology Laboratory at Woods Hole before moving to the University of Chicago, where he was much influenced by cytologist Bradley Moore Davis (1905–1907). He moved to England and there married Marie Stopes, then a palaeobotanist. The marriage was unconsummated and led to a sensational divorce trial in the House of Lords where it was proven that Marie Stopes' hymen was intact! Both Gates and Stopes testified that they didn't know that sexual intercourse was needed to produce

children because they grew up in Victorian households and led sheltered lives. Marie Stopes went on to become an advocate of the birth control movement after reading about Margaret Sanger's efforts in the United States. Gates moved from botany to human genetics (especially in the British eugenics movement) and anthropology. His views were usually opinionated and much ridiculed. When his death was announced at a national meeting of American anthropologists, Mrs. Garret Hardin told me that the audience cheered!

[19] Gates R.R. 1909. The stature and chromosomes of *Oenothera gigas*, de Vries. *Archiv für Zellforschung* **3**: 525–552.

[20] Gates R.R. 1909. op. cit. p. 549.

[21] Gregory R.P. 1909. Note on the histology of the giant and ordinary forms of *Primula sinensis. Proceedings of the Cambridge Philosophical Society* **15**: 243–246.

[22] Gregory R.P. 1909. The forms of flowers in *Valeriana dioica, L. Linnaean Society Journal—Botany* **39**: 91–104; see p. 104.

[23] Johannsen W. 1903. *Ueber Erblichkeit in Populationen und in Reinen Linien*. G. Fischer, Jena. English translation in 1955. *Selected Readings in Biology for Natural Sciences* (ed. Harold Gall and Elga Putschar), volume 3, pp. 172–215. Chicago University Press.

[24] Johannsen W. 1909. *Elemente der Exakten Erblichkeitslehre*. G. Fischer, Jena.

[25] Nilsson-Ehle H. 1909. Kreuzungsuntersuchungen an Häfer und Weizen. *Lunds Universitets Arsskrift* **5**: 2.

[26] For background on the early careers of plant geneticists in the first generation (about 1900–1910), see Castle W.E. 1951. The beginning of Mendelism in America. In *Genetics in the 20th Century* (ed. L.C. Dunn), pp. 59–76. Macmillan, New York. Also see Rhoades M.M. 1949. Biographical Memoir of Rollins Adams Emerson (1873–1947). *Biographical Memoirs 1949*, pp. 313–323. National Academy of Sciences, Washington.

[27] Emerson R.A. 1904. Heredity in bean hybrids. *Agricultural Experimental Station in Nebraska, 17th Annual Report*, pp. 33–68; see p. 33.

[28] Emerson R.A. 1904. op. cit. p. 45.

[29] They produced a 1 yellow:1 purple ratio, where the yellows were of two homozygous kinds and the purple was heterozygous for the two yellow factors (nonallelic to each other) that represented separate color factors. The cross of F_1 purple $\times$ F_1 purple would be represented today as $F_1Y_1Y_2 \times F_1Y_1Y_2$ yields F_2 $1Y_1Y_1:2Y_1Y_2:1Y_2Y_2$ or, phenotypically, 2 yellow (the homozygotes) to 2 purple (the heterozygotes).

[30] Emerson R.A. 1909. Factors for mottling in beans. *American Breeders Association* **5**: 368–376.

[31] Emerson R.A. 1909. op. cit. p. 375.

[32] For an account of East's life, see Jones D.F. 1945. Biographical Memoir of Edward Murray East (1879–1938). *Biographical Memoirs*, volume 23, pp. 217–242, National Academy Press, Washington D.C.

[33] East E.M. 1909. The transmission of variations in the potato in asexual reproduction. *Connecticut Experimental Station Report 1909–1910*, pp. 120–160.

[34] East E.M. 1910. Notes on an experiment concerning the nature of unit characters. *Science* **32**: 93–95.

[35] East E.M. 1910. A Mendelian interpretation of variation that is apparently continuous. *American Naturalist* **44**: 65–82.

[36] For an account of Davis' career and life, see Cleland R. 1957. Bradley Moore Davis (1871–1957). *Yearbook of the American Philosophical Society 1957*, pp. 113–117.

[37] Davis B.M. 1911. Cytological studies on *Oenothera*. III. A comparison of the reduction divisions of *Oenothera lamarckiana* and *O. gigas*. *Annals of Botany* **25**: 941–974.

Maize Genetics and
the Popularization of Genetics

None of the recorded Mendelian variations in maize is of a nature that would be advantageous to a wild plant and most of them are obviously detrimental.

G.N. COLLINS*

...Maintaining otherwise useless inbred lines of corn solely for the purpose of utilizing the heterosis resulting from their hybridization was revolutionary as a method of corn breeding.

PAUL MANGELSDORF[†]

AN EXPANDING INTEREST IN PLANT BREEDING in continental Europe, Great Britain, and the United States resulted after the rediscovery of Mendelism in 1900. The next few years saw the emergence of three widely different approaches. In continental Europe, de Vries lost interest in following up Mendelism and hoped that his work with the evening primrose, *Oenothera*, would turn out to be the basis of a long-awaited experimental evolution. In addition, Correns shifted his attention from Mendelism to more complex problems that emerged from his crosses in flowering plants. He encountered a maternal inheritance in plant leaf variegation that shifted him into studies of nucleo-cytoplasmic relations. In Great Britain, William Bateson seized on Mendelism as a possible inroad to an alternative to natural selection, working on infinitesimal differences. He sought discontinuities in evolution, but he was not convinced that de Vries had found the right system or model. He wanted a Mendelian model of evolution and began a lengthy study of character traits that yielded many new phenomena that he happily reported to the Evolution Committee of the Royal Society set up by Francis Galton. In the United States, the experimental agricultural stations were geared for an infusion of Mendelian ideas to be applied to crop plants on a large scale. Beans, wheat, cotton, maize, tobacco, and potatoes were among the first to be explored by the new breeding analysis imported from Europe. This brief

---

*1921. In Dominance and the vigor of first generation hybrids. *American Naturalist* **55:** 116–133; see p. 119.
†1951. Hybrid corn: Its genetic basis and its significance in human affairs. In *Genetics in the 20th Century* (ed. L.C. Dunn), pp. 555–577; see p. 562. Macmillan, New York.

overview suggests that a major motivating interest among breeders (1900–1910) in the newly rediscovered Mendelism was a desire to bring together the tributaries of breeding analysis and evolution.[1] This motivation was as true for American breeders as it was for their European colleagues. A major difference between the two schools was the heavy reliance among American geneticists to use plants of agricultural interest. European studies were more varied in their choice of organisms.

Not much cytology was applied to the new genetics. This was not primarily an indifference or antipathy to the Sutton-Cannon-Boveri-Wilson chromosome theory, although both attitudes existed among the early Mendelians. The relationship between cytology and genetics was still hypothetical. No one knew that there would be polyploidy, aneuploidy, or chromosome rearrangements to study; these would largely emerge after the first decade. The first direct ties of genes to chromosomes would require the work of the second decade, when fruit flies in Morgan's hands would work like an Aladdin's lamp, pouring out bountiful connections of genes to chromosomes. Instead, the plant breeders, who dominated the field of breeding analysis in this first decade, looked for species that would make breeding analysis an exact science. The organism that emerged as the most favorable was maize, *Zea mays*.

Maize Has Advantages for Genetic Studies

Maize, or corn, as it is called in North America, was used by both de Vries and Correns to augment the confirmation of pea (*Pisum sativum*) results in 1899 and 1900. The plant is usually large (it grows to be six to ten feet or two to three meters tall), and it produces large distinct male and female organs that can be cut off after being fertilized (Figure 1). Most of the strains are self-pollinating, but cross-pollinating is an easy procedure. The kernels in the ears of corn have three tissues that could be examined. The outer pericarp is a diploid maternal tissue. The endosperm is a triploid tissue, composed of one pollen nucleus and two embryo sac nuclei (the latter is roughly the equivalent of an egg nucleus in the complex double fertilizations that lead to seed formation in flowering plants). The third component, the embryo, is the diploid tissue that brings together a pollen nucleus and an egg nucleus. Additional advantages to using corn for genetic studies included the many available commercial varieties: podded (covered kernels), flint (a hardened outer endosperm), pop (a small soft corn used for making pop corn), dent (a soft endosperm that shrinks and wrinkles the kernel as it dries), floury (a soft, eating corn), sweet (a translucent eating corn), and waxy (an animal chow corn that uses erythrodextrin instead of starch in its endosperm). The kernels were numerous from a single plant, and the ears could be stored to be analyzed during the winter season. The kernels are arranged in parallel rows on the ear and thus, each ear from a plant could have its kernels counted for discontinuous traits that segregated according to Mendelian laws. In most varieties, the kernels could be stored for years before being planted, and their longevity allowed time to go back to material that needed testing as new phenomena were found.[2]

Maize was also useful because character traits other than the ears and their kernels could be studied, including the seedlings, adult leaves, and stems.

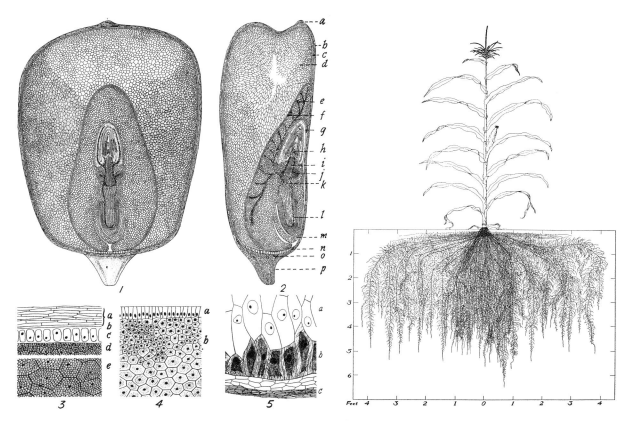

Figure 1. *The maize kernel, ear, and plant were well suited for genetic studies because of the ease of securing financial support for research and classifying genetic traits among the rows of kernels, as well as the large number of varieties available from commercial sources.* (Reprinted, with permission, from Kiesselbach T.A. 1999. The Structure and Reproduction of Corn. *Cold Spring Harbor Laboratory Press, Cold Spring Harbor, New York.*)

As a major crop plant, it was a favorite for support at the experimental agricultural stations in the land-grant colleges. Eating corn was picked up quickly by the settlers of the colonies, who learned to cultivate it shortly after landing in Massachusetts (they were taught by Native Americans to fertilize the ground with dead fish when planting the kernels). Maize for the European dinner plate was never popular and it was considered strictly animal chow. Maize did arise in the Western hemisphere, although at the beginning of the twentieth century, it was not known if it originated from Mexico, Central America, or the northwest of South America (especially Peru).[3] No wild varieties of maize were known.

Evolution Becomes an Early Reason for Studying Maize

Although maize turned out to be the most successful plant genetic organism in use for most of the twentieth century, several years elapsed before American breeders realized its advantages. The three pioneers in this effort were Edward Murray East, George Harrison Shull, and Rollins Adams Emerson. Like their European counterparts, American breeders first applied Mendelism mostly to smaller flowering plants. As the work from Bateson and his students reached American shores, the work of East, Shull, and Emerson began to center on maize as the best system to confirm and extend the European findings, but several issues needed resolution. For example, how many traits among known varieties would yield Mendelian ratios? Although many traits were clearly discontinuous, quite a few, such as row number on the ears or height of the adult plants, were not. Also at issue for farmers in the 1890s was the advantage of using inbred or highly outbred strains of corn. Some theories favored hybrids as superior planting material and others argued that a better effort would be to isolate inbred strains selected for higher yields.[4] Most of the traits studied in the European fields and greenhouses seemed to support a pessimistic finding that newly found mutations were losses of functions or monstrosities of little commercial value to cash crops, although they might prove interesting for horticultural varieties in ornamental plants.

The first generation of plant breeders was also less attracted to the chromosome theory of heredity, a reasonable reaction because few plants turned out to have sex chromosomes, and the existence of so many monoecious (hermaphroditic) crop plants made the question of sex determination more remote for botanists, who were very much interested in evolution. Maize lent itself to such studies even before the rediscovery of Mendelism, with papers on the evolution of maize, from teosinte (*Euchloena*) or from Gama grass (*Tripsacum*), dating from 1896 (Figure 2).[5] de Vries, of course, had aroused great interest in plant evolution through his studies of *Oenothera*, in the early 1900s, although, unfortunately for him, this turned out to be an aberrant system that offered far less than he and his school had hoped on the genetic relationship to the origin of species.

Fortunately, the practical and the pure science aspects often went together, which was important for those working in land-grant colleges. Despite the practical aspect always lingering in the background, all three of the first pioneers in maize genetics were motivated by questions of pure science. They also could trace the origin of their interest to Darwin. Darwin had studied inbreeding and experimented with plants (including maize) to see the effects of

Figure 2. *Maize and teosinte. Shown* (left to right): (first pair) *ear and cob of primitive corn;* (second pair) F_1 *hybrid ear and cob of cross of primitive corn and teosinte;* (third pair) F_2 *ear and cob of that self-fertilized* F_1 *hybrid; and* (far right) *an ear of teosinte. Teosinte is the most likely contributor to the bulk of the genome of maize. Note the small size of these ancestral components compared to the domesticated corn ear* (horizontal, top). (*Reprinted, with permission, from Beadle G.W. 1978. Teosinte and the origin of maize. In* Maize Breeding and Genetics [*ed. D.B. Walden*], *pp. 113–128. Wiley, New York.*)

repeated inbreeding.[6] At issue was the unresolved question of whether inbreeding intrinsically led to degeneration (a belief associated with the taboo on human incest) or if the inbreeding somehow brought unknown factors together and that the process was independent of the outcome. Many plant and animal breeders used inbreeding to obtain or fix novelties of horticultural or hobbiest interest.

Darwin communicated his results to Asa Gray at Cornell and it was Gray's student, William Beal, who started a study of inbreeding and outbreeding using maize, beginning in 1876 at Michigan State and later at the University of Illinois. That branch of corn genetics eventually led to East's contributions.[7] Independent of this line of investigation was the work of Shull, who was interested in quantitative traits, also a Darwinian legacy, but from a different slant. Darwin argued that changes on which evolution acted were of a fluctuating nature, usually minor, and that numerous generations of natural selection were required for such changes to accumulate.

Darwin was convinced that sports or qualitative mutations were pathologies that had little bearing on progressive evolution, the changes that led to new species. Instead, they were the losers in evolution and did not get transmitted. Some spectacular work was appearing in Denmark through the studies of Wilhelm Johannsen on quantitative traits. Johannsen studied bean size and weight and showed that inbred lines eventually led to pure lines, whose distribution curves remained constant regardless of the bean size used in a pure line. Outbred lines were quite different. In such lines, small beans produced a range of small beans and large beans produced a range of large beans; selection made a difference. It was the constancy of the pure lines that intrigued Johannsen and those who followed his work. Shull was enamored with Johanssen's work and sought to look for the effects of inbreeding and outbreeding in maize using the kernel row number in ears of corn. He began these experiments in 1905, but did not publish his first findings until 1908.

Emerson Makes Many Contributions to Maize Genetics

Emerson introduced the third branch of early maize genetics. Before this, he had bred beans (*Phaseolus*) for their color variations and demonstrated that color inheritance was complex, with many genes involved in flower color. He also worked with gourds to study the inheritance of their unusual shapes. He confirmed the relative constancy of the F_1 generation when a cross was made between contrasting traits, and the much greater variability of the F_2 when the F_1 was self-fertilized or crossed to the F_1. He added maize genetics to this roster of studying individual character traits and chose height because some corn plants were relatively short and others quite tall. Unlike most of his color crosses where the results tended to be discontinuous

following Mendelian patterns of inheritance, he found the maize results more complex, with incomplete dominance in some traits and unexpected gradations of size in other cases.

Although maize and plant genetics had earlier starts than fruit fly genetics, the findings that Morgan and his students reaped in abundance between 1910 and 1916 were either not yet found or could not be found in plant material. There are many reasons for this difference in success. Fruit flies have about 25 generations per year. An experimental cross can be carried out within two weeks with abundant F_1 progeny at the end of that time. They also have a small number of chromosomes: four in the gametic state. Maize has one (occasionally two) growing season. Its chromosome number is 10 in the gametic state, but that was not resolved for some time.

Yoshinari Kuwada was the first (1911) to attempt to work out the maize cytology.[8] He thought that he had found a haploid number of 12 (he used an unusual strain, black Mexican, with a variable number of chromosomes), but later, cytologists (especially Theodore A. Kiesselbach) reported 10 in a number of varieties of maize, a number Kuwada confirmed in 1925. Kuwada had studied cytology in New York and returned to Kyoto to do the bulk of his cytological studies. The larger the number of chromosomes in a species, the more difficult it is to find linked traits or to encounter what Bateson and his school referred to as coupling and repulsion. Plants usually lack sex chromosomes, and it was these chromosomes that made the genetic implications of nondisjunction accessible to Bridges using fruit flies. It is not that nondisjunction is rarer in plants. Blakeslee was fortunate that his aneuploids in jimsonweed, *Datura*, each gave unique anomalies that he could identify. Not all aneuploids provide such opportunities and in animals they are almost always fatal to early development (usually in organogenesis) unless found among the sex chromosomes. The sex chromosomes also make it easier to identify linked genes and map them. Without such a gift from the sex chromosomes, plant geneticists had to use indirect ways to identify linked genes, and even more difficult procedures had to be developed to map those factors to specific chromosomes.

Shull's Use of Hybrid Corn
Leads to an Agricultural Revolution

The two most significant findings associated with corn are the early triumphant analysis and application of hybrid corn to agriculture and the cytogenetic findings associated with the work of Barbara McClintock (1902–1992). McClintock's work belongs to the third generation of maize workers (most of her work was done in the 1930s to 1950s) and is a capstone of classical genetics (Figure 3). Shull's work resulted in one of the most important contributions to genetics because it eclipsed the approaches of Luther Burbank (a gifted selectionist who made no contributions to basic genetics and whose work required no knowledge of Mendelian genetics or cytology). Its economic and beneficial social success made genetics a much-admired science and one that attracted many outstanding students. It fulfilled the predictions of Bateson on how genetics would transform science and society beyond his own imagination.

In 1908, when Shull presented his work in "The composition of a field of maize," he showed that a field is genetically complex.[9] While self-fertilization leads to homozygosity and loss of variation, outbred corn shows enormous variation. The hybrids also fare better during the growing season and produce better yields. Shull (Figure 4) did his work at Cold Spring Harbor, the genetic experimental station that Davenport had organized with the hope that it would lead to experimental evolution. Shull expressed his motivation in a more comprehensive paper on maize in 1911: "In 1905 I undertook a rather extensive series of comparisons between cross-bred and self-fertilized strains of Indian corn

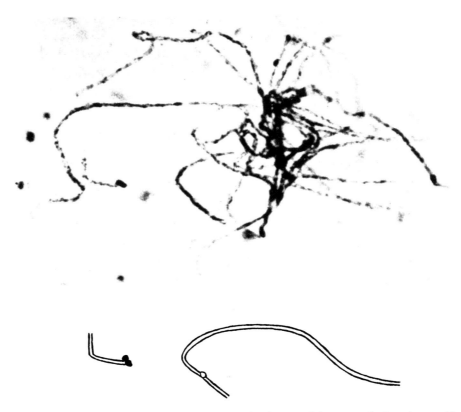

Figure 3. *Maize chromosomes. Barbara McClintock proved to be one of the most gifted cytologists. Her careful preparations and interpretations gave corn cytogeneticists an ease in identifying genetic phenomena and locating genes that fruit flies lacked until the discovery of salivary gland chromosomes. The chromosomes of maize were useful for studying cytogenetics, especially through the careful work of McClintock, whose contributions were numerous and profoundly influenced this field. Shown here is her analysis of chromosome 9. (Reprinted, with permission, from McClintock B. 1952. Chromosome organization and genic expression. Cold Spring Harbor Symposia on Quantitative Biology 16: 13–47.)*

for the purpose of discovering the effects of these methods of breeding upon variability, and these investigations have been continued each year since that time."[10] In 1909, Shull presented a remarkable finding that he presented under the title, "A pure line method in corn breeding."[11] He prepared two inbred strains, A and B, that were self-fertilized for three years, and then he crossed these to each other. When self-fertilized, strains A and B produced typical yields of about 12 to 13 pounds of

Figure 4. *George Harrison Shull was a plant breeder whose major contributions to maize genetics were carried out at Cold Spring Harbor. His discovery of heterosis (hybrid vigor) from producing hybrid strains of corn became an icon of the commercial worth of genetics. Shull's use of inbred strains produced dramatic increases in bushels of corn. (Courtesy of the Cold Spring Harbor Laboratory Archives.)*

corn per row of planted seeds, in contrast to the hybrid of a cross between these strains, whose kernels were planted in the alternate rows. This alternate row method served as an environmental control for soil and weather exposure differences. These hybrids yielded a range of 48 to 55 pounds per row. Shull quickly recognized the value that this would have to farming if hybrid seed were prepared commercially and sold to farmers for planting: "I am not prepared at present to say what will be the probable cost of seed-corn when produced by this method, but have reason to suppose that it would be more expensive than by the present method; nor can I surmise what relation this increased cost will bear to the increased yield that will be produced. These are practical questions which lie wholly outside my own field of experimentation, but I am hoping that the Agricultural Experimental Stations in the corn-belt will undertake some experiments calculated to test the practical value of the pure-line method here outlined."[12]

East Improves Hybrid Corn

East worked on tobacco and maize at the Connecticut Agricultural Station. His interest in Darwin's views on inbreeding and hybridization initiated his own crosses, and he prepared a review of the literature known to his day.[13] He pointed out that Darwin crossed *Ipomoea purpurea* for ten generations using outbred and inbred plots. Although the outbred plots maintained an average but consistent variability throughout, the inbred plot of plants plateaued instead of continuing a trend to shorter plants. Darwin recognized that an inherent physiological pathology associated with inbreeding is false. East agreed: "Some species thrive under inbreeding while others appear

to deteriorate."[14] For maize, he noted that even a single generation of inbreeding leads to diminished vigor among the progeny. Some species are perpetually self-fertilized (when Mendel did his studies on peas, he had to prevent this natural tendency of self-fertilization). Flowering plants like dandelions develop without fertilization; they are parthenogenetic. This is comparable to being in a perpetually inbred state because no new combinations of variations from crossing are possible.

East's work on tobacco and maize supported Darwin; no continued degeneracy from inbreeding occurred in these plants. East speculated that the

Figure 5. *East's double-hybrid method. East improved on Shull's original method for generating hybrid corn, and his proved more practical for farmers. He took a lifelong interest in the problem of hybrid vigor. (Reprinted from Sturtevant A.H. and Beadle G.W. 1939. An Introduction to Genetics. W.B. Saunders, Philadelphia.)*

difference between animal incest and plant inbreeding might be a developmental problem, such as the frequency of cell division. A "decrease in vigor and not degeneration of characters is usually the sole effect of inbreeding."[15] He rarely encountered new pathologies from inbreeding plants; instead, he got a loss of vigor. East based his argument on 30 crosses he had carried out in maize in 1908. He confirmed what Shull was finding in maize and praised his work. He noted in his own work how "every possible combination of dent, flint, and sweet maize was grown

and in every case an increase in vigor over the parents was shown by the crosses."[16] East later developed a superior method for hybrid corn production using a "double cross method" in which strains A x B produce hybrid C, and strains D and E produce hybrids of type F. When the F_1 hybrids C and F are crossed, the F_2 results in adult plant size, ear row number, and other commercially useful traits were dramatically increased. The popularization of East's work (mostly done in association with his colleague D.F. Jones) led to an agricultural revolution (Figure 5). Dozens of seed companies (including the one headed by Henry Wallace, the first geneticist of national prominence, who became Vice President to Franklin Roosevelt) supplied hybrid seed to farmers, who quickly realized its superiority to the yields they had obtained before commercial hybrid corn based on the Shull-East model was available. East was an extremely effective speaker and writer. His book *Inbreeding and Outbreeding* was a scientific best seller.[17]

East also used his plant studies to study quantitative traits. In 1910, he published an article called "A Mendelian interpretation of variation that is apparently continuous."[18] He had used maize and found that his own work confirmed the work of Nilsson-Ehle, who had used cereal grain coat color to work out bell-like curves from modified 9:3:3:1 ratios (yielding 1:4:6:4:1 distributions by intensity or hue and 15:1 by color to noncolor) to modified 63:1 ratios in wheat, oats, and barley. In Nilsson-Ehle's more complex ratio, three factors were involved in the red cereal coat color. In maize, East reported two factors for yellow color in the endosperm, three (possibly four) in the pericarp, and two factors for the aleurone. In each case, he obtained the modified ratios found by Nilsson-Ehle. East claimed that he had actually obtained the results when Nilsson-Ehle's work came out and thus, he had independently discovered this phenomenon.

Emerson's Maize School Dominates Plant Genetics

Emerson is best known for the school of maize he developed at Cornell University. Among the best-known students in his Botany Department were Milislav Demerec, W.H. Eyster, George Wells Beadle, Marcus Rhoades, and Barbara McClintock (Figure 6). Beadle, who later shifted to fruit flies and then to fungi, went on to pioneer studies in the genetic control of biochemical pathways, a brilliant analysis of gene function referred to as the one-gene–one-enzyme theory. It is frequently acknowledged as a founding branch of molecular biology. McClintock stayed with maize and worked out its complex cytogenetics mostly in the 1930s; this included the remarkable study of chromosome breakage, its effects on cell division, and the discovery of transposable genetic elements that move from chromosome to chromosome. Both Beadle and McClintock ended up with Nobel Prizes. Demerec became a pioneer of gene structure in bacterial chromosomes. These second- and third-generation maize investigators worked out cytogenetics in maize and competed with *Drosophila* genetics in generating new ideas and launching new fields of genetics. Emerson obtained numerous single-gene mutations (most of them pathologies), which he and his students used to study the number of factors in different character traits. He demonstrated cytoplasmic inheritance in maize, and he developed methods to map maize genes on their chromosomes.[19]

Emerson had doubts about the interpretation of traits in coupling or repulsion. Bateson encountered traits that stayed together (coupling) in a gamete or that segregated into equal numbers of gametes (repulsion). He could not see the logic of a mecha-

Figure 6. *In the 1920s, R.A. Emerson established a school of maize genetics at Cornell. This photo includes some of his major students: Charles Burnham, Marcus Rhoades, Rollins Emerson, Barbara McClintock, and George Beadle (with dog). (Courtesy of the California Institute of Technology Archives.)*

nism for the same traits to keep them in repulsion in one parent and in a state of coupling in an altogether different parent. Emerson proposed an explanation: "Although this is pure speculation, it is worthwhile to note that if genes were definitely located in chromosomes and that if parental chromosomes separated bodily at the reduction division we should have an 'explanation' not only of perfect genetic correlation and of allomorphism—spurious or otherwise—but of independent inheritance as well." He put a footnote after this remark and stated that he wrote it just before reading Morgan's paper of 1911, which introduced the idea of shifting genes in homologous chromosomes.[20] Emerson did not specifically state that there was an exchange between paired linked genes; Morgan was the first to make that proposal but Emerson's concept was virtually there. He solved the problems of there being more genes than chromosomes and how they could account for their linked (coupled) or repulsed (on opposite chromosomes in a pair of homologs) states. Emerson was and should be given credit for being the first to accept Morgan's results and for seeing the relationship of genes to chromosomes in maize.

Wilson and his students coupled Mendel's laws with meiosis through the chromosome theory of inheritance. On his own, Morgan attempted (but failed) to establish a discontinuous origin of species by "sudden mutations" in fruit flies, thereby uniting breeding analysis with evolution. Instead, Morgan and his students established a union of cytology and genetics that went far beyond Sutton's chromosome theory of inheritance. They introduced linkage groups, mapping, sex-linked inheritance, and nondisjunction, among many other findings that fruit fly genetics revealed. Morgan's group even made sense of Darwinian evolution by natural selection, despite Morgan's antipathy to that model. The first generation of corn geneticists also united breeding analysis with evolution. Although Emerson accepted the

chromosomal model of mapped genes in fruit flies, the cytology and the genetic tools to provide this in maize would take many more years. But maize offered brilliant insights into unresolved issues of the nineteenth century. The work on hybrid corn dispelled simplistic theories of the evils of inbreeding and firmly established the role of Mendelian genes in producing quantitative traits. Where fruit flies supported natural selection through the analysis of chief genes and modifiers, maize genetics supported natural selection through its ease of relating specific genes to quantitative traits.

For both maize genetics and the genetics of smaller plants in both Europe and the United States, the studies were carried out without a sense of competition across the Atlantic. With the exception of maize breeding, Europe led the way in novelties discovered. de Vries, Correns, and Bateson were the acknowledged leaders in the European community of geneticists; their work was much admired by their American counterparts. Bateson, in particular, used his findings to introduce a flood of new terms into the field of genetics. But Bateson had a handicap that American geneticists did not encounter: He had to fight off a fierce resistance to genetics by those Darwinists who believed that Mendelism was of no value to natural selection and that it was misguided in seeking applications to evolution.

End Notes and References

1 Mangelsdorf P.C. 1951. Hybrid corn: Its genetic basis and its significance. In *Genetics in the 20th Century: Essays on the Progress of Genetics During its First 50 Years* (ed. L.C. Dunn), Chapter 24, pp. 555–572. Macmillan, New York.

2 Eyster W.H. 1934. Genetics of *Zea mays*. *Bibliographia Genetica* **11:** 187–392.

3 The most favored interpretation today is that maize arose by mutations occurring in teosinte in southern Mexico; the greatest number of varieties of ancient corn originates from that region. In the 1920s, the Russian geneticist, N.I. Vavilov, proposed that centers of origin for domesticated plants and animals are most likely to have the greatest number of varieties.

4 In pre-Mendelian breeding studies, en masse cross-fertilization among several varieties was tried as an alternative to the farmer's traditional method of saving the healthiest and largest ears for the next year's planting. See Shull G.H. 1909. A pure line method in corn breeding. *Proceedings of the American Breeder's Association* **5:** 51–59.

5 Beal's work is cited in Shull G.H. 1911. The genotypes of maize. *American Naturalist* **45:** 234–252 (Beal W.J. 1876. *Reports to the Michigan Board of Agriculture* 1876–1881).

6 Darwin C. 1876. *Effects of Cross and Self Fertilization in the Vegetable Kingdom*. Murray, London

7 Mangelsdorf P.C. 1951. op. cit. p. 563.

8 Kuwada Y. 1911. Meiosis in the pollen mother cells of *Zea mays* L. *Botanical Magazine, Tokyo* **25**. In Kuwada Y. 1925. On the number of chromosomes in maize. *Botanical Magazine, Tokyo* **39:** 227–234.

9 Shull G.H. 1908. The composition of a field of maize. *Proceedings of the American Breeders Association* **4:** 296–301.

10 Shull G.H. 1911. The genotypes of maize. *American Naturalist* **45:** 234–252; see p. 239.

11 Shull G.H. 1909. A pure line method in corn breeding. *Proceedings of the American Breeders Association* **5:** 51–59.

12 Shull G.H. 1909. op. cit. p. 59.

13 East E.M. 1909. The distinction between development and heredity in inbreeding. *American Naturalist* **43:** 173–181.

14 East E.M. 1909. op. cit. p. 173.

15 East E.M. 1909. op. cit. p. 177.

16 East E.M. 1909. op. cit. p. 179.

17 East E.M. and Jones D.F. 1919. *Inbreeding and Outbreeding: Their Genetic and Sociological Significance*. Lippincott, Philadelphia.
     Unfortunately, East was also enamored with the eugenics movement, and some of his racist remarks, in an era when racism was not likely to raise protests, are embarrassing to read: "The Negro is a happy-go-lucky child, naturally expansive under simple conditions; oppressed by the restrictions of civilization, and unable to assume the white man's burdens. He accepts his limitations; indeed, he is rather glad to have them. Only when there is white blood in his veins does he cry out against the supposed injustice of his condition. White germplasm in a Negro complex spurns its hopeless situation... ." The reference to "white blood in his veins" probably reflects the popular view of the mulatto as degenerate and evil, a portrayal seen in D.W.

Griffith's film, *The Birth of a Nation*. East's quote is from Population. *Scientific Month*ly. June 1920, pp. 603– 624; see p. 621.

[18] East E.M. 1910. A Mendelian interpretation of variation that is apparently continuous. *American Naturalist* **44:** 65–82.

[19] For a more detailed account of R.A. Emerson's career and his influence on the maize group at Cornell, see Berg P. and Singer M. 2003. *George Beadle: An Uncommon Farmer*. Cold Spring Harbor Laboratory Press, Cold Spring Harbor, New York.

[20] Emerson R.A. 1911. Genetic correlation and spurious allelomorphism in maize. *Annual Report of the Nebraska Agricultural Experimental Station*, pp. 59–90; see p. 80.

W. Bateson, C.C. Hurst, and C. Davenport established independent assortment of two or more genes associated with the comb morphology of poultry. The extension of Mendelism to animals was a major reason for the rapid growth of genetics as a science. (Reprinted from Bateson W. 1909. Mendel's Principles of Heredity. Cambridge University Press, United Kingdom.)

Animal Genetics in the First Decade of the Twentieth Century

From this survey of evidence mostly already published, it is clear that Mendelian analysis provides a means of elucidating a large part of the phenomena. The majority of observations are in accord with the Mendelian hypothesis in a simple form. The true solution of several subordinate problems still remains obscure. The value of the Mendelian analysis will be the more appreciated when it is remembered that previously the whole body of facts must have been regarded as a hopeless entanglement of contradictions, as reference to any non-Mendelian discussion even of these very phenomena will show.

WILLIAM BATESON*

IN 1906 LONDON, JULY 30 TO AUGUST 3, what should have been the third international conference on plant breeding was renamed, at the urging of William Bateson, the Third International Conference 1906 on Genetics.[1] The Royal Horticultural Society, then in its 103rd year, hosted it. The first conference on plant breeding took place in 1899 (about 30 were in attendance, also in London), and the second conference was held in New York in 1902. At this second conference, Bateson promoted Mendelism, and the message was carried away with great enthusiasm by American plant and animal breeders in attendance, who had already begun their first confirmations and exten-

sions of Mendel's law of segregation. At the third conference, Bateson formally introduced his new name, genetics, for the field (he had used it a year earlier in a letter for the Quick professorship that he did not get) that he considered a new branch of physiology emphasizing breeding, reproduction, and evolution. Note that Bateson did not mention cytology, although the chromosome theory was first introduced in 1902. Bateson was also confident in the future of genetics as a science. He told his audience, "A knowledge—a precise knowledge—of the laws of heredity will give man a power over his future that no other science has yet endowed him with. I am not going to say that this knowledge is going to create the millennium of the human race; I can only say it will change man's destinies profoundly—whether for good or evil the future alone will show!"[2]

*1903. The present state of knowledge of colour heredity in mice and rats. *Proceedings of the Zoological Society of London* **2**: 71–99; see p. 95.

It was unusual for a plant breeding conference to also sponsor papers on animal genetics, but Bateson felt that all genetics work in progress should be represented. Bateson had been working with his students using poultry and found some new genetic phenomena demonstrating that complex characters, such as the shape of combs in poultry, could be analyzed by Mendelian breeding crosses with three major factors involved. In plants, he had earlier shown modified ratios of 9:7 and 9:4:3, as well as the complex three-factor ratio of 27:9:28. He had now extended this success of Mendelism to poultry.[3] Bateson was not alone in presenting work on animal genetics: C.C. Hurst (England), C.B. Davenport (U.S.), A.D. Darbishire (England), L. Cuénot (France), and W.E. Castle (U.S.) had already published results of animal crosses confirming Mendelism, and they were there to present their work or their work was discussed at the meeting.

Early Studies of Mouse Genetics Confirm Mendelism

The first studies of mouse genetics were among the first to yield important findings. Lucien Cuénot in France (Figure 1), W.E. Castle in the United States, and A.D. Darbishire in England were the first to begin breeding mice, with Mendelism clearly motivating their experimental designs.[4] Castle worked on albinism in mice (and extended it to humans), Darbishire studied a behavioral trait called waltzing, and Cuénot worked on coat colors with color dominant to albinism and normal locomotion dominant to the neurological waltzing impairment. Surprisingly, Cuénot found that one color, yellow, did not breed true[5]; it seemed to be a perpetual hybrid. Cuénot's 1903 findings stimulated several interpretations. He suggested that the two yellow gametes are incompatible and do not fertilize, thus arguing from the prevailing plant model of self-sterility. Bateson thought that it might be due to the rarity of the homozygote, assuming quite a few factors were involved in the formation of yellow. Yellow was dominant to the normal gray or agouti color, and it was also dominant to black in the heterozygotes. But the dominance was slightly incomplete and hints of the gray or black could be seen in the hybrids. The debate was not resolved until 1910 when Castle and C.C. Little showed that the precise ratio in the yellow mouse cross was a 2 yellow to 1 white, black, brown, or agouti, depending on the colored parent

Figure 1. *Using mice, Lucien Cuénot demonstrated that coat color factors in mice were Mendelian, with the exception of yellow mice. These were perpetual hybrids and the mechanism for their heterozygosity remained a puzzle for several years. Cuénot was one of the few French biologists to embrace Mendelism, which was vigorously resisted by those who favored Lamarck's model of acquired characteristics. (Reprinted from Vilmorin P.D., ed. 1913. IV Conférence Internationale de Génétique. Libraries de l'Académie de Médicine, France.)*

Ledger No.	Yellow Young	Non-yellow Young	Total Young	Per Cent. Yellow
1–5,400	423	238	661	63.99
5,401–5,514	22	11	33	66.66
5,515–5,824	97	45	142	68.30
5,825–6,437	184	110	294	62.58
6,438–6,621	74	31	105	70.47
Total	800	435	1,235	64.77
Cuénot's results......	*263*	*100*	*363*	*72.45*
Grand total.	1,063	535	1,598	66.52

Figure 2. *Mouse yellow × yellow cross. The perpetual hybrid state of yellow mice was successfully explained by Castle whose larger series showed a 2:1 (and not 3:1) ratio. The modified ratio arises because one category, homozygous yellow zygotes, abort. (Reprinted from Castle W.E. and Little C.C. 1910. On a modified Mendelian ratio among yellow mice. Science 32: 868–870.)*

used in the cross (Figure 2).[6] Every yellow mouse tested (263 of them) was heterozygous. Castle and Little argued, "Now the evidence which will presently be offered shows that, contrary to the idea of Cuénot as well as to the suggestion of Bateson and Punnett, the yellow egg which by chance has met a yellow sperm has its career ended thereby."[7] This implied a lethality of the zygote or its embryo, but the stage of the life cycle was not quite clear. Castle and Little were the first to demonstrate that the ratio was 2:1 for the yellow mice, but they were not the first to obtain a 2:1 ratio. Erwin Baur found this in 1907 in *Antirrhinum*, where an "aurea" race, involving two hybrid parents, produces one fourth of its seeds that germinate and die as seedlings, in contrast to the hybrid parents, which survive with a yellow-green color and give a ratio to the surviving progeny of 2 yellow-green to 1 green.[8] Fruit fly geneticists would later describe Baur's chlorophyll mutation as semilethal. The death of the seedlings suggested to Castle and Little that a similar "...physiological inability to develop may permanently modify a Mendelian ratio, causing the loss of an entire class."[9] The term lethal again was not applied. Two years later, Morgan would make this designation for an X-linked lethal that he encountered in *Drosophila*, and Castle's prediction of a missing class of expected progeny (in Morgan's case, of males) was vindicated.[10]

Castle Makes Many Contributions to Mammalian Genetics

William Ernest Castle (1867–1962) was born on a farm in Ohio and died in California after a long academic career, spent mostly at Harvard.[11] His father was a farmer and school teacher and his mother an M.D. He received a B.A. in classics in 1889 at Dennison University and taught classics for three years in Kansas. He returned for a second B.A. at Harvard, this time majoring in natural history. He worked with C.B. Davenport on a master's degree and with E.L. Marks on his Ph.D. (received in 1895) in developmental biology. At that time, heredity was still a philosophic idea and not yet a science. His dissertation was on *Ciona intestinalis* (a tunicate), and he showed that its mesoderm in embryos was developed from pouches in the endoderm, a classic developmental problem. Of more interest, sensitizing him to his future interest in heredity, was his finding that *Ciona* was a hermaphrodite and its gametes were self-sterile. Self-sterility was known in plants, but it was not yet known to occur in animals. Castle taught at Wisconsin and in Illinois before returning to Harvard as an instructor in 1897, and he remained there with a distinguished career until his retirement in 1936, when he moved to the University of California at Berkeley. Castle's first genetic studies were with fruit flies, but that work, which studied inbreeding effects for 60 generations, had little Mendelian interest.[12] He shifted to mammalian genetics and remained in that field for the rest of his career.

From 1900 on, Castle became the chief advocate of Mendelism in the United States. He extended Mendelism to guinea pigs and rabbits, independently finding the aberrant 9:3:3:1 ratios and the existence of two or more genes for a given character, such as coat color. He wanted to determine the validity of Mendel's laws for contrasting traits, with most of those studies being carried out in 1903–1907. He also ran into a problem when he studied a spotting pattern in rabbits or a hooded pattern in rats (1906–1919) (Figure 3). Here, the trait behaved like a typical unit factor, giving a 3:1 ratio as expected, but among the spotted rabbits or the hooded rats a range of variation occurred.[13] Some rabbits were white with a few spots, and at the other extreme, they were so filled with black spots that the white fur was a small minority of the pelt. Most showed mixtures of spots and thus there was a graded series.

What was the source of this variation? Castle erred in attributing it to the gene itself. He believed heterozygosity was akin to mixing two layers of differently colored modeling clay or wax, with the subsequent separation yanking off impurities. This theory of the impurity of the gametes (originally proposed by Morgan) was at odds with Mendel's original claim that the hybrid separates the two parental types without contamination.[14] Mendel did not obtain chartreuse or variegated green peas in the F_2 generation. Castle went one step further. He argued that he could select for hoodedness and shift the pattern to more extreme pigmented hoods or to much smaller hoods. If the homozygous hooded rats could be selected, then this might be a basis for Darwinian

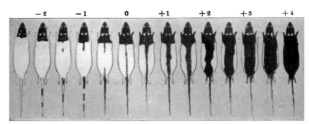

Figure 3. *Castle's hooded rats. Castle criticized the Morgan school for its alleged invention of modifier genes to explain variable traits. Castle believed that it was the gene itself that fluctuated, tossing off Darwinian plus-and-minus changes. Eventually, Castle realized that Morgan and his students were right after all. (Reprinted, with permission, from Castle W.E. 1940. Mammalian Genetics. Harvard University Press, Cambridge, Massachusetts © 1940 by the President and Fellows of Harvard University.)*

plus-and-minus variations, and an unstable gene would be the basis for Darwinian small variations. It was logical and appealing, but as the work of

Johannsen and Nilsson-Ehle was showing in plants, variable traits were frequently associated with multiple factor inheritance.

After 1912, Morgan and his students got into the fray, and they believed that there were categories of genes that Muller called modifier genes, which could influence the expression of a chief gene for a trait. After 13 years of fighting his cause, Castle finally did the essential crosses and confirmed what Morgan's group had pointed out. What they called "residual inheritance" served as a background to intensify or diminish the activity of the chief gene, *hooded*. The *hooded* gene did not vary; its expression varied. In Castle's own words, "My own early observations indicated that they were modifiable, and to this end I stubbornly adhered, like Morgan in his early opposition to Mendelism, until the contrary view was established by a crucial experiment."[15] That was a key difference between Morgan's school and Castle's. Castle's error arose because he believed, as did Bateson, that the unit character (later called a gene) produced the character in some sort of direct way. In Morgan's group, the influence of Johannsen and Nilsson-Ehle's work was significant. They saw the gene as relatively stable, but they found evidence for modifier genes.

Despite his failings in interpreting some of the results of his early experiments and observations, Castle made some lasting contributions. One of the most powerful experiments he conducted was with his student John C. Phillips; it surpassed the tail-cutting experiments of Weismann as a test of Lamarckism.[16] Castle and Phillips used guinea pigs (Figure 4). They substituted the ovaries of a black guinea pig for the albino ovaries of an albino female. They then had the surgically altered albino mate with an albino male. The three litters of offspring from repeated crosses with albino males were all black and showed not a speck of white in the fur. Castle and Phillips argued that the long presence of the black ovaries in the albino body of the host

guinea pig ruled out any physiological route of transmission of the albino heredity into the black ovaries. They believed that they had confirmed Weismann's theory of the germ plasm—soma did not influence the hereditary potential of the reproductive tissue.

Castle found multiple allelism in the English rabbit and contrasted these fixed alleles with the fluctuating types derived from a single gamete. He used an evolutionary argument to support the importance of his model. "If unit characters are not constant, selection requires much of the importance which it was regarded as possessing in Darwin's scheme of evolution, an importance which many have recently denied it." Castle was right about the basis for the evolution but wrong about the basis for the fluctuation.[17] Since he was weak in mathematics, Castle avoided studies involving complex statistics or quantitative traits requiring a lot of calculation. He preferred simple experiments to execute. With his students, he was formal, courteous, and reserved, but, like Morgan, he

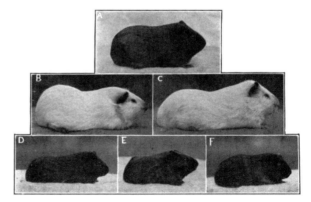

Figure 4. *A test for Lamarckism by ovary transplants. Castle and Phillips used guinea pigs and showed no influence of the albino mother on the pigmented strain of ovaries that replaced her albino ovaries. When mated with an albino male, her offspring were all heterozygous for the dominant black color of the ovary donor. (Reprinted from Whitney L.F. 1933. The Basis of Breeding. Earle C. Fowler, New Haven, Connecticut.)*

treated them as colleagues. Unlike Morgan, he found no new phenomena to compensate for his published errors. His major contribution was in establishing a flourishing school of geneticists, who made mammalian genetics, from mice to horses, the passion of their careers in genetics.

Hurst Analyzes Animal Coat Colors and Other Traits

Castle's counterpart in England was C.C. Hurst, who had a similar range of interest in mammalian genetics (Figure 5). He studied the coat color of mice and showed that there were many genes involved. He studied rabbits, poultry, and rats and like Bateson, he also studied plants, especially the inheritance of color in tomatoes, peas, and orchids.[18] Hurst had obtained a 9 gray:3 black:4 white ratio for fur color in the Belgian hare. He assigned roles to the two genes involved, one allowing for color and the other the specific color. He hinted that a third color factor may also be involved, based on similar findings that Bateson had reported. In his poultry analysis, he studied three factors for comb development, one for feathering of the shanks, one for extra toes, one for the color of the shanks (yellow or white), and two for plumage. In his work on thoroughbred horses, he found chestnut to be recessive to bay and brown. Altogether, he surveyed 43 characters in a variety of plants and animals and concluded that 41 of them involved Bateson's model of a simple presence and absence of a factor. He said there were three possibilities for the meaning of "absence." It could be dormant or latent but still there, it could be a real factor representing an absence of the trait, or it could be a literal absence of the factor itself. Bateson favored the latter, but Hurst pointed out that no pairing of factors was possible in such a scheme. Pairing of factors was implied for an orderly separation of the many factors into the gametes. He concluded that "All three views are possible, and all are open to some objection; in the present state of knowledge, it is difficult to say which of the three is the most reasonable. ...On this view mutational variations may consist simply of the addition of new unit-characters and the subtraction of old ones." He concluded with a broad view of the unit character: "The biological problem of the future will be not so much the origin of species as the origin of unit characters."[19]

Figure 5. *C.C. Hurst studied numerous traits in animals and extended Mendelism to a sufficient variety of traits to convince Bateson that Mendelism did not just deal with trivial traits but rather extended to the most fundamental tissues and organs of animal life. Bateson admired Hurst's work. (Reprinted from Wilks W. 1906. Report of the Third International Conference 1906 on Genetics. Royal Horticultural Society/Spottiswoode and Company, London.)*

Animal Crosses Reveal Sex-associated Traits

The mammals and birds used in these studies were male and female. Most of the plants were monoecious or hermaphrodites, where sex is not an issue in a cross. Bateson was concerned about several traits in animals that showed an effect in males but not in females. For example, Dorset lambs showed horns in the males but not in the females, as if horns were dominant in males and recessive in females. He noted that the same was true for tortoise shell pattern in cats, which always occurs in females. He included in this sexual dimorphism the well-known observation that it is human males who are color blind and not females, but it is the females who pass it on, not the unaffected males in a family. Color-blind women were rare, but from a study of seven such women, he noted that all 17 of their sons were color blind. This fit his belief that this trait is dominant in males and recessive in females.[20] After Morgan's 1910 finding of X-linked traits in fruit flies, that interpretation would change.

It is noteworthy that the chromosome theory was proposed in 1902, and the debates about the "accessory chromosome" of C.E. McClung with respect to sex determination were in force throughout this period, but for the next eight years, no one inferred a sex chromosome explanation. One reason for this reticence might have been the complexity of sexual dimorphism in relation to traits. L. Doncaster and G.H. Raynor studied the currant moth, *Abraxas*, and two of its species, *A. grossulariata* and *A. lacticolor*.[21] *lacticolor* females crossed with *grossulariata* males gave F$_1$ offspring that were both *grossulariata* (Figure 6). The F$_1$ × F$_1$ of this cross in the F$_2$ gave *grossulariata* males and females and *lacticolor* females, but no *lacticolor* males. This made sense on a Mendelian scheme if special assumptions were made, coupling the male sex factor with the *grossulariata* trait. No

Figure 6. *The moth* Abraxas *revealed a sex-associated trait that did not fit into the Mendelian model. It was the female* Abraxas *that exhibited the mutant trait and the heterozygous male that transmitted it to the daughters. Doncaster, unlike Wilson and Morgan, did not attribute these findings to a chromosomal mechanism. Because the Bateson school rejected cytological approaches, they failed to identify the ZW mechanism of sex inheritance (which is like the XY, but the male is homogametic and the female heterogametic, using Wilson's terminology). (Reprinted from Doncaster L. 1914. The Determination of Sex. Cambridge University Press, United Kingdom.)*

such coupling was present in the *lacticolor* trait. Morgan's work in 1910 offered a non-Mendelian interpretation, assigning the traits to one of the sex chromosomes.

Actually, two differences existed between Doncaster's moths and Morgan's fruit flies. In the moths, an attempt was made to explain the events without reference to chromosomes. In fruit flies, that assignment was made once Morgan found several such factors showing sex-limited ratios. The second difference was the constant frustration that biologists encounter when universal interpretations encounter biological diversity. As Wilson and Nettie Stevens demonstrated about the same time that Doncaster was publishing his results, fruit flies had **XY** (or **XO**) = male and **XX** = female. In *Abraxas*, the situation with respect to sex was reversed: **WW** = male and **WZ** = female. The use of the **WZ** nomenclature was introduced after

Morgan and Wilson introduced the **XY** system. Just as Mendel was frustrated by the nonuniversality of Mendelism that *Hieracium* presented, so too was Bateson frustrated by this lack of agreement between sex-limited traits and the sex in which the traits were expressed. If anything, the British were on stronger ground for considering the *Abraxas* case the more likely, because Bateson's school quickly encountered sex-limited traits in poultry, which also turned out to have the **WZ** rather than the **XY** mode of inheritance, as these systems were to be designated in the 1910s.

Mendelism Becomes the Predominating Model of Heredity for Plants and Animals

As the first decade of animal genetics drew to a close, most breeders agreed that there were no fundamental differences between plant breeding and animal breeding with respect to Mendelism. Many traits were involved; most showed a simple dominance for contrasting traits, and many unit characters (genes) were involved in character formation, such as color or organ shape. Both had quantitative traits or continuous traits that could not be resolved by a simple Mendelian experiment and thus required more attention. Animals with separated sexes revealed many traits with a sexual dimorphism, and special assumptions had to be made to reconcile their inheritance with typical Mendelian ratios. The extension of Mendelism to behavior, as in the waltzing mouse analysis, made it probable that unit characters were not limited to trivial traits, but affected the most fundamental features of life.

Bateson offered this thought to those assembled at the Third International Conference 1906 on Genetics: "Colour, shape, habit, power of resistance to disease, and many another property that might be named have one by one been analyzed and shown to be alike in the laws of their transmission, owing their excitation or extinction to the presence or absence of such units or factors. Upon them the success or failure of every living thing depends. How the pack is shuffled and dealt we begin to perceive: but what are they—the cards? Wild and inscrutable the question sounds, but genetic research may answer it yet. Substances which excite disease or confer resistance, which preserve health or produce deformity, have been extracted, and it may not be more difficult to determine the nature of those critical factors than those which excite hoariness or colour in a plant."[22]

End Notes and References

[1] Wilks Rev. W. 1906. *Report of the Third International Conference 1906 on Genetics.* Royal Horticultural Society/Spottiswoode and Company, London.

[2] Bateson W. In Wilks Rev. W. 1906. op. cit. p. 76.

[3] Bateson W. and Saunders E.R. 1902. Experimental studies in the physiology of heredity. *Reports to the Evolution Society* **I:** 1–160.

[4] Cuénot L. 1903. l'Hérédité de pigmentation chez souris. *Archives de Zoologie Experimentale and Generale* **1:** 33–41; Darbishire A.D. 1906. Recent advances in animal breeding and their bearing on our knowledge of heredity. In Wilks Rev. W. 1906. op. cit. pp. 130–137; and Castle W.E. 1905. Recent discoveries in heredity and their bearing on animal breeding. *Popular Science Monthly* **66:** 193–208.

[5] In 1904, Castle's student, Glover M. Allen, attempted to explain the inability to render yellow homozygous in terms not too different from Kölreuter's. He states that Cuénot's "...result indicates a strong individuality for the yellow character and strengthens the belief that the yellow type of mouse is in origin different from the black, chocolate, and golden-agouti types and may, as Bateson suggests, be derived from another species, perhaps *Mus sylvaticus*." In The heredity of coat color in mice. *Proceedings of the American Academy of Arts and Sciences* **40:** 61–163; see p. 158. Cuénot's paper appeared in 1905: Les races pures et leur combinations chez les souris. *Archives de Zoologie Experimentale et Generale* **3:** 123.

   Although he gives his numerical findings as 73% yellow to 27% gray in the progeny from yellow x yellow pairs, he saw this as close enough to a 3:1 ratio (instead of the expected 1**YY**:2**YG**:1**GG**). He should have had a 2 yellow:1 gray (67% yellow to 33% gray) ratio. He could not explain the absence of the **YY** mice among his 73% yellow mice.

[6] Castle W.E. and Little C.C. 1910. On a modified Mendelian ratio among yellow mice. *Science* **32:** 868–870.

[7] Castle W.E. and Little C.C. 1910. op cit. p. 868.

[8] Baur E. 1907. Unters über die erblickeitsverhaltnisse einer nur in bastardform lebensfahigen sippe von *Antirrhinum majus. Berichte Deutsche Botanische Gesellschaft* **25:** 442.

[9] Castle W.E. and Little C.C. 1910. op. cit. p. 869.

[10] Morgan T.H. 1914. Two sex-linked lethal factors in *Drosophila* and their influence on the sex ratio. *Journal of Experimental Zoology* **17:** 81–122. See also Rawls E. 1913. Sex ratios in *Drosophila ampelophila. Biological Bulletin of Woods Hole* **24:** 115–124; and Morgan T.H. 1912. The explanation of a new sex ratio in *Drosophila. Science* **36:** 718–719.

[11] For an account of Castle's life and career, see Dunn L.C. 1962. William Ernest Castle (1867–1962). *Biographical Memoirs* **38:** 33–80. National Academy Press, Washington D.C.

[12] Castle W.E., Carpenter F.W., Clark A.H., Mast S.O., and Barrows W.M. 1906. The effects of inbreeding, cross-breeding, and selection upon the fertility and variability of *Drosophila. Proceedings of the American Academy of Arts and Sciences* **41:** 729–786.

[13] Castle W.E. 1912. The inconstancy of unit-characters. *American Naturalist* **46:** 352–362.

[14] Morgan had originally suggested a similar model of impurity, but abandoned it when he and his students ran into modifying factors. Muller responded to Castle's claims in Muller H.J. 1914. The bearing of the selection experiments of Castle and Phillips on the variability of genes. *American Naturalist* **48:** 567–576.

[15] Castle W.E. 1919. Piebald rats and the theory of genes. *Proceedings of the National Academy of Sciences* **5:** 126–130.

[16] Castle W.E. and Phillips J.C. 1909. A successful ovarian transplantation in the guinea pig and its bearing on the problems of genetics. *Science* **30:** 312–314.

[17] Castle W.E. 1915. The English rabbit and the question of Mendelian unit-character constancy. *Proceedings of the National Academy of Sciences (U.S.)* **1:** 39–42.

[18] Hurst C.C. 1906. Mendelian characters in plants and animals. *Report of the Third International Conference 1906 in Genetics,* op. cit. pp. 114–129.

[19] Hurst C.C. 1906. op. cit. p. 128.

[20] Bateson W. 1909. Heredity and Sex. In *Mendel's Principles of Heredity,* pp. 172–173. Cambridge University Press. United Kingdom.

[21] Doncaster L. and Raynor G.H. 1906. Breeding experiments with Lepidoptera. *Proceedings of the Zoological Society of London* **1:** 125.

[22] Bateson W. 1909. op. cit. pp. 90–96; see p. 96.

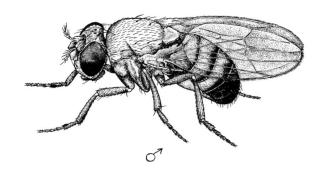

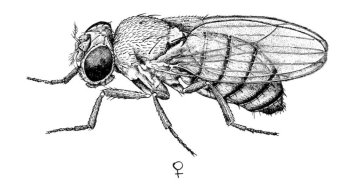

Work sheet used in Muller's laboratory at the University of Texas (1922–1932). The male (top) shows a fusion of bands in the lower abdomen and sex combs on the front legs. The female (bottom) shows unfused banding of the abdomen and an absence of sex combs on the front legs. The two karyotypes (below) show the three paired autosomes, the female XX, and the male XY. As a convention to represent the Y, Bridges used the bent sex chromosome. (Reprinted from Bridges C.B. 1916. Non-disjunction as proof of the chromosome theory of heredity. Genetics 1: 1–52.)

FEMALE   MALE

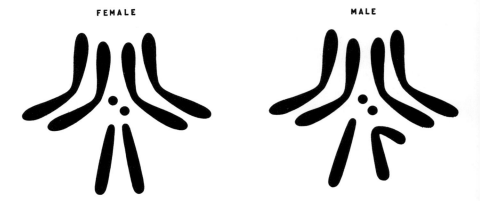

Morgan and Fruit Fly Genetics

Morgan was too good a scientist to hold to a conclusion after he believed it had been clearly disproved.

WILLIAM E. CASTLE*

The results are a simple mechanical result of the location of the materials in the chromosomes, and of the method of union of homologous chromosomes and the proportions that result are not so much the expression of a numerical system as of the relative location of the factors in the chromosomes. Instead of random segregation in Mendel's sense we find "associations of factors" that are located near together in the chromosomes. Cytology furnishes the mechanism that the experimental evidence demands.

T.H. MORGAN†

THOSE SCIENTISTS ORIGINALLY INVOLVED in breeding analysis had used plants or animals that were convenient or useful. As Mendel and his rediscoverers first demonstrated, horticultural and agricultural plants could be relatively easily and inexpensively used to work out the laws of heredity. Animals were more expensive to breed and maintain and were messy, requiring cleaning of cages. They also produced far fewer progeny than flowering plants. The plants were better in that respect, but they usually had only one or two growing seasons. Investigators might try to get around this by planting several experiments at once, as Bateson certainly did, but this would require a lot of luck in finding new phenomena that could be resolved within a year or two of breeding. Mouse, rabbit, guinea pig, or hamster studies would be even more difficult to use to find new topics to explore, because it would take years to get enough unexpected events and data to analyze them. What was needed to open up genetics to new phenomena was an organism that bred rapidly, produced a lot of progeny, and was inexpensive to maintain. This organism turned out to be the fruit fly *Drosophila melanogaster* (originally called *Drosophila ampelophila*) and Thomas Hunt Morgan was the person who would use it so effectively. However, he did so for an entirely different reason than our logic of hindsight: He was trying to find new species of flies.

---

*1951. The beginnings of Mendelism in America. In *Genetics in the 20th Century* (ed. L.C. Dunn), pp. 59–76; see p. 65. Macmillan, New York.
†1911. Random segregation versus coupling in Mendelian inheritance. *Science* 34: 384.

Morgan Was a Complex Person

It is difficult to discuss Morgan without encountering contradictions and the unexpected. Morgan was a complex person, but liked to seek simplicity in his work. Fruit flies were used as a genetic organism before Morgan began using them. He came to genetics in his middle age as an established and successful developmental biologist. Morgan's fruit fly laboratory was home to one of the most successful scientific teams ever assembled. It was also an unhappy laboratory, divided from the beginning into two rival groups driven apart by their mutual distrust. Morgan was both admired and resented by his students; he led his students to greatness often as much as they led him.

Morgan began his plunge into genetics with skepticism of Darwin's theory of natural selection, skepticism about the chromosome theory, and a conviction that Hugo de Vries had found a theory of evolution by saltations that he hoped to confirm using animals. Morgan treated his students as colleagues, but he did play favorites. He came from a distinguished family of patriots, and he held ancestral pride in contempt. Morgan was a supreme skeptic, sometimes brilliant and sometimes careless in his speculations; he was a sloppy dresser and an indifferent and disorganized teacher who sometimes forgot to go to his classes. He was generous to those he admired. He was a committed reductionist who believed all scientific problems could be resolved through experimentation. Morgan cast aside failed hypotheses and foot-in-the-mouth speeches, and he published articles without embarrassment or apology and barreled ahead with new hypotheses when he, his students, or other geneticists proved him wrong.[1]

Morgan's Developmental Biology Leads Him to Genetics

Morgan was an embryologist who was enamored by the German educational program stressing Entwicklungsmechanik (the reductionist, experimental approach to living cells and embryos). His reputation was established at Bryn Mawr and at Columbia for studies of parthenogenesis, regeneration, twinning, and embryo fusion. Morgan enjoyed dipping into the past and went back to the eighteenth-century work of Charles Bonnet (1720–1793) on preformation in aphids and Abraham Trembley (1710–1784) on regeneration in hydra to assess how far Entwicklungsmechanik had altered thinking on these problems. By 1898, he was fairly confident that the nucleus was essential to regenerating a cell because cytoplasmic parts failed to thrive as Boveri had demonstrated. For vertebrates, he was puzzled why an amputated limb formed a blastema of cells at the stump and produced an arm from that undifferentiated mass of cells; for plants, however, lopping off a stem or branch did not lead to blastema formation, but it did lead to bud formation farther down the amputated stump. He was puzzled about polarity. Lopping off the head of an earthworm led to a new head, and lopping off a tail led to a new tail. In addition, he wondered why a whole animal regenerated a part, but the part did not regenerate the whole animal

Figure 1. *Lobsters and crabs can lose a claw and regenerate a new one, sometimes smaller and sometimes of the other sex. (Reprinted from Morgan T.H. 1927. Experimental Embryology. Columbia University Press, New York.)*

(Figure 1).[2] These questions forced some to think of vitalist explanations and others to think of reductionist explanations, but the evidence for either view was not very good. Morgan said, "We err, I think, in going at present to either extreme, i.e., either in ignoring this something that has been called a vital force and pretending that physics and chemistry will soon make everything clear, or, on the other hand, in calling the unknown a vital force and pretending to explain results as the outcome of its action."[3] As the century came to a close, skepticism was Morgan's strong suit. Years later, in his obituary for Morgan, Muller said of him, "he doubted the doubt until he doubted it out."[4]

The turning point of Morgan's life was meeting Hugo de Vries in 1900 and seeing for himself the numerous species that de Vries claimed had arisen in his garden and in the surrounding fields in which they grew as weeds. An experimental approach to evolution was exactly what Morgan thought that the field of evolution needed. He abhorred theory that led nowhere but to debate and that could not be put to an experimental test. In 1900, he rejected Weismann's feeble attempt to use his germ plasm theory to explain regeneration. He saw in "determinants," "idioplasm," and similar inferred hereditary units a backdoor reentry of preformationism. "This host of divisible germs, moving at the command of Weismann's imagination, is supposed to carry out the process of regeneration. No one can fail to see that the difficulty is only shifted into a region where fancy can have free play and a scientific

experimental test cannot be applied."[5] Instead, Morgan looked with favor on Julius Sachs' theory that plants regenerate when they produce substances for root formation or for bud formation and that these substances flow toward or away from the wound during regeneration. Unfortunately, his own work with planaria, inspired by Sachs' model, did not work out as he had hoped. Morgan slit planaria below the head into symmetric longitudinal halves, but at the junction below the head, they produced two smaller heads (Figure 2). If the head was intact, why produce two more? The puzzle remained.[6]

Morgan felt uncomfortable with the newly discovered Mendelism and the attempts by the German cytologists to make a case for a hereditary role for the chromosomes. A great deal was focused on the reduction of chromosome number, the casting out of polar bodies during oogenesis, and a parallel reduction in the production of mature sperm. Morgan was not convinced: "Strasburger's idea that the essential element of the nucleus, the idioplasm, is reduced to one half, and must be renewed by the material received

Figure 2. *Morgan was disappointed that regeneration did not follow a principle of regulatory feedback. In his experiments with partially cleaved planaria, he expected a simple symmetric restitution, but instead found that new partial heads appeared. (Reprinted from Morgan T.H. 1899. Twelfth lecture. Regeneration: Old and new interpretations. In Biological Lectures from the Marine Biological Laboratory of Woods Holl. Ginn & Company, Boston.)*

from the spermatozoon, is also undoubtedly a wrong interpretation, as subsequent work has shown; for we now know that the female and the male nucleus if supplied with protoplasm are each capable of forming a complete embryo."[7] Morgan and his era did not know that signals and materials for developmental processes were stored in the cytoplasm of the egg and could be stimulated into development by partheno-genetic processes. He, and they, knew parthenogene-sis occurred, but they had no clue why the cytoplasm responded. Here, Morgan erred by assuming that the alleged genetic reduction would make such an event impossible and since it occurred, the nuclear theory of reduction had to be false.

Morgan took a dim view of natural selection and the broad theories stimulated by it. He mentioned Huxley's 1849 observation of the dual tissue layers of coelenterates and his speculation that there might be a tie to the presence of such simple layers in more advanced animals at the start of their organogenesis, an idea explored more fully by Ernest Haeckel and known variously as the biogenetic law or through the slogan "ontogeny recapitulates phylogeny." He ridiculed the effort to make such a connection: "We find in this series of affirmations a confusion of ideas that will appall anyone who will try to think out how the phy-logeny can be the mechanical cause of the ontogeny."[8] Morgan preferred a physiological model of a retained structure, not a phylogenetic memory, which is passed on. At least that model "we may call an explanation, since it removes the element of chance and of mysti-cism from the phenomenon."[9] Avoiding explanations on the basis of teleology, mysticism, and chance was key to Morgan's thinking. He found what could be put to experimental test, looked for causes for everything, and kept a tight rein on speculation.

Natural selection was almost on the verge of mys-ticism to him. He did not know how such small or infinitesimal variations arose and did not believe there was evidence that these either occurred or were selected by natural forces. Natural selection was built on speculation alone, and however logical it was, Morgan rejected it, as he did all scientific claims that could not be put to experimental test. He had no doubts about evolution; the fossil record and comparative anatomy provide more than sufficient evidence that life had changed and that all life is related through descent. But Morgan quickly learned that finding a suitable animal with which to work was not an easy task. He tried chickens, mice, and rats, but they were expensive, time consuming, and did not produce sufficient amounts of offspring to look for new species among them. He also doubted whether this would occur with larger animals. There were many hobbyists and breeders, as well as vast numbers of domesticated animals headed to the din-ner table, and at least among these animals, no de Vriesian "mutating period" had been observed.

As the chromosome theory entered the picture, bringing with it an amalgam of cytology and Mendelism, Morgan shook his head in disbelief. In 1903, he worked out an overview of sex determina-tion. He pointed out that Cuénot's experiments showed that simple feeding did not change the sex of maggots destined to be males or females. Larger mag-gots produced females and smaller ones produced males, not because they could be made larger or smaller by feeding or starvation, but because their sexes were fixed earlier. Large maggots made smaller by starving still produced females and small maggots fattened by excess food still produced males.[10] This was not true for Morgan's phyloxerans. In good times, these plant lice were wingless females, and they pro-duced parthenogenetic wingless females. But when conditions dried out, they produced half-winged males and half-winged females and after overwinter-ing the fertilized eggs would produce a crop of wing-less females once more capable of parthenogenetic development. In contrast, "The discoveries of McClung, Montgomery, and Sutton in this connection indicate that there are two kinds of spermatozoa, and McClung has argued that this difference is connected with the determination of sex; but there is nothing more than the supposition that this may be so to go on at present."[11] He came to the conclusion that environment can alter sex in many different ways, and even when the sex is fixed, the mechanism may have nothing to do with sex chromosomes, as studies of bees suggested, where both nutrition (for queens) and ploidy level (for sterile worker females or for fertile drones) were involved. "It may be a futile attempt to try to discover any one influence that has a deciding influence on all kinds of eggs."[12] He rejected Castle's attempt to render sex a Mendelian trait, and called it a hypothesis filled with contradictions. For Morgan, whatever merits there might be to the chromosome theory, they certainly had nothing to do with universality.[13]

Morgan was enamored with de Vries' mutation theory;[14] for him, it represented an advance over Darwinism. In 1905, he contrasted the two models. He denied that selection had anything to do with the origin of species and believed that adaptationist arguments were no better than mysticism. "It seems that the method of the Darwinian school of looking upon each particular function or structure of the individual as capable of indefinite control through selection is fundamentally wrong."[15] In contrast, de Vries offered two classes of changes: those that came about as losses, such as an albino instead of a pigmented form, and that followed a Mendelian ratio, and those that de Vries called "elementary species," in which something new had arisen. Darwin called those sports and dismissed them as monstrosities; like de Vries, Morgan rejected that pessimistic assessment. Certainly, the elementary species arising in *Oenothera* were not monstrosities; they were new varieties or combinations of traits not seen in the parental plant before. Morgan asked for a change in terminology, with "mutation" to be reserved for something new and "retrogression" to be used for the loss of a trait. Then evolution would make sense: "So far as a phrase may sum up the difference, it appears that new species are *born*; they are not *made* by Darwinian methods, and the theory of natural selection has nothing to do with the origin of species, but with the survival of already formed species."[16] That sounded good to Morgan. Experimental evolution had arrived, and he had seen it with his own eyes in 1900 when he visited de Vries' laboratory in Hilversum, near Amsterdam. What he hoped to do now was to look for it in animals.

Fruit Flies Come to Columbia University

Exactly how Morgan began working with fruit flies is almost as mired in contradictions as the rediscovery (or confirmation) of Mendelism. Castle used fruit flies since 1901.[17] With his students, he published papers (beginning in 1906) (Figure 3) on selection effects on fertility and vigor. He was testing the well-known Darwinian problem (and dubious status) of inbreeding as a cause of degeneracy. Castle used fruit flies to study this problem on the recommendation of a colleague at Harvard, Charles William Woodworth, an entomologist, because Woodworth had bred them on grapes and bananas and found them to be versatile. According to Davenport's account, about a year after Castle began breeding flies at Harvard, Frank E. Lutz began breed-

INTRODUCTION.

PHYSIOLOGICAL characters are inherited no less than morphological ones. In each case there doubtless is in the germ a structural basis on which the development of the peculiarity in question rests. In no instance as yet have we been able to identify beyond question the physical basis of any particular character, but a first step in that direction has been taken in the discovery of specific morphogenic substances in the animal egg and of specific differences among the chromosomes of the germinal nuclei in both sexes. While cytologists attack the problem of heredity from the side of the structure of the germ-cells, it is important that their labors be supplemented by a study of the heritable characters themselves, so that the mutual relations of characters and the modifications which they undergo from generation to generation may be better understood.

Among physiological characters which beyond question are heritable may be mentioned fertility, i. e. the capacity for reproduction. This varies among individuals and among races, as every experienced breeder knows, but the conditions upon which it depends are somewhat uncertain. In some cases external conditions are supposed to induce sterility, as, for example, abundant nutrition and lack of exercise, resulting in excessive vegetative growth without reproductive activity. In other cases inbreeding is assigned as a cause of sterility, or sterility may occur spontaneously without any assignable cause. The relation of inbreeding to sterility has been studied experimentally in mammals by Crampe ('83), Bos ('94), and Guaita ('98) ; and in birds by Fabre-Domengue ('98). They all find the relation to be a causal one, continuous inbreeding, as of brothers and sisters, resulting in decreased fertility, attended more or less commonly by lack of vigor, diminution in size, partial or complete sterility, and pathological malformations. It was our expectation that similar effects would be observed in the fly, Drosophila, when

Figure 3. The first paper on fruit fly genetics. Castle and his students worked on an evolutionary problem, selecting a continuous trait in fruit flies. Their work showed the utility of flies for producing large numbers of progeny in a relatively short generation time of less than two weeks. (Reprinted from Castle W.E., Carpenter F.W., Clark A.H., Mast S.O., and Barrows W.M. 1906. The effects of inbreeding, cross-breeding, and selection upon the fertility and variability of Drosophila. *Proceedings of the American Academy of Arts and Sciences* 41: 729–786.)

ing them at Cold Spring Harbor, on Castle's recommendation.[18] Lutz (who later became curator of insects at the American Museum of Natural History) tested

Lamarckism by starving flies to determine whether that brought about a change in wing size. He also studied correlation of wing size and femur size using a variety of influences, including the presence or absence of sexual activity, feeding, and temperature changes. Most of Lutz's findings were published after 1910, because he carried these studies on for many years of selection.[19] Davenport also claimed that Castle suggested to Morgan that he switch to fruit flies, after Morgan complained about his poor progress with larger animals, and not too long after that conversation, Davenport claimed, Morgan did. Lutz gave this account to Davenport: Morgan went to Cold Spring Harbor to visit Lutz and was impressed by the flies' rapid life cycle and abundant progeny. He asked Lutz for a sample to show his classes. Lutz allegedly gave the culture to Morgan, from which he derived his white eye mutation. Davenport claims that he gave his article on Lutz's role to Castle, who confirmed this version.[20]

Blakeslee wrote to Morgan on May 22, 1935, claiming that Woodworth was first to use fruit flies for experimental studies while at Harvard. Woodworth was studying the embryology of *Drosophila*, which he grew on Concord grapes, but Castle substituted bananas when grapes were not available. Castle also found that too many of the flies drowned in the grape juice. Morgan replied to Blakeslee on May 27, 1935, confirming that Castle's laboratory had taken up fruit fly breeding through Woodworth. He said that he was familiar with both Woodworth's and Castle's work, so fruit flies were not unknown to him. He was not sure if Lutz had given him his first batch of fruit flies; "...if so I have forgotten it."[21] In this letter, Morgan implies that it was his knowledge of the fruit fly work at Harvard and Cold Spring Harbor that made him think of using fruit flies for testing de Vries' mutation theory.

This last interpretation may be consistent with Fernandus Payne's version. I interviewed Payne on August 14, 1970, in his office at Jordan Hall, Indiana University.[22] Payne had received his B.A. and M.A.

Figure 4. *Fernandus Payne in 1949 at Indiana University. Payne gave Morgan his first batch of fruit flies to use for experiments. He also recruited Muller to Indiana University in 1945. (Reprinted, with permission, from Payne F. ca. 1995. Memories and Reflections. Indiana University, Bloomington.)*

from Indiana University and arrived at Columbia in 1907 (Figure 4). For his M.A., he had worked on blind cave fish with Carl Eigenmann. He also served as a teaching assistant to A. Petrunkevitch, a Yale professor and cytologist who agreed to teach introductory zoology for Eigenmann while Eigenmann was on sabbatical leave. It was Petrunkevitch who wrote the letter of recommendation that secured a scholarship to Columbia for Payne. Payne worked for both Wilson and Morgan. Morgan suggested to Payne that since he had experience working with blind cave fish, he might try using fruit flies raised in the dark as a test of Lamarckism. At that time, Morgan was not using fruit flies but probably knew, from Castle's papers, that they were useful for laboratory work. Payne placed some bananas on the windowsill of the graduate laboratory in Schermerhorn Hall (room 613 then) and collected the first *Drosophila* cultures at Columbia. Later, Morgan used these cultures for his *Drosophila* work (beginning in 1908). Payne raised his cultures for 69 generations in the dark and failed to find any loss of visual ability or change in eye morphology (Figure 5).[23] He estimated that he looked at as many flies in those two years as did Morgan from 1908 to

---

FORTY-NINE GENERATIONS IN THE DARK.[1]

FERNANDUS PAYNE.

Since the time Lamarck put forward the well-known theory of the transmission of acquired characters, this subject has been discussed pro and con by many writers; some believing that it is one of the guiding principles in the evolution of species; others that the transmission of an acquired character is an utter impossibility. I shall not attempt a review of the literature, as every book on heredity and evolution has its chapter devoted to that. In fact, it seems to me that there has been too much discussion and not enough experimental work. I believe it can justly be said that there has not been a single decisive experiment which proves or disproves the theory. They are all open to criticism in one way or another. The main argument which the opponents of the theory advance is that there has been no proof brought forward, and further there is no conceivable way in which an acquired character could be transmitted. On the other hand, the supporters believe that this is one of the easiest ways of explaining evolution and as it helps us out of many a difficulty it must be true.

Much interesting data has been collected, but none of it is conclusive. What we need at the present time is more experimental work to test the validity of the theory. With this attitude toward the subject, I started an experiment October 21, 1907, while at Columbia University to test the effect of darkness upon the common fruit fly, *Drosophila ampelophora*, and, if any effect was noticeable, to test its transmissibility. This seemed to be a suitable experiment as the present condition of cave animals is easiest explained by the assumption of the transmission of acquired characters.

The paper is not a finished report, but it may be of interest to scientific men to learn that such an experiment is in progress, and that this fly has been bred in the dark for forty-nine generations. Most certainly the length of time is rather short, but the

[1] Contribution from the Zoölogical Laboratory of Indiana University, No. 112.

Figure 5. *Payne's test of Lamarckism. Payne used fruit flies as a project for Morgan before Morgan began working with fruit flies for mutation studies. Payne attempted to grow flies in the dark for 49 generations. A later paper extended it to 69 generations. Payne was inspired by studies of his teachers at Indiana University, who tried to interpret the evolution of character changes in blind albino cave fish found in nearby limestone caverns in southern Indiana. Payne gave Morgan his own cultures to use for mutation studies. (Reprinted from Payne F. 1910. Forty-nine generations in the dark. Biological Bulletin 18: 188–190.)*

1910 and "it was a matter of luck that white eyes turned up in his cultures rather than mine."[24] H.J.

Muller backed Payne's version because, in his speech for Payne's retirement (in an undated 1948 note), Muller mentions "...my own flies, which I got from Dr. Morgan, who got them from not yet quite Dr. Payne... ."[25] Payne told me that when he first came to Columbia, Morgan was using chickens, rats, and mice. He also mentioned that at Indiana University, W.J. Moenkhaus was studying inbreeding effects on fruit flies and had actually begun the work before Castle started using fruit flies, but that when Moenkhaus was scooped by Castle's publication he set aside the project and delayed publication of his own studies. At the time, Morgan did not know of Moenkhaus' work on fruit flies.[26]

Morgan Works Alone with Fruit Flies in a Search for a "Mutating Period"

Morgan's motivation for using the flies was born out of frustration working with the chickens, rats, and mice that Payne described Morgan as using. He was using them in the hope that a "mutating period" might arise in one of those species, as it had in de Vries' *Oenothera*. He was frustrated by the cost, the small yields of progeny, and the amount of time required to maintain these animals. Exactly when he had his discussion with Castle is unclear, but it was probably in late 1907 or early 1908. We know that by comparing Payne's entry into Columbia for his graduate work and the fact that Morgan had been working on fruit flies for at least two years before he found his most significant mutation, a white-eyed fly.

For this new approach, Morgan was his own first student. He bred the flies for two years (1908–1910) without assistance. His method was very simple: He grew them in any smaller-sized milk or cream bottle that he collected from his breakfast table and used the banana method that Castle and Lutz had employed. Morgan liked to work standing up, and used a jeweler's loupe to examine his cultures (Figure 6). To isolate a fly, he used a "halving method," in which he would pull the plug of a bottle that contained the fly he wanted to isolate, place that bottle neck-to-neck with an empty bottle, let half the flies crawl in, separate the two bottles, and replug them (quick wrist reflex action is needed for this). He would then choose the bottle with the desired fly, take another empty bottle, and repeat the process until he ended up with a bottle containing only the fly that he sought. If he wanted to get rid of flies he was not interested in using, he mashed them with his thumb.[27]

Figure 6. *Morgan liked to work standing up and he used a jeweler's loupe to look at the flies in his bottles. He improvised crude techniques to isolate flies, but abandoned these when Bridges made systematic improvements that greatly reduced both labor and error. (Courtesy of the American Philosophical Society, Curt Stern Papers.)*

He was getting frustrated in the winter of 1909 when Ross Harrison visited; he pointed to the shelves with flies and told Harrison that he had wasted two years and had gotten nothing for his work.[28] Fortunately for Morgan, his luck changed considerably by the spring. In the fall of 1909, Morgan was attempting to induce mutations with acids, bases, salts, sugars, and different food media. That too was profitless. But in January 1910, he noticed a fly with a trident pattern on its thorax. He isolated it and attempted to establish a stock of it; he was only partially successful. The trait was variable, and Morgan was not sure if it Mendelized. It was recessive when crossed to the normal flies lacking the trident. He called the mutant strain "with" (Figure 7) and he

called the normal allele "without."[29] Morgan thus started the trend to name mutants in cute ways that sometimes became inside jokes. Morgan's difficulty was that he was not quite sure how to guarantee virginity (at room temperature, usually about 75°F [26°C], flies remain virgin for about 12 hours), and he was learning how to carry out his crosses by trial and error.

It would be several years before more efficient methods (etherization, binocular microscopes, and pushing flies on porcelain plates with Winsor & Newton watercolor brushes) came into play. Morgan's second mutation was not much better. He called it olive, and the body color seemed a bit darker than normal. In March, he also found his third mutant,

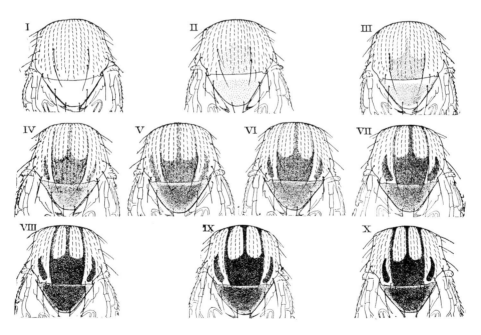

Figure 7. *Morgan's first mutant, "with." In I, no pigmentation is apparent in the scutellar or dorsocentral region of the thorax. The most extreme gradation is in X with a characteristic trident in the dorsocentral region and a heavily melanized scutellum. (Reprinted from Bridges C.B. and Morgan T.H. 1923. The third chromosome group of mutant characters of* Drosophila melanogaster. *Carnegie Institution of Washington Publication No. 327, pp. 1–257.)*

Figure 8. *Morgan's third mutation, "speck" wings, showing the melanized pigment associated with the juncture of the wings to the thorax. (Reprinted from Bridges C.B. and Morgan T.H. 1919. The second chromosome group of mutant characters. In Contributions to the Genetics of Drosophila melanogaster. Carnegie Institution of Washington Publication No. 278, pp. 123–304.)*

again using a clever name, "speck," because the flies showed a puddling of melanin in the axial region where the wing joins the thorax (looking like a miniature hairy armpit in appearance) (Figure 8). This too was a difficult trait to work out, but it also appeared to be recessive.

May was when the revolution began. Morgan found a white-eyed male running around in one bottle. He mated it with several of its sisters and obtained some white-eyed sons and daughters to establish a white-eyed stock.[30] He then did reciprocal crosses and found a surprising departure from Mendelism. He called this "crisscross inheritance" because a red-eyed male crossed to white-eyed females produced white-eyed sons and red-eyed daughters (Figure 9). The crisscross relationship did not extend to the reciprocal cross: A white-eyed male crossed to red-eyed females produced only red-eyed F_1 offspring. But when he crossed the F_1 to one another, he obtained a modified 3 red:1 white ratio with all the white-eyed flies male and two thirds of the red-eyed flies female. This was new, and he hurried his first *Drosophila* paper to *Science*.[31] In the paper, Morgan makes no mention of Wilson's sex chromosomes. For Morgan, this was, at best, a correlation, and he did

not want to enter into the still unresolved issue of sex determination and sex chromosomes.

Morgan was also using radium to treat the flies after his chemical treatments failed, and he obtained a fly, "Beaded" wings, with scalloping of the wing edges. He was not sure if the radium was the basis or not, but this fly did come from a line that had been treated. *Beaded* acted as a dominant mutation. To top off his merry month of May, Morgan found yet a third mutation, another olive body color, and this one was easier to demonstrate as being a typical Mendelian recessive. Morgan wasted no time sending in a paper describing this breakthrough as a "mutating period in *Drosophila*."[32] Morgan thought that he was on his way to experimental evolution, the possible emergence of elementary species, and a confirmation of de Vries' findings in *Oenothera*.

The white eye case, however, piqued Wilson's interest. He told Morgan that the crisscross inheritance and modified sex ratio for his white- and red-eyed flies made good sense if he assumed that the Mendelian factor involved was on or associated with the X chromosome. Still skeptical about the chromosome theory, Morgan was not willing to go that far. However, Wilson liked the model so much that he even proposed that it could be used to explain human color blindness as being a sex-limited trait on or associated with a human X chromosome. Morgan dismissed the idea as speculation, but Wilson did not hesitate to publish the idea in 1911.[33]

In June 1910, another sex-limited trait appeared. Morgan called it *"rudimentary"* wings, the wings being truncated and about half the normal length. He found a pink-eyed mutation in July; it was an autosomal recessive trait. In August, two more mutations were added: a third wing mutation, which he called "miniature" wings, because the wings looked proportionate but compressed, and a fourth wing mutation that looked superficially like *rudimentary* wings but

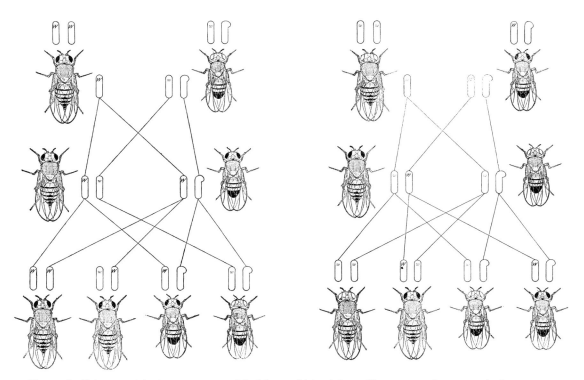

Figure 9. *Crisscross inheritance as a modified form of Mendelism. The reciprocal crosses of white eyes × red eyes gave different outcomes. Morgan successfully interpreted these by assuming that a gene for the white eye color was associated with the X chromosome. White eyes was the third mutation that Morgan found. (Reprinted from Morgan T.H. 1919. The Physical Basis of Heredity. J.B. Lippincott, Philadelphia.)*

which he called "Truncate." *Truncate* acted like a peculiar dominant mutation and, like *Beaded* wings, it came from one of the radium-exposed lines. The rest did not. With so many more spontaneous mutations than allegedly induced ones during this "mutating period," Morgan played it safe and dismissed the role of radium as an agent inducing mutations.[34]

What caused this sudden appearance of so many mutations after a nearly blank period of two years? If it was a mutating period, it did not happen to Payne's flies, still growing in the dark, but with samples being hauled out every generation to look for signs of degeneration or loss of vision or shape or change in color of the eyes. More likely, it occurred as a result of gaining familiarity with the flies. Muller once told me that looking for subtle changes, such as bent bristles or a slightly distorted vein in a fruit fly's wing, did not require a special gift of vision. He said it was like being in a room with students, with one of the students lacking a nose—it would just jump right out at you because you are so used to looking at people's faces.

The summer of 1910 found Morgan with six mutations to study and lots of cultures to pursue. He wanted to cross the sex-limited flies to each other to see the results. He also wanted to test independent assortment for the various autosomal traits that he found and to cross sex-limited and autosomal traits to see if the sex factor stayed connected to the sex-limited traits. He also wanted to keep an eye out for the emergence of elementary species. Morgan was a workaholic and carried out many projects at once. He had to write up his papers on aphids and phylloxerans, and he wanted to publish several papers on these new mutations.

Morgan's Unassisted Contributions Extend the Chromosome Theory

During the years 1910–1916, Morgan's unassisted findings were largely confined to the first two years; after that he worked with his students or allowed them to find interesting problems that he left them to work out. He established *Drosophila* as a choice organism for genetic research. He found X-linked inheritance, the linkage of traits on the **X** chromosome, and a chromosomal interpretation of exchange of linked genes by crossing-over.[35] Morgan reasoned that the degree of linkage was associated with the distance that genes were apart on the chromosome. He found linkage for the second group of autosomal traits[36] and showed that there was no crossing-over among autosomal linked traits in the male.[37] He proved that an environmental influence was at work in the expression of a mutant trait when he analyzed *Abnormal* abdomen.[38] Morgan worked out the eye color interactions of different allelic and nonallelic combinations of eye mutations.[39] He discovered the existence of a recessive lethal factor that he found on the **X** chromosome[40] and found that eosin eye color was a partial reversion of white eyes, thus disproving that white was a loss of the normal gene.[41] This disproved the universality of Bateson's "presence-and-absence" model of allelism.

Of these findings, two were world class and earned him a Nobel Prize some 20 years later. He found both of these before he worked with his most famous students—Bridges, Sturtevant, and Muller. Morgan's reluctance to accept the chromosome theory shifted when he found more than one sex-limited trait. It was a theory on which he relied when he came up with crossing-over. Morgan's conversion was not through arguments with Wilson, although they proved stimulating and sharpened his thinking. What convinced him was the shower of mutations that allowed him to carry out the crucial experiments. When he crossed *white* with *rudimentary*, he thought he was getting independent assortment with two sex-limited traits. *white* and *rudimentary* did not appear to be linked, as he thought they were if they were both on the same **X** chromosome (Figure 10). Morgan did not then know what we now know—that they were more than 50 map units apart. Thus, they appeared to assort independently.[42] Yellow body color was the fifth sex-limited mutation (the fourth was vermilion eye color).

His illustrator and assistant, Elizabeth M. Wallace, found a yellow male in a culture in January 1911. The addition of yellow body color clinched the story.[43] Morgan used three independent body traits to provide unambiguous combinations—white eyes, yellow body color, and rudimentary wings (vermilion could not be seen if white eye color was present, and

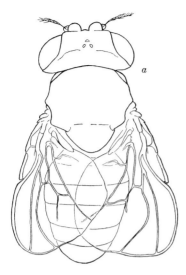

matids among the tetrads that we today associate with the pachytene and diplotene stages of reduction division. Janssens called this "chiasmatypie," and cytologists today refer to the twists as chiasmata (Figure 11). Morgan recognized that the chiasmata had to break and unwind for the paired homologs to

Figure 10. rudimentary wings is an X-linked recessive trait, almost at the other end of the X chromosome from the white eye mutation. This beautiful line drawing by Elizabeth Wallace was characteristic of her careful work describing the mutants and preparing them for publication. (Reprinted from Morgan T.H. and Bridges C.B. 1916. Sex-linked inheritance in Drosophila. Carnegie Institution of Washington.)

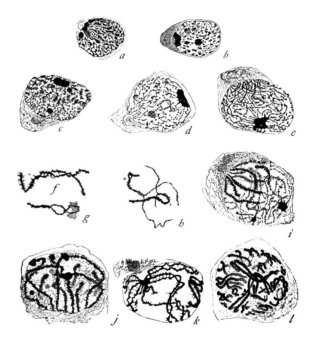

Figure 11. The Belgian cytologist, Janssens, used the salamander, Batrachoseps attenuatus, to illustrate chiasma formation in meiosis (note f–h). It was this observation that suggested to Morgan the possibility that chiasmata may represent exchanges of maternal and paternal chromatids in paired homologs and thus lead to recombination (loss of coupling or repulsion) for the genes involved. What is twisted has to come apart, and Morgan believed that a simple model of breakage and reunion of the broken ends was sufficient. (Reprinted from Morgan T.H., Sturtevant A.H., Muller H.J., and Bridges C.B. 1915. The Mechanism of Mendelian Heredity. Henry Holt, New York.)

a lopping off of the wing tip could not be seen when the fly also had miniature wings). As we now know for more than 85 years, *white* was at 1.5, *yellow* was at 0.0, and *rudimentary* was at 54.5 on the X chromosome map. Although Morgan did not recognize that he could map his genes, he knew from his data that white eyes and yellow body color were tightly linked, and both of those traits were a good distance from *rudimentary* wings. Morgan also connected this theory to the chromosome theory of heredity by relying on F.A. Janssens's article in *La Cellule*.[44] Studying salamander chromosomes (in *Batrachoseps attenuatus*), Janssens showed a twisting of chro-

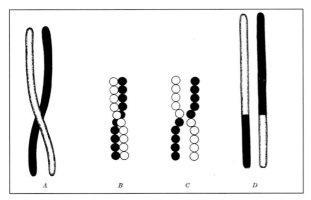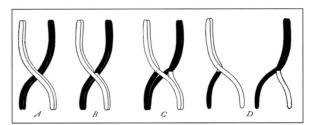

Figure 12. *Crossing-over and chromosome strands. (Left) Morgan's representation of crossing-over. In A, a chiasma forms between two chromatids. In B, these are represented as a string of beads (a simplified teaching aid for genes on a string). In C, the exchange has taken place, and in D, the two recombinant chromatids are represented. (Right) A shows two homologs before breakage and reunion. The chromosomes in B show crossing-over with two of the four strands undergoing an exchange. In C, the recombinant strands and the two nonrecombinant strands may be seen as the homologs begin to separate. In D, the homologs have separated and each cell gets one nonrecombinant and one recombinant strand. These will separate later during equation division. (Reprinted from Morgan T.H., Sturtevant A.H., Muller H.J., and Bridges C.B. 1915. The Mechanism of Mendelian Heredity. Henry Holt, New York.)*

separate. It was this breakage-and-reunion event that Morgan saw as consistent with the shifting of segments of chromosomes between homologs (Figure 12). Muller said that Morgan described it to his class using the analogy of the shifting blocks of normal and boldface type seen on railroad timetables to designate A.M. and P.M. schedules.[45]

The laboratory was not yet a team; this emerged between 1911 and 1913 when Morgan's experiments began to overwhelm him, and he knew that he had to find a way to expand his work. What was clear by 1912 was revolutionary: Morgan had found specific genes associated with a specific chromosome, the **X**. He had provided the first experimental evidence for the chromosome theory of heredity. Moreover, he had found and demonstrated by experimental crosses that coupling and repulsion could be interpreted through crossing-over. The same chromosome theory of heredity incorporated linked genes. Morgan quickly promoted these new findings in the context of Wilson's chromosome theory and quietly dropped his search for new species occurring in a "mutating period." He had now cast his lot with Wilson and abandoned de Vries. Classical genetics was now launched.

End Notes and References

[1] There are two biographies of Morgan: Allen G. 1978. Thomas *Hunt Morgan: The Man and his Science.* Princeton University Press, New Jersey; and Shine I. and Wrobel S. 1976. *Thomas Hunt Morgan, Pioneer of Genetics.* University of Kentucky Press, Lexington.

Allen's book covers more detailed accounts of Morgan's scientific contributions. The Shine–Wrobel account stresses Morgan's personality and career. Many obituaries were written, the most personal and informative from his closest student, Sturtevant A.H. 1959. Thomas Hunt Morgan. *Biographical Memoirs of the National Academy of Sciences* **33**: 283–325.

2   Morgan T.H. 1899. Some problems of regeneration. *Biological Lectures for 1897–1898 at the Marine Biological Laboratory of Woods Holl*, pp. 193–207. Ginn and Company, Boston.

Trembley's work was performed in the 1740s. Note that in the nineteenth century, Woods Hole was still designated as Woods Holl. The archaism of the geographical term, holl, led to the name change by the village and laboratory. For an account of the preformationist period and the experiments conducted in the mid-1700s, see Baker J.R. 1952. *Abraham Trembley, Scientist and Philosopher, 1710–1784*. Edward Arnold & Co., London. For an overview of the connections of developmental biology to genetics in Morgan's generation, see Gilbert S., ed. 1991. *A Conceptual History of Modern Embryology*. Plenum, New York, particularly the articles by Churchill F.B., Fischer J.L., Maienschein J., and Oppenheimer J.M.

3   Morgan T.H. 1899. op. cit. p. 206.

4   Muller H.J. 1946. Thomas Hunt Morgan. *Science* **103**: 550–551.

5   Morgan T.H. 1900 Regeneration: Old and new interpretations. *Biological Lectures for 1899 at the Marine Biological Laboratory of Woods Holl*, pp. 190–201; see p. 191. Ginn and Company, Boston.

6   Morgan noted that "...The material of the body is almost as plastic as that of an undivided or dividing egg." In his article, Experimental studies of the regeneration of *Planaria macolata,* in 1898, *Archiv für Entwicklungsmechanik der Organismus* **7**: 364–297, Morgan reasoned that the head should have been full with "head stuff" and thus there was no reason for the appended body parts to produce new heads when the body halves were not detached from the original head.

7   Morgan T.H. 1901. The problem of development. *International Monthly,* March 1901, volume 3, pp. 1–47; see p. 7.

8   Morgan T.H. 1901. op. cit. p. 16.

Note that Morgan distorted the meaning of Haeckel's phrase and reasoned that "phylogeny causes ontogeny," because he asked the reader to see the absurdity of phylogeny (i.e., phylogenetic memory) as a *cause* of ontogeny. Haeckel's view is not as silly as Morgan claims. In a rough way, there is a parallel between the events of a vertebrate's development (one-celled ball, multicelled ball, hollowed ball, two layers, three layers, segmentation, etc.) and the phylogenetic history (single-celled protozoa, colonial protozoa, coelenterates, mesozoa, coelomates, metameric organisms, etc.). If evolution builds the new on the old, such a "biogenetic law" would not be illogical or naive. It is not the extinct or ancestral *species* that are being "recapitulated," but the ancestral *body plans* that had undergone successive evolutionary changes.

9   Morgan T.H. 1901. op. cit. p. 16.

10   Morgan T.H. 1903. Recent theories in regard to the determination of sex. *Scientific Monthly* **64**: 97–116.

Morgan's point in this essay is a good one. Some species have environmental factors determining sex and some do not. If sex determination is so varied, how can one have a universal determining mechanism?

11   Morgan T.H. 1903. op. cit. pp. 107–108.

12   Morgan T.H. 1903. op. cit. p. 116.

13   Castle W.E. 1903. Mendel's law of heredity. *Proceedings of the American Academy of Arts and Sciences* **38**: 535–548.

At best, as Morgan pointed out, one sex would have to be heterozygous, and each homozygote would form the other sex. There was no logic for such a scheme.

14   de Vries H. 1901–1903. *The Mutation Theory* (1910 English translation by J.C. Farmer and A.D. Darbishire). Open Court, New York.

15   Morgan T.H. 1905. The origin of species through selection contrasted with their origin through the appearances of definite variations. *Scientific Monthly* **67**: 54–65; see p. 57.

16   Morgan T.H. 1905. op. cit. p. 64.

17   Castle W.E., Capenter F.W., Clark A.H., Mast S.O., and Barrows W.M. 1906. The effects of inbreeding, cross-breeding, and selection upon the fertility and variability of *Drosophila. Proceedings of the American Academy of Arts and Sciences* **41**: 729–786. Castle briefly describes his intentions in Castle W.E. 1951. The beginnings of Mendelism in America. In *Genetics in the 20th Century* (ed. L.C. Dunn), Chapter 4, pp. 59–76; see p. 73. Macmillan, New York.

He does not say that he gave Morgan his fruit flies; instead, he states, "...It called to Morgan's attention a new source of material for experimental study, not subject to the slow-breeding laboratory mammals." The account involving Castle and Lutz in Morgan's use of flies comes from Davenport C.B. 1941. The early history of research with *Drosophila. Science* **93**: 305–306.

18 Frank E. Lutz worked at Cold Spring Harbor and with G.H. Shull and Anne Lutz constituted the original permanent staff appointed by Davenport to begin research at Cold Spring Harbor Laboratory in 1904. Shull did plant breeding and Anne Lutz worked on cytological problems (especially with *Oenothera*, toward the end of that decade). Davenport had no fixed plans for his staff other than to make their work relevant in some way to the broad theme of experimental evolution. See MacDougal E.C. 1946. Charles Benedict Davenport, 1866–1944: A study of conflicting influences. *Bios* **17**: 1–50.

MacDougal, an associate staff member in that 1904 start, wrote a very harsh judgment (to some) of Davenport not only for his eugenics work, but also because he ran the Cold Spring Harbor Laboratory in an autocratic way.

19 Lutz F.E. 1907. Inheritance of abnormal wing-venation in *Drosophila. Proceedings of the Seventh International Zoological Congress, Boston*, pp. 411–419 (the volume came out in 1912); Lutz F.E. 1913. Experiments concerning the sexual difference in the wing length of *Drosophila ampelophila. Journal of Experimental Zoology* **14**: 267–273; and Lutz F.E. 1915. Experiments with *Drosophila ampelophila* concerning natural selection. *Bulletin of the American Museum of Natural History* **34**: 605–624; article XXI; December 10, 1915.

20 Davenport C.B. 1941. op. cit. p. 305–306.

21 The Morgan–Blakeslee letters are in the Millikan Library, the California Institute of Technology, Pasadena.

22 Fernandus Payne. Interview with Carlson E.A., Bloomington, Indiana. August 14, 1970. Unpublished, six pages.

23 Payne F. 1910. Forty-nine generations in the dark. *Biological Bulletin* **18**: 188–190; and Payne F. 1911. *Drosophila ampelophila* bred in the dark for sixty-nine generations. *Biological Bulletin* **21**: 297–301.

24 Carlson E.A. 1970. op. cit. p. 1.

25 Muller H.J. 1948. Unpublished notes, Farewell Upon Dean Payne's Retirement. Muller archive, Lilly Library, Indiana University, Bloomington.

26 W.J. Moenkhaus' work on fruit flies is briefly mentioned in 1907 as an untitled report, *Yearbook of the Carnegie Institution of Washington, 1906*, No. 5, Annual Report of the Department of Experimental Evolution, p. 105; and Moenkhaus W.J. 1911. The effects of inbreeding and selection on the fertility, vigor, and sex ratio of *Drosophila ampelophila. Journal of Morphology* **22**: 123–154.

27 I learned much of Morgan's personal habits from Muller, Sturtevant, and Altenburg. In addition, Sturtevant's memoir provides many insights into his work habits.

28 Harrison R.G. 1937. Embryology and its relations. *Science* **85**: 369–374.

29 Morgan describes his difficulties in 1919. The second chromosome group of mutant characters. In *Contributions to the Genetics of* Drosophila melanogaster. Carnegie Institution of Washington Publication No. 278, p. 128.

He writes that "In the fourth generation of selection for a race 'without' such a trident, there appeared (in March 1910) a few individuals with a tiny black speck at the juncture of each wing with the thorax... At first the breeding results obtained with this character were irregular... Some of this irregularity may have been due to nonvirgin females (24 hour females were used) and to the practice of using mass cultures, though probably more was due to the difficulty of classification before familiarity with the characteristics of the mutation had been acquired."

30 Although Morgan listed it as having arisen in May, he also cited April as the month of discovery. The April date is in Morgan T.H. 1919. op. cit. p. 128.

Note that the white must have been present in several females of the culture, but in only a tiny percent of the entire bottle's population, or there would have been more males with white eyes. The fact that some of the females were heterozygous rules out its origin as a solitary mutation in an egg.

31 Morgan T.H. 1910. Sex limited inheritance in *Drosophila. Science* **32**: 120–122.

Morgan's symbolism for white eyes varied with his models of the genes involved. He tried associating both sexes (**X** or **O**) with the eye colors (W or R). To make his crosses work, he had to assume that the two traits were absolutely coupled. He ran into contradictions if he used **O** for white (a presence-and-absence situation), because there was no effective way that he could represent the males with a single **X** (at this time, there was no **Y** chromosome in Morgan's mind) (See Carlson E.A. 1966. *The Gene: A Critical History*, pp. 43–45. W.B. Saunders, Philadelphia.)

[32] Morgan T.H. 1910. Hybridization in a mutating period in *Drosophila. Proceedings of the Society for Experimental Biology and Medicine* **7**: 160–162.

[33] Wilson E.B. 1911. The sex chromosomes. *Archiv für Mikroskopische Anatomie* **77**: 249–271; see p. 265.

Wilson states, "Color-blindness being a recessive character, should appear in neither daughters nor granddaughters, but in half the sons and grandsons, as seems to be actually the case. The same interpretation will apply equally to the heredity of white eye-color in *Drosophila*, as observed by Morgan."

[34] Within a year of Roentgen's discovery of X rays in 1895, many people began applying radiation to living material to study its effects. Blakeslee, like Morgan, could not really tell if he had induced any mutations. Over the years, the contradictory results would not be resolved until Muller (and independently, a year later, L.J. Stadler) showed that X rays were indeed mutagenic. What was original was not the idea of trying it, but of proving it.

[35] Morgan's papers on crossing-over include 1911. Random segregation versus coupling in Mendelian inheritance. *Science* **34**: 384; and 1911. Chromosomes and associative inheritance. *Science* **34**: 636–638. Morgan's work on the second chromosome appeared in 1912. Complete linkage in the second chromosome of the male. *Science* **36**: 719–720.

[36] Morgan T.H. and Lynch C.J. 1912. The linkage of two factors in *Drosophila* that are not sex linked. *Biological Bulletin of Woods Hole* **23**: 33–42.

[37] Morgan T.H. 1914. No crossing over in the male of *Drosophila* of genes in the second and third pairs of chromosomes. *Biological Bulletin of Woods Hole* **26**: 195–204.

[38] Morgan T.H. 1912. The masking of a Mendelian result by the influence of the environment. *Proceedings of the Society for Experimental Biology and Medicine* **9**: 73–74.

[39] Morgan T.H. 1911. The origin of five mutations in eye color in *Drosophila* and their modes of inheritance. *Science* **33**: 534–537.

[40] Morgan T.H. 1912. The explanation of a new sex ratio in *Drosophila. Science* **36**: 718–719.

[41] Morgan T.H. 1912. Further experiments with mutations in eye color in *Drosophila:* The loss of the orange factor. *Journal of the Academy of Natural Sciences and Philosophy, Philadelphia* **15**: 321–346.

[42] Morgan T.H. 1910. The method of inheritance of two sex-limited characters in the same animal. *Proceedings of the Society for Experimental Biology* **8**: 17–19.

[43] Elizabeth Wallace prepared realistic illustrations for most of the new mutations found by Morgan and his students, as well as mutations obtained from other laboratories. She also drew the pathologies of flies, including mosaics and gynadromorphs. Her style was impeccable, meeting the highest standards of museum or encyclopedia illustration. Chromosome maps predict the frequency of linkage between any two genes. Tightly linked genes, such as *yellow* and *white*, show very little (1.5%) recombination based on their map locations. But with *white* and *rudimentary*, the distance is so far that there are just as many recombinants as nonrecombinants. Despite having drawn hundreds of illustrations of fruit flies, Wallace had no publications of her own nor any with Morgan.

[44] Janssens F.A. 1909. La théorie de la chiasmatypie, nouvelle interpretation des cinèses de maturation. *La Cellule* **25**: 387–411.

[45] Muller H.J. Mutation and The Gene, course, Indiana University, fall, 1954.

Forming the Fly Lab: Contributions of A.H. Sturtevant and C.B. Bridges

The difference between normal red eyes and colorless (white) ones in Drosophila is due to a difference in a single gene. Yet red is a very complex color, requiring the interaction of at least five (and probably of very many more) different genes for its production. And these genes are quite independent, each chromosome bearing some of them.

A.H. Sturtevant[*]

The improvements that have gradually evolved have been to a large extent the outcome of Bridge's unusual inventive faculty. The hand lens was replaced by a binocular microscope of his designing; the wall cases have been supplemented by incubators with fans and expensive regulators; the banana has given place to a synthetic medium consisting of agar-agar, corn-meal, molasses and yeast.

T.H. Morgan[†]

ORGAN WAS NOT INNOVATIVE in using or designing new equipment and so he needed help. Like most of his generation in 1911, he worked without grant support, doing his own research out of pocket and by himself, i.e., he bought his own bananas, used his household glassware, and prepared and cleaned the glassware for the cultures of flies he needed. His first additions to the laboratory were recruited from one of his undergraduate classes. At the time, other students were in his graduate laboratory, but they were working on developmental problems. As the fly work grew in importance, Morgan set aside these other projects for future work or he abandoned them. This led to what is often called the "fly lab," a group of major students—Alfred Henry Sturtevant, Calvin Blackman Bridges, and Hermann Joseph Muller—centered around Morgan. There were many other students who were their contemporaries in the laboratory, but they are seldom mentioned or remembered when the term "fly lab" is used (see page 310 for some of the fates of these lesser known students).

---

[*]1915. The behavior of the chromosomes as studied through their linkage. *Zeitschrift für Induktive und Abstammungs und Vererbungslehre* **13**: 234–287; see p. 265.
[†]1939. Personal recollections of Calvin B. Bridges. *National Academy of Sciences Biographical Memoir* **22**: 31–48; see p. 34.

Morgan Treats His Students as Co-workers

Both Sturtevant and Bridges took Morgan's introductory zoology course at the same time.[1] Actually, it was pure luck that benefited them because Morgan was covering for James Howard MacGregor, who was on sabbatical leave. Morgan did not mention his fly work in that course, but he did mention some of the problems in heredity that were occupying his thoughts. Sturtevant had an independent interest in heredity as a consequence of his family circumstances. The Sturtevants were an academic family and had moved from Illinois to Alabama to become farmers. Alfred Henry Sturtevant (1891–1970), the youngest of seven children, got out of farming quickly after he tried picking strawberries and came back almost empty-handed compared to his siblings. Sturtevant had a severe red-green color deficiency and thus could not readily distinguish a berry from a leaf. But he did take an interest in the horses on the farm and followed up on them when he came to New York (he stayed with his brother's family; his brother taught philology at Barnard College). At the New York Public Library, he read Punnett's text in genetics and applied Mendelism to coat color inheritance of his family's horses, looking up their pedigree records and applying symbols to the various traits. His brother suggested that he show it to Morgan; he did. Morgan was impressed by this undergraduate who could do independent research, and he encouraged him to pursue a career in genetics. When Sturtevant ran short of tuition money a year later, without Sturtevant's knowledge, Morgan arranged

for him to receive a scholarship that he had secretly funded for him.[2]

This illustrates several important features of the emerging fly lab. Morgan was generous, but kept his generosity secret, which is understandable because he did not want Sturtevant beholden to him. Morgan played favorites with those he liked. I say this not as a criticism, but as a fact of life in relationships of graduate students and their sponsors. When faculty sponsors take a liking to students, they often shower them with opportunities. Issues that can lead to sibling rivalries in a family can also lead to equivalent rivalries among graduate students vying for the affection and attention of their sponsors. It is often not deliberate, and the sponsor may not be aware of the favoritism doled out to some members of a laboratory. The ambivalence works both ways, because a sponsor may try to be fair to all students in the laboratory, just as parents claim to love their children equally, but one child (or student) may feel less favored. The relationship of graduate students and sponsors in the sciences is often an intense and intimate one. Daily conversations and long hours occur in a setting where, for committed graduate students with a passion for science, the laboratory becomes a home, and the members in it a family that may remain intact for several years while each student passes through the rites of passage to a Ph.D. What was unusual about the Morgan laboratory was his collegiality. He treated his students as co-workers and he promoted their work.[3]

Controversy Arises Over Assigning Credit for Work Done

Morgan abided by some of the conventions of his era, conventions that were still largely in force when I began my graduate training in 1953, but which

began to disappear when I got my Ph.D. in 1958. This was the year after Sputnik and the year when large amounts of money became available to students for predoctoral fellowships from the National Science Foundation and the National Institutes of Health. This policy is largely forgotten today. The rule sponsors applied was to put students on papers as co-authors if they were supported by fellowships (that made them students) but to leave them off as co-workers when they were paid from departmental research assistantships or the sponsor's grants (that made them employees). Employees who had no part in designing the experiment were not usually co-authors; they were acknowledged in a footnote at the end of the article. I remember Irving Tallan lamenting the amount of work he did for T.M. Sonneborn on *Paramecium*, only to see his name as a footnote thanking him for his assistance. Muller broke that tradition: It did not matter to him whether the student was on fellowship or Muller's grants and the following story explains why this was so. Morgan was careless or inconsistent in the way in which he assigned joint authorship. None of his best-known students (Sturtevant, Bridges, and Muller) published articles before 1913. Some were working in his laboratory from 1911 on. Morgan co-authored papers largely with Cattell, Lynch, and Tice (1912–1914); Bridges entered as a co-author in 1913–1919.

There is a second component to this story of graduate school sponsorship. For at least the first half of the twentieth century, possibly a decade longer, it was a standard rule that a student's dissertation work was only in the name of the graduate student (the professor received the footnote acknowledgment for advice and support). Muller and Sonneborn both abided by that rule at Indiana University in the 1950s. Thus, Sturtevant published his own work as sole author in the formative period of 1913–1916. The one exception to these rules was reserved for broad review articles or books. Here, Morgan was quite generous in his text, *The Mechanism of Mendelian Heredity* (Figure 1), to list all four names as co-authors. He would have added a fifth, Edgar Altenburg, who was in and out of the laboratory, but Altenburg was already

drafting his own textbook of genetics (it took him 30 more years to finish it), and he declined Morgan's offer to help write some chapters.[4] At the same time, Morgan could be selective in mentioning Bridges and ignoring Muller and Sturtevant, or mentioning Sturtevant and Bridges and ignoring Muller. Sonneborn thought that this was simply a matter of Morgan being dazzled by mapping and nondisjunction, in contrast to what Morgan may have believed were not very shining events in Muller's papers during that formative period.[5] Whatever the reasons for Morgan's selective or casual

THE MECHANISM
OF
MENDELIAN HEREDITY

BY

T. H. MORGAN
PROFESSOR OF EXPERIMENTAL ZOOLOGY
COLUMBIA UNIVERSITY

A. H. STURTEVANT
CUTTING FELLOW, COLUMBIA UNIVERSITY

H. J. MULLER
ASSISTANT IN ZOOLOGY, COLUMBIA UNIVERSITY

C. B. BRIDGES
FELLOW IN ZOOLOGY, COLUMBIA UNIVERSITY

NEW YORK
HENRY HOLT AND COMPANY

Figure 1. *Morgan and his students published their work in 1915 as a monograph that shaped genetics textbooks for a generation to come. This is the title page of the first edition. Each of his students wrote several chapters that Morgan assigned to them. (Reprinted from Morgan T.H., Sturtevant A.H., Muller H.J., and Bridges C.B. 1915.* The Mechanism of Mendelian Heredity. *Henry Holt, New York.)*

way of assigning credits, it rankled Muller, who allied himself with Altenburg (an outsider who was not formally accepted as a student of Morgan's), and this coalition split the students into two camps.

This tension, however, did not prevent the students from sharing data, providing interpretations, and enjoying the excitement of their discoveries. They admired and recognized each fellow students' talents; resentment was reserved for acknowledging ideas and interpretations. Morgan had an important part in this feature of the fly lab. He was indifferent to speculation and ideas, and with his long history of skepticism, he believed that what mattered was the executed experiment with the positive or negative results to test a theory. Morgan, Payne told me, threw out dozens of ideas in the laboratory, and most of them were worthless. He did not care for ideas that he felt were plentiful and that almost anyone could propose. Payne told me that what mattered to Morgan was doing the experiment.[6]

I cannot claim to be able to tease apart all of the individual and joint contributions of Sturtevant, Bridges, Muller, and other students and employees in Morgan's laboratory during this period (roughly, 1910–1920).[7] No laboratory is so efficient that all of its records remain and all of its data recorded for easy access. Morgan jotted down notes on scraps of paper, as did Muller. Morgan threw these out after approximately five years (he cleaned out his files this way, to make room for his present and future activities). Morgan had a distaste for showing off honors and historical self-congratulation. Muller saved everything like a pack rat, but unfortunately, most of the early work (up to 1932) was kept in boxes at the University of Texas while he was in Europe for seven years, and J.T. Patterson (then not on good terms with Muller) threw them out as junk.[8] Sturtevant was meticulous in his note keeping and numbered every cross he carried out; his records are available at the California Institute of Technology archives. Because Bridges led an unconventional life, Morgan (or Bridges' family) may have secreted or burned Bridges' personal papers after his death.[9] Fortunately, whatever one's habits, if letters are sent out, enough of them will be saved for partial reconstruction, and certainly most of the published information pertinent to these formative years was written during the era of first-person articles, where the editor's blue pencil did not chop away some of the motivation and personality involved in the production of the article.

Sturtevant Works Out the First Genetic Map

Sturtevant entered Morgan's laboratory as an undergraduate in August 1910, three months after Morgan had found his white-eyed mutation (Figure 2). Morgan liked Sturtevant's new interest in heredity. Originally, Sturtevant was considering a major in chemistry. He took Morgan's 1909–1910 zoology course at the urging of a friend of his brother, the pathologist L.W. Famulener.[10] This turned out to be the only time in his 24 years at Columbia that Morgan taught the introductory course. That fall (1910), because he was impressed by Sturtevant's initiative in working out the genetics of horse pedigrees, Morgan gave Sturtevant a desk in the fly lab and some flies with which to work. In his 1911 course on experimental zoology, Morgan mentioned the work he had been doing with his sex-limited traits, and Sturtevant asked Morgan if he could study the actual data. Muller's account is filled with admiration for

Sturtevant's accomplishment: "...I saw a 19-year-old student, Sturtevant, who was the first person in the world to strike upon the idea that it is possible to make a real map of the components of a chromosome and who actually constructed it in the course of only one year. One young man in one year's time could do that! What will not be done by a friendly joint-effort of scientists in the course of time?"[11] Muller's unpublished article, "Results of a decade of research in *Drosophila*," was presented as a talk in 1921 at Cold Spring Harbor. It illustrates that the sibling rivalry in 1911–1913 was not severe, that Muller and Sturtevant got along then, and that Sturtevant's impressions are compatible with Muller's. During those first few years, Muller stayed isolated, whatever resentment he had of Morgan's favoritism.

The achievement is described somewhat differently by Sturtevant in 1965: "In the latter part of 1911, in conversation with Morgan about this attempt—which we agreed had nothing in its favor—I suddenly realized that the variations in strength of linkage, already attributed by Morgan to differences in the spatial separation of the genes, offered the possibility of determining the sequences in the linear dimension of the chromosome. I went home and spent most of the night (to the neglect of my undergraduate homework) in producing the first chromosome map, which included the sex-linked genes y, w, v, m, and r in the order and approximately the relative spacing that they still appear on the standard map."[12] In his 1913 paper (dissertation) on this finding, Sturtevant acknowledges the discussions with his fellow students: "I have been greatly helped by numerous discussions of the theoretical side of the matter with Messrs. H.J. Muller, E. Altenburg, C.B. Bridges, and others. Mr. Muller's suggestions have been especially helpful during the actual preparation of the paper."[13] Sturtevant introduced (at Muller's suggestion) the idea of using "percent of crossing-over" as the "map distance" (later called a cMo or centiMorgan by J.B.S. Haldane). Morgan's nomenclature was unusual then, with uppercase letter symbols representing a trait in its generic form (B = body color for the yellow and its normal amber color). Thus, B–C/O–P–R–M would be the sequence for yellow–white eyes or its allele eosin–vermilion eye

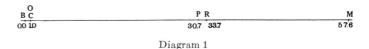

Diagram 1

Figure 2. *Sturtevant maps Morgan's genes. (Left) Sturtevant, as a youth, about 1920 at Woods Hole. (Above) Sturtevant's map of six X-linked genes. The terminology reflects Morgan's nomenclature: B = body color (now yellow is y and its normal allele is y⁺), O = eosin eye color (now eosin is wᵉ and its allele is wᵉ⁺); C = eye color (now w is white and its red eye color allele is w⁺), P = pink eye color (now vermilion is v and its allele is v⁺), R is miniature wings (now miniature wings is m and its allele is m⁺), and M = rudimentary wings (now rudimentary wings is r and its allele is r⁺). The distances also differ somewhat: yellow = 0.0, white = 1.5, vermilion = 33.0, miniature = 35.5, and rudimentary = 55.0. ([Left] Courtesy of the Cold Spring Harbor Laboratory Archives. [Above] Reprinted from Sturtevant A.H. 1913. The linear arrangement of six sex-linked factors in* Drosophila, *as shown by their mode of association.* Journal of Experimental Zoology **14**: 43–59.)

color–miniature–*rudimentary* (see Figure 2). Note that at that time, the name of the mutant was different from its symbol when two or more traits affected the same characteristic. Morgan's thinking was that if miniature is mutant, it is still nonrudimentary and thus R should represent the normal gene for miniature, the mutant. Later, the use of lowercase letters for recessive mutations eliminated this attempt to use the generic normal uppercase letter method of describing genes by their alleged normal functions. In addition, in the nomenclature, the white eyes region was considered compound in some way, with the normal eye color (designated as C) associated with another normal component designated as O (a presumed orange color).[14] Since eosin arose from white eyes, white could not be a simple loss of a gene for red eye color. Of course, Sturtevant used Morgan's nomenclature.

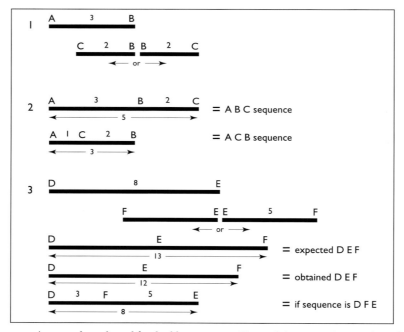

Figure 3. *Sturtevant's sum rule and need for double crossovers. He used the sum rule to work out linkage. If segment AB is 3 map units and segment BC is 2 map units, then the length AC must be the sum or subtraction of the two, either 5 map units (if the gene order is ABC) or 3 map units (if the gene order is ACB). For smaller distances, this model would work as predicted; for larger distances, it would fail. Thus, if DE is 8 map units and EF is 5 map units, the predicted length of DF for the sequence DEF will be 13 map units. In reality, somewhat less is consistently found (in this case, it is listed as 12 map units). The predicted 8 map units will be realized for the length DE in the sequence DFE, because it is a shorter total length. Muller interpreted the difference between the expected and the obtained frequencies of crossing-over as a consequence of what he called interference. He provided a mathematical measure of coincidence and interference in crossing-over for his Ph.D. dissertation, and he showed that Sturtevant's predicted distance is obtained if the longer distance is constructed by using several contiguous smaller segments of gene distances. (Figure drawn by Claudia Carlson.)*

In Sturtevant's paper, Muller claimed that he was acknowledged in general, but not in specific detail, for introducing the idea of calculating crossovers to total progeny and not crossovers to noncrossovers as Sturtevant had originally done. Muller also contributed to the discussions by pointing out that double crossovers in longer spaces would shorten the map, and thus, for accurate representation of the map by Sturtevant's Euclidian "sum rule" of AB + BC = AC, one had to use mutants between these outlying genes for revealing these double crossovers that would count otherwise and erroneously as noncrossovers (Figure 3). It was this exchange of ideas and interpretations and a constant back and forth to the blackboard to describe various results and ideas for experiments that exhilarated all involved.[15] Whatever envy, insecurity, or resentment may have rancored in their minds, there was little to be seen or felt in the honest exchange of criticism and enthusiasm for each other's work that made the fly lab such a productive and enjoyable part of all their lives and memories (Figure 4).

Figure 4. *The fly lab at Schermerhorn Hall, Columbia University, shows the ever-present bunch of bananas used to feed fruit flies (and hungry staff). Renovations after 1920 eliminated the original room and floor of the laboratory. (Courtesy of the American Philosophical Society, Curt Stern papers.)*

Sturtevant Interprets Multiple Allelism

Sturtevant's other major contribution in 1913 was introducing the idea of "multiple alleles" to explain some unusual findings by animal breeders. Castle, Hurst, and Punnett found that rabbits have three distinctive color varieties, one called self-color (such as gray), one called Himalayan (with a booted or variegated pattern), and albino. In their interpretation, based on the idea of one character unit to one trait and a concept of coupled traits, they listed self as CC, Himalayan as Cs, and albino as cs. The absence of the expected recombinant from Mendel's law of independent assortment, cS, sug-

gested an absolute linkage of these two factors for each of these three color patterns. Sturtevant suggested an alternative, with S, H, and A representing the three traits as alleles or variants of one common gene. As support for this model, he cited the red eye color of the normal fly, the white eye mutant that arose from it, and the eosin eye color recessive to red, which arose in a stock of white-eyed flies as a partial reversion to the normal eye color (Figure 5). As Sturtevant succinctly expressed it, "It will be seen that triple allelomorphs may be substituted for complete coupling as an explanation of any case where only three of the four combinations possible on the complete coupling scheme are known."[16] This was one of the many assaults on Bateson's model of coupling and repulsion. To Morgan's group, the eosin reversion from white also nullified the legitimacy of the presence-and-absence model. In addition, Sturtevant recognized that multiple allelism had already been established for two mutant factors by Emerson in beans and by Baur in columbines.

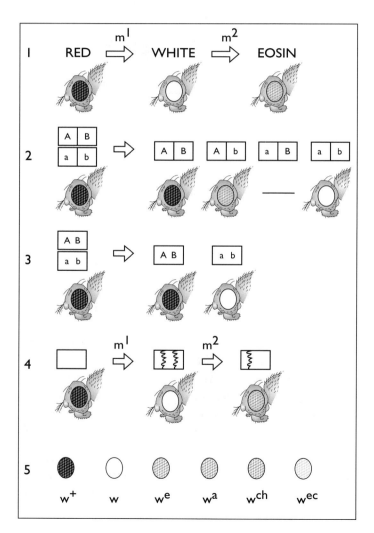

Figure 5. *The white eye series of alleles (1) provided the fly group with a model of multiple allelism interpreted by Sturtevant. Two problems were at issue. The mutant white eyes arose from red-eyed flies. The mutant eosin arose from a stock of white-eyed flies. To Morgan's group, this implied the death of Bateson's presence-and absence-theory of mutations—you can't get something (eosin) from nothing (white). But it also forced Sturtevant to reject a digenic model of white eyes. In such a model (2), there should be four possible combinations, two common (red and white) and the other two relatively rare (eosin and some unknown other color) if the two factors were coupled. The absence of this inferred fourth combination suggested a permanent coupling (3) in which mutation of one of the factors (B to b if coming from the AB parent, or a to A if coming from the ab parent) leads to eosin (aB). The inferred fourth factor, looked for but not found, from crosses Morgan and his students carried out, was called "simple white" (in case it was a single lesion resulting in white eye color) or some color similar to eosin. Sturtevant rejected the coupling of two genes and instead argued (4) that there was one intact gene with two lesions for white eyes and one lesion for eosin. Because he assumed no intragenic recombination, it was impossible to extract eosin or the other color from white. To represent this unitary status of a normal gene and its family of alleles, Muller proposed a nomenclature using superscripts that the fly lab adopted. Thus, (5) w^+ = red, w = white, w^e = eosin, w^a = apricot, w^{ch} = cherry, and w^{ec} = ecru. (Figure drawn by Claudia Carlson.)*

Sturtevant Has Multiple
Interests in Genetics and Evolution

In this formative period, Sturtevant also contributed insights into variations in map length. He then shifted to a variety of other problems. Sturtevant thought that Diptera would provide a good system for studying species and in 1915, he obtained funds to go on expeditions in the United States and Central America to find different species of *Drosophila*. He had a brief stint in the U.S. Army during World War I in 1918. In 1920–1921, he joined Morgan and Bridges to work on fruit fly projects at the University of California (Berkeley) and the Hopkins Marine Station. His evolutionary studies in the 1920s led to many important findings, the most important of which was the similarity in chromosome maps of *D. melanogaster* and *D. simulans*.[17] Sturtevant also obtained hybrids of these two species (a difficult job) and stimulated an interest in using genetic similarities and differences tested through hybridization of species.

As an undergraduate, Sturtevant did not do well in Wilson's course, because his color blindness prevented him from discerning slides stained with acetocarmine and other red dyes. He never used a compound microscope for research after that experience. Thus, Sturtevant (and Morgan) left to Bridges the cytological work on fruit flies. This visual impairment limited Sturtevant in working with eye color mutations or in finding those that did not show a change in intensity. I became aware of the extent of Sturtevant's difficulty when I interviewed him in 1967. He took me to his experimental garden of irises (on the California Institute of Technology campus). He loved breeding them and waved his arm over the many varieties he produced in his hybridization studies. "Aren't they beautiful!" he proclaimed. I was polite and nodded, but to me

they were a forlorn and funereal combination of mustards, puce, and odd variants of mauves, umbers, and ochres. Despite the visual drawback, Sturtevant had a brilliant mind to compensate, and he was able to infer that the C factors (crossover reducers or inhibitors) were actually due to a chromosomal rearrangement to which he gave the name "inversion."[18]

Sturtevant was regarded as the sage of Morgan's laboratory. He read the literature extensively and did not limit his readings to genetics. He had a near-photographic memory for details, and he frequently wrote his papers in his head and then typed first drafts that were virtually ready to send out for review. He regularly did the Sunday *New York Times* crossword puzzles and the far more fiendish *Manchester Guardian* puzzles. Like Morgan, he was a sloppy dresser and visitors would frequently encounter Sturtevant through the soles of his shoes, his feet propped up on his desk when they entered to see him. Sturtevant was friendly, relaxed, and always ready to talk. However, he was a terrible correspondent, and if a letter meant work to answer it or if it dealt with anything that was unpleasant, he usually ignored it.[19]

I believe that Sturtevant's loyalty to Morgan cost him a Nobel Prize. I once asked Åke Gustafson at a genetics congress about the awards to Morgan (1933) and Muller (1946) and not to Sturtevant or Bridges. He said the "rule of three" was already established in the 1930s when the fly lab was considered, and the Karolinska Institute reasoned that "each of Morgan's students was good enough to win a prize on his own." Muller charted an independent path and fulfilled that prediction; Bridges and Sturtevant did not. There is a lack of stimulation and an unacknowledged deference to working all one's life for the same mentor.

Morgan Adds Bridges to the Laboratory

Calvin Blackman Bridges (1889–1938) was raised by his grandparents in upstate New York, both his parents dying young. He was a talented student but his grandparents were poor, and Bridges had to make do with clothing that was constantly mended. He was too ashamed to go to social activities in high school because of his ragged appearance. He received a scholarship to attend Columbia University, but he had to support himself with part-time work. Bridges took the same introductory biology course as Sturtevant, and Morgan, who learned of Bridges' circumstances, asked him to be a part-time bottle-washer and food preparator for the fly work that was gaining momentum in Morgan's laboratory. Bridges would eavesdrop on the conversations in the laboratory and he developed a friendship with Sturtevant.[20] Muller and Altenburg were also part of his circle, although they were not part of the laboratory in 1911. They collectively formed a biology club where some of the new work coming from Morgan's laboratory would be hotly debated. Muller was away from Columbia University most of that year pursuing a master's degree in nerve physiology at Cornell University, and Altenburg was still an undergraduate like Sturtevant and Bridges.[21]

Bridges Introduces the Technology for Fly Work

Bridges' circumstances changed approximately a year after he began working for Morgan (Figure 6). He showed Morgan a bottle that contained a fly whose eye color seemed to be brighter than usual. Morgan isolated the fly, showed that it carried another X-linked trait, and called that trait vermilion. He also assigned Bridges to a desk and told him to look for more mutations. Bridges was very gifted mechanically. He introduced the use of ether to knock out the flies, porcelain plates on which to line flies up with watercolor brushes, and good lighting (cooled and diffused in its passage to the plate by a flask of water). Bridges also standardized the food medium, and the lab shifted from just using a wedge of mashed bananas in the bottom of a jar or vial to a cooked liquid medium with molasses, cornmeal, and agar (later, they added mold preservatives) that could be squirted into vials or bottles so that several trays of gelled cultures could be prepared daily.

Figure 6. *This photo of Calvin Bridges was taken at Woods Hole in 1922. While Morgan was at Columbia University, his laboratory frequently spent summers at Woods Hole or Cold Spring Harbor. (Courtesy of Mrs. H.J. Muller to Elof Carlson.)*

Bridges' Personality Limits his Academic Career

Bridges had a child-like quality to his personality, which made him engaging. He also had an unconventional reputation as a womanizer. This repelled both Morgan and Sturtevant, who were more conservative.[22] Bridges liked the Bohemian life and frequented Greenwich Village to rub shoulders with the creative and socialist groups there; one of his friends was author Theodore Dreiser. Some ambiguity exists in the accounts of Bridges' life regarding his socialist interests. Muller felt close to him in those days because they were labor class socialists—those supporters who liked to read *The Masses*, support the Industrial Workers of the World (IWW), and look upon capitalism as an oppressor of the working class. This was before the Russian Revolution and the establishment of a Bolshevik state. After the Bolshevik victory, Muller was a communist in everything but membership. Like Muller, Bridges had no strong commitment to a political change in the world that would establish either socialism or communism, and he did not consider himself an activist as was Muller from 1910 to 1936.

Morgan was concerned about his students' political outlook and their moral reputation, not because he was uptight in his own attitude, but because he knew that most colleges had "moral turpitude" covenants and even tenured teachers could be fired for living with someone without being married, having been treated for a venereal disease (reported by physicians then in public documents from the public health authorities), or advocating birth control. Although colleges were shifting to research and providing utilitarian majors to their students, they still had strong *in loco parentis* obligations for faculty to ensure that students remained celibate while living on campus. Those laws were still in force when I was a graduate student at Indiana University in the 1950s. For Morgan, this presented a dilemma. He liked Bridges' work and supported him as a scientist. But he believed Bridges would never be able to get or hold an academic position because of his loose reputation. It was one thing to be a creative artist or writer in the Village; it was altogether different in the staid moral atmosphere of a college or university. This made Bridges dependent on Morgan for grant money support. Morgan kept that lifetime obligation and made sure Bridges would work wherever Morgan went.[23]

Bridges Discovers Nondisjunction but Encounters Two Priority Disputes

Bridges may not have been the first to find aneuploidy (abnormalities in chromosomes numbers), and Reginald Ruggles Gates claimed that he had priority for that discovery, which Bridges interpreted as arising by nondisjunction. It is true that Gates had found irregular chromosome numbers in some species of *Oenothera*. But what Bridges found was far more significant, because he related his findings to the genes carried by those chromosomes. In addition, Gates used cytology to *report* irregular chromosome numbers, but Bridges used breeding analysis to *infer* (and later confirm cytologically) the abnormal chromosome numbers. Specifically, Bridges used the white eye series. His classic example was associated with the white-eyed crisscross expectations and the reciprocal cross. He found occasional unexpected flies. Thus, if a white-eyed virgin female is mated to a red-eyed male, the expectation is for all white-eyed sons and all red-eyed daughters. But Bridges found rare instances (about 1 in 1000) in which a red-eyed male would appear among the progeny. Such males were always sterile. In the reciprocal cross of white-eyed males with red-eyed females, he found occasional white-eyed males among otherwise red-eyed flies. These white-eyed males were also sterile. Most interesting were the crisscross results that produced an occasional white-eyed female. These females, Bridges' analysis revealed, provided a quite different outcome. When crossed to a red-eyed male, they produced red- and white-eyed males and red- and white-eyed daughters. The ratios always had the unexpected classes at a lower frequency (about 4%). In addition, the sterilization of exceptional male flies seemed to have disappeared.

Although Bridges carried out these experiments and eventually wrote up the results as his own dissertation work under his own name, the entire laboratory was in on the discussions and planning of the experiments, not because Bridges was lacking the skills to do so, but because the results were so exciting that everyone wanted to have a go at interpreting the new findings. At least, that is my interpretation of the comments that Muller, Altenburg, and (indirectly to others) Morgan or Sturtevant might have felt.[24] In the terminology used after 1915, when the Y chromosome that Nettie Stevens had reported (her h2 heterochromosome) and Wilson named returned to fruit fly genetics

and cytology, the exceptional white-eyed female from P_1 **WW × RY** was actually **WWY**. When crossed to **RY** males, she produced eggs with haploid **W** (the majority of her X-bearing eggs) and the unexpected **WW**, **WY**, and **Y** eggs. Fertilization of these eggs by the half of the sperm bearing the **R** produced daughters that were **WR** (red-eyed and fertile females), **WWR** (semi-lethal red-eyed and sterile females), **WRY** (red-eyed and fertile females), and **RY** (red-eyed and fertile males). When crossed to the Y-bearing sperm, these same eggs produced **WY** white-eyed sons who were fertile, **WWY** fertile white-eyed females, **WYY** fertile white-eyed males, and **YY** zygotes that aborted.

After consulting with Wilson, Bridges chose the term nondisjunction to make a parallel comparison with de Vries' phrase for the law of disjunction of hybrids (Figure 7). In this chromosomal model, nondisjunction was an error in which one gamete received two sex chromosomes or no sex chromosomes, when it would normally receive just one X chromosome. Bridges adopted the term secondary nondisjunction to describe the distributive process of chromosomes from a female with an **XXY** constitution. If all three paired randomly (X with X, or X with Y), four types of eggs should be produced in equal amounts (X, XY, XX, and Y). In fact, the two X chromosomes pair more often than does one of the X chromosomes with a Y.

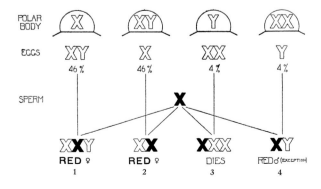

Figure 7. *Bridges' use of nondisjunction to prove the chromosome theory. Bridges found the exceptional flies and carried out the experiments to test their predicted chromosomal composition. All members of the lab enjoyed discussing the experiments to be carried out, but Bridges had the priority, and the desire, to use this for his dissertation. (Reprinted from Morgan T.H. 1919. The Physical Basis of Heredity. J.B. Lippincott, Philadelphia.)*

Bridges wrote his first paper on nondisjunction in 1913. But at that time his analysis was incomplete. When he had it in polished form with the loose ends neatly accounted for, his paper was sent to the *Journal of Genetics* (Bateson's journal), only to be rejected. This galvanized Morgan to join with Shull and other geneticists to establish a new journal that they named *Genetics*. Bridges' article was given the honor of starting on page one of the first issue in 1916 (Figure 8). Bridges was not as prolific as Sturtevant, Muller, or Morgan in publications; a major reason for this was his role in the fly lab. He mapped every new mutation and kept meticulous records of all the crosses so that they could be written up. He supervised the construction and maintenance of all the fly stocks as the mutants and their combinations for different stocks emerged over the years. Later, when Painter described giant salivary chromosomes, it was Bridges who established the most accurate depiction of the bands and correlated the genetic maps with the cytological maps.[25]

NON-DISJUNCTION AS PROOF OF THE CHROMOSOME THEORY OF HEREDITY[1]

CALVIN B. BRIDGES
Columbia University, New York City

[Received October 21, 1915]

TABLE OF CONTENTS

I

INTRODUCTION

There has been a long series of observations and experiments which has led more and more definitely to the conclusion that the chromosomes are the bearers of the hereditary materials.[2] It was observed that male and female contribute equally to the inheritance of the offspring, and yet the contribution of the male consists of little more than a nucleus. That

[1] Contribution from the Zoological Laboratory of Columbia University.
[2] For a fuller discussion of the steps in the accumulation of this evidence see "The mechanism of Mendelian heredity" by MORGAN, STURTEVANT, MULLER, and BRIDGES.

GENETICS 1: Ja 1916

Figure 8. *Bridges published his work on nondisjunction in the new journal,* Genetics. *Bridges' dissertation, which a British journal had turned down, was published as the first article in this new journal. This is the title page; note its citation as volume 1, page 1. (Reprinted from Bridges C.B. 1916. Non-disjunction as proof of the chromosome theory of heredity.* Genetics **1**: *1–52.)*

Chromosome Rearrangements Are Explained in Bridges' Papers

Bridges recognized that some mutations of that time were unusual, and he devoted much effort to working out their solution. They were like complex puzzles; the first of these was the deficiency (Figure 9). That is a name Bridges assigned in 1914, but he did not publish the concept until 1917.[26] Morgan noted in his obituary for Bridges that although Bridges wrote clearly, he had difficulty getting around to writ-

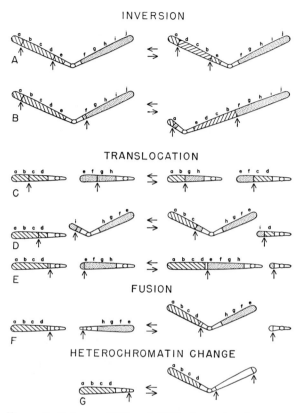

INVERSION

TRANSLOCATION

FUSION

HETEROCHROMATIN CHANGE

Figure 9. *Between 1915 and 1930, the major chromosomal rearrangements—inversions, deletions, duplications, and translocations—had been worked out. Shown here are some of the rearrangements as depicted in genetics texts of the mid-twentieth century. (Reprinted, with permission, from Patterson J.T. and Stone W.S. 1952. Evolution in the Genus Drosophila. Macmillan, New York.)*

ing up his abundant notes and experiments. At his death, he left a sizable amount of unpublished data on mapping relations between crossover maps and salivary gland chromosome maps. In fact, Bridges would wait to the last day before writing a paper for a meeting. None of the other fly lab workers had this writer's block problem.

The concept of genetic deficiency arose when a mutation arose, showing vermilion in the female whose other X chromosome contained that mutant in the stock being used. This new vermilion was also a recessive lethal but was so closely linked to it that it did not separate. It was unclear whether there were two adjacent genes that had independently mutated (what was then called a "line mutation") or whether there was an inactivation, in which the genes were shut off but still present. Bridges successfully argued that this deficiency was the most likely explanation, because the genes included in the deficiency did not undergo recombination, and they showed a "pseudodominance" in the heteroduplex condition, i.e., one chromosome contained the recessive viable mutation (e.g., v) and the other homolog contained the deficiency. Similar deficiencies were all found to show these properties, and somewhat larger deficiencies also showed a reduced frequency of crossing-over for genes that were present to the left or right of the deficiency, which is what Bridges expected for a chromosome segment slightly reduced in length between those genes.

Priority Over the Bar Eyes Duplication Is Disputed

The second finding of chromosomal rearrangements was the duplication.[27] Although the most common duplications are tandem, a few are larger, and the

segment might be shifted elsewhere on that same chromosome or to an entirely different chromosome into which it is inserted. Duplications were of interest because they provided a basis for more genes to accumulate in the evolution of the entire genome or set of chromosomes. They also could act as apparent suppressors of what should have been a homozygous recessive mutation sought in a cross. Bridges labeled that in 1916 and published his idea with some genetic evidence in 1918. In the mid-1920s, Morgan and Sturtevant worked out a gene duplication that turned out to be the origin of a dominant X-linked mutation called Bar eyes (found by Sabra Tice in 1913). It was difficult to keep a pure stock of Bar eyes because occasional revertants of Bar eyes to round eyes occurred (the frequency being about 1 in 1000 flies). Morgan and Sturtevant interpreted the Bar phenotype not as a gene mutation but as a new phenomenon they called position effect. They did so because if the duplicated genes are shifted by crossing-over back to the single-gene state, Bar disappeared.[28]

When salivary chromosome analysis opened up in the mid-1930s after T.S. Painter described their correlation to the much smaller somatic or meiotic chromosomes in *D. melanogaster*, Bridges at the California Institute of Technology and Muller (then in Leningrad working with a cytologist, Alexandra Prokofyeva-Belgovskaya) independently reported Bar eyes as a chromosomal segment that had duplicated (Figure 10).[29] A dispute on priorities erupted here. Bridges had been accused before of stealing a concept (nondisjunction) from Gates. But as we saw, Gates had no genetic evidence to correlate with his cytological observation of variations in chromosome numbers in certain species of *Oenothera*. Now it was Bridges whom Muller accused of swiping the idea when Muller's student, Carlos Offermann (accompanied by Muller's first wife), had returned to California from Leningrad. Bridges scooped Muller by quickly publishing in *Science*; Muller's Soviet paper took more time to get published, although it was submit-

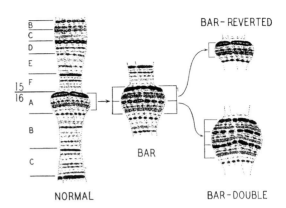

Figure 10. *Bridges drew the Bar duplication, showing the bands involved on giant salivary chromosomes. (Reprinted from Bridges C.B. 1936. The Bar "gene," a duplication. Science 83: 210–211.)*

ted earlier. I do not think that Muller was robbed any more than I think Gates was, because the weight of the evidence points to coincidence. When Painter's work appeared it was like the rediscovery papers of 1900, and it sent *Drosophila* workers scrambling to apply it to a number of problems. By the mid-1930s, all of the major chromosome rearrangements were already known from their genetic characteristics. The Bar case was unusual, and interest in position effect was high among geneticists. It is no surprise that Bridges would work on such a problem in Morgan's laboratory. It is also no surprise that Muller, intensely studying position effect at the time, would have been interested in looking at Bar (Figure 11). What is ironic is that Bridges (and hence the suspicion of duplicity), who was known for sitting on things before publishing them, worked so rapidly to publish the Bar case. I suspect that it was because he knew that Muller, Painter, or someone else would realize as he did that Bar was a natural for examining under the microscope.

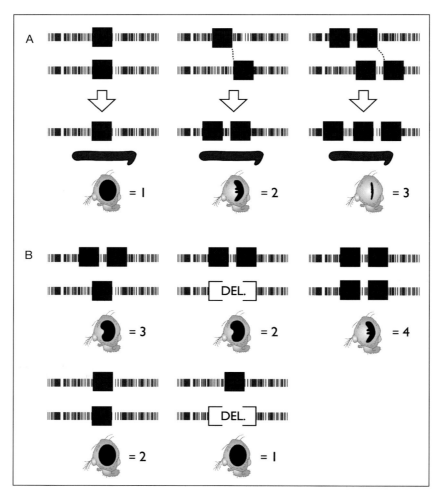

Figure 11. *A contested interpretation of Bar eyes. Interpreting the Bar case in 1935 reveals how Bridges and Muller were guided by two different models. Bridges clearly saw the Bar region as duplicated once Painter's techniques were applied to this region of the X chromosome (very near the centromere region). (A) Using males, Bridges found a graded series. One dose of the region resulted in round eyes, two doses tandemly duplicated through unequal crossing-over produced the original Bar eyes, and mispairing of the duplication during meiosis led to the "triple-Bar" condition, an even more reduced eye shape. Hence, to Bridges it was a matter of taste whether Bar arose as a position effect or a dosage effect. (B) Muller argued otherwise, using the data on females from his studies. Thus, round eyes could be one dose of the region (in a female, heterozygous for the deletion, but otherwise normal) or they could be two doses (the homozygous normal round-eyed female). The kidney-bean heterozygous form of the eye could arise from three doses of the region, as in the typical B/+ genotype or from two doses, as in the B/del(Bar) genotype. The homozygous four doses of Bar result from the B/B genotype. Muller argued that position effect was the mechanism for the altered function and not dosage of segments. (Figure drawn by Claudia Carlson.)*

Bridges Has a Role in Working Out Translocations

The third rearrangement tackled by Bridges was what came to be called translocations.[30] The first translocation (called Pale) arose in 1918 as a complex event that created a puzzle. Two pieces called PII and PIII (the Roman numerals are for the chromosomes in which these two pieces were found) seemed to be involved, and both pieces were needed for the fly to live. Homozygosity for either PII or PIII (and, of course, for both) was lethal to the embryo. Bridges believed that there was a resemblance between both the features of a deficiency (a tendency to be lethal when homozygous) and duplication (a tendency to repress the expression of a homozygous condition when present elsewhere). He then thought he would play a trick on Muller and sent him a postcard with a puzzle, giving the data but not the interpretation to determine whether Muller could figure it out. Muller did and when Bridges published on the Pale translocation, Muller was annoyed that he was not given credit for providing the solution to the puzzle. It is one of the realities of professional life (not only in science) that jokes, parodies, satires, and puzzles often backfire and lead to wrongful consequences. As noted earlier, the ovists were lampooned in the eighteenth century by at least one joker who submitted a fictional illustration of a homunculus in a sperm to ridicule the idea that a comparable encapsulation existed in eggs during the preformationist debates. That picture of a sperm homunculus is seen in almost every embryology textbook, with contemporary authors interpreting eighteenth-century scientists as projecting "the little man that wasn't there" into the sperm. Unfortunately for Bridges, his various personality features got him into more than one difficulty. He was not diabolical; he was guileless and enjoyed boyish pranks.

Much more important for the Bar case was the resolution of its interpretation. For Bridges, the position effect could be some sort of gene interaction or just a dosage effect "as a matter of taste." Not so, Muller argued: The position effect was real because a female who has one **X** chromosome with Bar and the homolog with a deficiency for that region shows no change in appearance (the same kidney-bean shape appears in both the heterozygote with the normal gene and the heteroduplex with the deficiency). Muller also argued that Bar was important because it extended Virchow's cell doctrine, and he argued that "all genes arise from preexisting genes" (except for the Ur gene that started it all).[31] The mechanism of tandem duplication was important because it also extended Bridges' idea that repeats had a role in evolution. To Muller, they arose from a "primary unequal crossing-over" and once established in a homozygous condition, the pairing could be askew and result more frequently in "secondary unequal crossing-over," leading to reverse mutations to the normal red eye or to triplications or nests of several cognate or duplicate genes, which could undergo independent evolutionary histories.

Bridges Analyzes Triploids and Gynandromorphs

Bridges made substantial contributions to the working out of triploids and to an understanding of gynandromorphs.[32] The gynandromorphs are also known as hermaphrodites, but in fruit flies, they show a segmental symmetry, often one side of the body being male and the other side being female. Bridges used these sex chromosome differences in gynandromorphs (the female side was **XX** and the male side was **XO**) (Figure 12) and his nondisjunction work (comparing **XXX** females with **XX** females, and with **X** males). To complete the story, he obtained a cross that produced a triploid female (**XXX** with three of each autosome). She was fertile and from her occasional triploid offspring, Bridges obtained triploid **XXY** individuals who were neither male nor female, not in a segmental manner as in gynandromorphs, but as sterile intersexes. He also obtained nondisjunction of chromosome IV (the dot-like small chromosome with very infrequent crossing-over). Such flies, being haplo-IV, also had disturbed sex development.[33] This reinforced an idea that went back to Wilson's original suggestion that sex determination in fruit flies might be quantitative instead of qualitative. Bridges called this the "balance theory" of sex determination.[34] In the diploid XX:AA = 1.0 = female, XO(or XY):AA = 0.5 = male, XXX:AAA = 1.0 = female, XXY:AAA = 0.67 = intersex, XXX:AA = 1.5 = superfemale (a sterile female), and XY:AAA = 0.33 = supermale (a sterile male).

In 1938, Bridges developed a streptococcal infection of the pericardium and died at the relatively young age of 49. Muller had heard of his illness while in Edinburgh and he wrote a comforting and hopeful letter. It was returned to Muller unopened. Bridges had already died.

Figure 12. *Hermaphroditic flies are called gynandromorphs. The top figure shows a sex comb in the right foreleg but not in the left foreleg. The loss of a chromosome with the normal allele for white eye color renders the male side partially white eyed. The wing is also smaller. However, the abdomen is almost exclusively male. On the lower left, the male side is on the left of the fly, with a partial white eye and a sex comb present on the foreleg. The right side shows a notching of the wing. The abdominal banding is mixed. The events for the third fly on the lower right show a largely female body (but note the male sex comb on the left front leg); the eyes are male (in the lost X, the normal allele, w⁺, is absent) and so is the normal wing on the left (the wing length is smaller than that of a female). (Reprinted from Morgan T.H. and Bridges C.B. 1919. Contributions to the genetics of Drosophila melanogaster. I. The origin of gynandromorphs. Carnegie Institution of Washington.)*

End Notes and References

1   Anonymous. 1970. Undated otherwise. *Alfred Henry Sturtevant 1891–1970*. Printed remarks and pamphlet from memorial service held at the California Institute of Technology, 13 pages.

   Also see *A.H. Sturtevant—Biographical Notes*, unpublished, undated, typescript with chronology, in the first person; it also contains inked marginalia by Sturtevant. I may have have gotten this Xerox copy from Sturtevant on one of my visits to the California Institute of Technology. The last written entry in his chronology is dated 1966 and the last typed entry is dated 1960.

2   Sturtevant A.H. 1965. *A History of Genetics*. Harper and Row, New York.

3   Altenburg E. 1946. *T.H. Morgan, Democrat*. Unpublished obituary. Lilly Library, Indiana University, Bloomington.

4   Edgar Altenburg. Interview with E.A. Carlson, Houston, Texas. 1967.

5   Cited in Allen G. 1978. *T.H. Morgan—The Man and His Work*. Princeton University Press, New Jersey.

6   Fernandus Payne. Interview with E.A. Carlson, Bloomington, Indiana. August 14, 1970.

7   Muller prepared an undated list of contributions of Morgan, Sturtevant, Bridges, and himself alone or in joint effort. I have included that in Chapter 15, in Muller's account of the early contributions (although I could not make out all of his handwriting).

8   I am not sure who told me of this discarding of Muller's early papers (I learned of it between 1965 and 1968), but it may have been one of Patterson's students who happened to be visiting UCLA. I made inquiries by mail and during a visit to the archivist at the University of Texas, but they had no record of Muller's papers.

9   Mark Graubard. Interview with E.A. Carlson, University of Minnesota. May 24, 1974. Carlson diary, volume 8, pp. 29–32. Graubard was Bridges' student at the California Institute of Technology.

10 Sturtevant, personal recollections. Undated op. cit. p. 2.

11 Muller H.J. 1920. *A Decade of Progress in Drosophila*. Speech given in English and later in Russian. Translated from the Russian by Joel Wilkinson.

12 Sturtevant A.H. 1965. op. cit. p. 47.

13 Sturtevant A.H. 1913. The linear arrangement of six sex-linked factors in *Drosophila*, as shown by their mode of association. *Journal of Experimental Zoology* **14**: 43–59.

14 Morgan and Sturtevant began an attempt to separate these alleged components of the white-eyed series of alleles, beginning with eosin and red. If red is CO and white is co, then crossing-over should produce two recombinants, one eosin Co and the other possibly another shade of diluted red. See Sturtevant A.H. 1913. op. cit. p. 46. In August 1911, eosin was found as a partial reverse mutation in a white-eyed stock. Unfortunately, their assumption was wrong. They would have found recombinants if they crossed white to eosin and sought red-eyed recombinants, but their logic at the time made it irrational to consider such a cross.

15 Edgar Altenburg. Interview with E.A. Carlson, Houston, Texas. 1968. I have the interview on tape (a Uher 2000 recorder was used) but never transcribed it. Altenburg did not want me to record him, so after the interview, I had to rush back to my hotel in Houston and dictate the interview while it was fresh in my mind. I found T.S. Painter to have the same reluctance to be interviewed on tape.

16 Sturtevant A.H. 1913. The Himalayan rabbit case, with some considerations on multiple allelomorphs. *American Naturalist* **47**: 235–238.

17 Sturtevant A.H. 1920. Genetic studies on *Drosophila simulans*. I. Introduction. Hybrids with *Drosophila melanogaster*. *Genetics* **5**: 488–500.

   A *D. m* female x *D. s* male produces only female offspring. The reciprocal cross of a *D. s* male and a *D. m* female only produces males. In both cases, the hybrids are sterile.

18 Sturtevant A.H. 1926. A crossover reducer in *Drosophila melanogaster* due to inversion of a section of

the third chromosome. *Biologischen Zentralblatt* **46:** 697–702.

19 Most of this I gathered from speaking to him at meetings, interviewing him, or from discussions with his colleagues and former students. In fact, when I entered Sturtevant's office to interview him, his feet were propped up on his desk.

20 Morgan T.H. 1939. Calvin Blackman Bridges. *Science* **89:** 118–119.

21 I knew Altenburg well and interviewed him when I wrote Muller's biography. Altenburg said that he and Muller "knew each other inside out" and they were close friends since their days at Morris High School in the Bronx. Muller (still a graduate student) served as an ex officio adviser for Altenburg's Ph.D. dissertation, a genetic map constructed from the data of *Primula* crosses. It was the first plant to be mapped. Morgan refused to let Altenburg work on fruit flies in his laboratory for his dissertation and this may have created the bitter (but ambivalent) feelings that he had towards Morgan. Curiously, his unpublished obituary, *T.H. Morgan, Democrat*, belies the advice that he gave to Muller (i.e., do not overly praise him and attribute to him what he did not do, out of respect for the positive attributes demanded for an obituary). It is a fine tribute to Morgan, and it implies that Morgan is a pioneer in founding team research in science.

22 Scientists, like most people, enjoy gossip. Bridges' love life was well known (he was open about his infidelities and love for sex). At Cold Spring Harbor, he named his boats the VIR and the GIN. He had an affair with the wife of a Soviet geneticist when he visited the USSR in 1933 or 1934. He lived with an attractive "gypsy," as he called her, who stole $5,000 from him when she dumped him. These are only a few of the rumors that I heard about him. I tried checking some out, to no avail. The only one that I wish were true was that Theodore Dreiser used him as a lead character in a novel he never finished. I called the archivist of the Dreiser papers, who knew of no such manuscript. Whether true or not, Bridges did nothing to discourage such stories. I was only seven years old when he died, so I did not have the opportunity to know him. As a graduate student, I naively asked Katherine Brehme Warren if Bridges died of a venereal

disease (yet another rumor); I was greeted by an icy "Calvin never had a venereal disease." My cheeks still turn red at the thought of my *faux pas*. Bridges also experienced tragedy in his life. He fell asleep while babysitting his two small children, and one of them fell into a fireplace and died from the severe burns. This haunted Bridges for the rest of his life.

23 I learned a lot about Bridges' personality, his work habits, and his relationships to Sturtevant and Morgan from conversations with Katherine Brehme Warren. Younger readers may not realize how rigid sexual behavior codes were for faculty before the 1960s. When I was a graduate student at Indiana University, graduate students were sexually segregated, lounges always had a matron checking for the number of feet seen under couches, and students (especially female) had to abide by the curfews in the dormitory.

24 Bridges C.B. 1913. Non-disjunction of the sex chromosomes of *Drosophila. Journal of Experimental Zoology* **15:** 587–606; Bridges C.B. 1916. Non-disjunction as proof of the chromosome theory of heredity (part 1). *Genetics* **1:** 1–52; and Bridges C.B. 1916. Non-disjunction as proof of the chromosome theory of heredity (part 2) *Genetics* **1:** 107–163. The second paper shows the detailed correlation of genetics and cytology. The first was done without knowledge that males had a **Y** chromosome.

25 See the foldout insert in the back of Bridges C.B. and Brehme K.S. 1944. *The Mutants of* Drosophila melanogaster. Carnegie Institution of Washington Publication No. 552, Washington D.C.

26 Bridges C.B. 1917. Deficiency. *Genetics* **2:** 445–465.

27 Bridges C.B. 1919. Duplication. *Anatomical Record* (abstract) **15:** 357–358.

28 Sturtevant A.H. and Morgan T.H. 1923. Reverse mutation of the bar gene correlated with crossing-over. *Science* **57:** 746–747.

29 Bridges C.B. 1936. The Bar "gene," a duplication. *Science* **83:** 210–211; and Muller H.J., Prokofyeva-Belgovskaya A., and Kossikov K.V. 1936. Unequal crossing over in the bar mutant as a result of duplication of a minute chromosome section. *Comptes Rendus (Doklady) de l'Academie des Sciences de l'URSS* **1:** 87–88.

30 I corresponded with Schultz on the Pale translocation. Schultz sent me a copy of Bridges' unpublished manu-

script (no title page, 20 typed pages). I replied to this in a four-page letter dated May 28, 1969.

31 Muller H.J., Prokofyeva-Belgovskaya A., and Kossikov K.V. 1936. op. cit. p. 88.

32 Bridges C.B. 1921. Triploid intersexes in *Drosophila melanogaster*. *Science* **54:** 241–254.

33 Bridges C.B. 1921. Proof of non-disjunction for the fourth chromosome of *Drosophila melanogaster*. *Science* **53:** 308.

34 Bridges C.B. 1926. The mechanism of sex-determination. *Proceedings of the Birth Control Congress, New York,* pp. 10–13.

This too was a disputed issue between Muller and Bridges, with Muller claiming that the idea of using the ploidy ratio of **X** chromosomes to autosomes was his.

Muller's detailed listing of the accomplishments of the members of the fly lab written about 1930. See Muller's version of the early contributions below (p. 209) for my interpretation of Muller's handwritten notes. (Courtesy of Mrs. H.J. Muller to Elof Carlson.)

Forming the Fly Lab:
Contributions of H.J. Muller

...In the past, a mutation was considered a windfall, and the expression "mutation frequency" would have seemed a contradiction in terms. To attempt to study it would have seemed as absurd as to study the conditions affecting the distribution of dollar bills on the side walk. You were simply fortunate to have found one.

H.J. MULLER*

ONLY ONE MEMBER OF THE FLY LAB, besides Morgan himself, went on to win a Nobel Prize. Hermann J. Muller (1890–1967) went beyond his dissertation triumphs to develop a new field of science: radiation genetics (Figure 1). Of Morgan's main members, in the formative years 1910–1915, there was no doubt that Sturtevant was looked upon as the most accomplished geneticist. Bridges recognized that he was not as much of an intellectual hotshot as Sturtevant, but that did not rankle him. He had found his own niche in carrying out experiments, mapping new mutations, and finding more mutations than any other fly worker. Muller was the new kid on the block who tried to fit in under difficult circumstances. He happened to be two years ahead of Sturtevant and Bridges and had received his B.A. before Morgan published his work on X-linked

inheritance and crossing-over. To Muller, Morgan was still an embryologist with a muddled sense of evolution when Muller took his classes in his sophomore and junior years. It was the Biology Club that began to attract Muller to Morgan through encounters with Sturtevant and Bridges. The relationship of Muller to Morgan and his fellow graduate students would prove to be difficult, each finding in the other some personal irritation, as well as a communal satisfaction that they were creating something new and important.

Morgan, Bridges, and Sturtevant descended from American settlers who arrived during the colonial period. Muller's paternal grandparents arrived after the 1848 revolution from Coblenz in Germany. Morgan, Bridges, and Sturtevant were American Protestants who had dropped their faith in favor of what would today be called a secular or "unchurched" lifestyle. Religion was not part of their lives. Muller's ancestry was Catholic on his grandfather's side and mixed Jewish and Anglican on his

---

*1922. Variation due to the change in the individual gene. *American Naturalist* **56:** 32–50.

Figure 1. *Muller at the time he completed his Ph.D. and about the time he joined the faculty at Rice University in 1915. (Courtesy of Mrs. H.J. Muller to Elof Carlson.)*

mother's side. His father raised Muller as a Unitarian, but Muller quickly dropped even that vestige of religion for a more aggressive humanism, in which he regarded most religion as superstition and hostile to science. He was a third-generation New Yorker with few traces of his grandparents' immigrant background left. He was also a third-generation Socialist, his grandfather being forced out of Germany after the 1848 revolution. Muller's father died when the young Muller was ten years old. His mother lived off a subsidy from her brother-in-law, a partner in the Muller's brass art works store on Canal Street. These circumstances, like those of Bridges, made it impossible for Muller to attend college without support. Fortunately, he was quite a good student in high school and won a full-tuition scholarship to Columbia.[1]

Muller Joins Morgan's Laboratory in 1912

Although H.J. Muller was a year younger than Bridges and a year older than Sturtevant, he began his college two years earlier than Sturtevant and Bridges because he had skipped grades in elementary school. He liked science and had originally thought he would become a school teacher, but changed his mind after taking Wilson's course and shifted to genetics after reading Lock's first text for that field, *Heredity, Variation, and Evolution*, published in 1906.[2] Muller completed his B.A. in 1910, before Morgan published or taught any of his work on fruit fly genetics. This may have motivated Muller to seek a master's degree at Cornell in neural physiology, which he completed in 1912. While at Cornell, he learned of Morgan's work from Sturtevant and Bridges, who had all been members of the Biology Club in Muller's senior year. Muller would regularly drop by for the discussions and in all likelihood,

Bridges and Sturtevant put in a good word for Muller's contributions to their discussions. In 1912, Morgan accepted Muller as the third of his great students. It was actually a larger laboratory complex with six members crowded into the fly lab and additional departmental graduate rooms known then as the "Boys'" or the "Girls'" graduate rooms. For a variety of reasons, the less well-known members of the laboratory did not take the same interest, enjoy the same discussions, or have the time to participate in doctoral programs.[3] Some, like Edgar Altenburg, Muller's high school chum, ached to work for Morgan but that was not to be (Figure 2). I suspect that this was because of Altenburg's loyalty to Muller. He adored him, and the two formed a close alliance that Morgan might have felt would be too much of a strain on the laboratory. Unlike Muller, Altenburg was not in financial need; his father made and sold pianos.[4]

Muller Does Not Publish Until 1914

Muller did not publish his first article in Morgan's laboratory until 1914. But for four years, he had taken an active part in the discussions through the Biology Club, the Peithologian Society, and most of all, through the visits to Sturtevant and Bridges in the fly lab. Altenburg was a frequent participant as well. Although Sturtevant acknowledged Muller's help as an aside in his major article on linkage, he did not specify Muller's actual contributions. Muller and Altenburg had a view of co-authorship or, at least, formal acknowledgment in an article, different from Morgan, Sturtevant, and Bridges. I do not believe that Sturtevant, Bridges, and Morgan were devious and wanted to strip Muller of credit; in Morgan's case, it was very clear that ideas meant little to him compared to the execution of worthwhile experiments. Sturtevant adored Morgan as much as

Figure 2. Muller's *lifetime friend from high school, Edgar Altenburg* (left), *was not accepted by Morgan as a graduate student. He worked with Muller on flies while completing his Ph.D. in botany. Altenburg collaborated with Muller on several projects while in Texas.* (Courtesy of Mrs. H.J. Muller to Elof Carlson.)

Altenburg adored Muller and that was the problem. It is not a surprise that a student mimics the habits and values of a mentor, and these relationships were very much in that vein. Unfortunately, that is not the way each group saw each other. The Muller-Altenburg group saw an injustice to Muller, who did not have the financial support to do fly work the way Sturtevant and Bridges had through Morgan's largesse. The Morgan-Sturtevant-Bridges axis saw the resentment as one of jealousy and a sour grapes attitude on the part of Muller.

If Muller is to be believed, his word is best supported by Altenburg in their correspondence at the time of Morgan's death.[5] Each was asked to write an obituary, and Altenburg, who seemed like a scolding Iago in this exchange of views, urged Muller not to heap praise on Morgan unless he also cited his faults, the most serious of which was his failure to acknowledge Muller's role in the formative years of 1910–1915. This was when Muller's ideas shaped the interpretation of the data and the design of experiments for the exceptional flies that emerged in the laboratory. Muller claimed to have suggested to Bridges the way to design an experiment to distinguish a deficiency from a line mutation. He advised Bridges on the experimental design that demonstrated the presence of a **Y** chromosome in the **XXY** and the need to test exceptional flies for fertility. He suggested to Morgan and Bridges the use of balance ratios for sex determination, relying on Wilson's general theory that sex determination in *Drosophila* was quantitative, not qualitative. He suggested to Sturtevant (as I mentioned in the account of Sturtevant's contributions) that it is crossovers to total progeny, not crossovers to noncrossovers, that formed a basis for mapping and that double crossovers had to be discussed and demonstrated to account for the failure to obtain a linear map based on Sturtevant's sum rule. He also introduced the symbolism used today that replaced Morgan's cumbersome method of inferred uppercase letters for remnant functions. He suggested to Morgan that somatic cell nondisjunction was the explanation for gynandromorph formation.[6]

Why Did Muller Have a Priority Complex?

This is a heady performance, if true (see Muller Gives his Version of the Early Contributions, p. 209). What was the reason that Muller did not write his own articles or carry out the experiments in the way that Bridges and Sturtevant had? Was Muller a theoretician who lacked experimental skills? According to both Muller and Altenburg, the explanation was economic. Muller had to work as a Wall Street clerk, rush to work eating a sandwich on the subway to teach classes at night in English as a second language for immigrants, and tuck in the time to go to his classes and study at Cornell for his master's degree. Whatever free time he had was spent arguing and suggesting ways in which Bridges and Sturtevant could improve their interpretations or test Muller's ideas with experiments. As for Muller, his commitment to working himself to a frazzle was one of filial duty; he had to support his mother. The workload proved damaging to Muller, and he had his first mental breakdown in 1912.[7] He does not describe what actually occurred, but in his occasional autobiographical accounts, he refers to it as leaving him with reduced physical energy for the rest of his life. It also cost Muller dearly, because the resentments that built up over these years of seeing his ideas unacknowledged and assigned to others gave him the reputation of having a priority complex. Later, when he was quite famous on his own, these youthful hurts lingered and his friends could not understand why he did not just drop these grievances. He had glory enough on his own, so what did it matter who was assigned credit for those ideas? To Muller, it mattered. They were the fruits of his creativity at a time when he was left out of the glory, which went to Morgan, Sturtevant, and Bridges.[8] All sibling rivalry may be wrong, but it can have an underlying truth for its existence.

Even if Muller's argument is correct, I cannot fully fault Morgan, Sturtevant, and Bridges. All three believed that sooner or later the interpretations and the experimental designs would have come to them. Just because Muller was faster at coming up with ideas and experimental tests did not mean that they were unaware (in the broad sense) of the same interpretations or incapable of formulating them on their own.[9] Muller may have had a more complete and polished way of going about the work, but they were convinced that step by step, the facts would have forced them to come to the same conclusions. I do not doubt that this is correct, but Muller did not believe this. I believe that the Muller-Altenburg error was a desire for Muller to be fully acknowledged, preferably as a co-author, whereas the Morgan-Sturtevant-Bridges error was one of minimizing that acknowledgment to such a generality that it did not give an adequate picture of Muller's contributions to the formative years of the fly lab. To Altenburg, they were out-and-out "steals."

Muller Begins to Publish, Then Leaves Morgan's Laboratory

Muller entered the laboratory in 1912—this time with departmental support to do research in Morgan's laboratory. He chose two areas to explore. The first was an analysis of linkage and events associated with variations in linkage strength or distance. The second was a pursuit of a project that Morgan had worked on and then assigned to one of his lesser students, John S. Dexter, and the project was turning out to be complex.[10] This study involved the inconstancy of the *Truncate* wing mutation (Figure 3). It could not be rendered homozygous, and its effect on wing shape was variable with ratios that did not at all conform to Mendelian expectations. For the first project, which became his Ph.D. dissertation, Muller introduced the idea of coincidence and interference in crossing-over, as well as a quantitative measure to explore them. When genes are close together, fewer than predicted double crossovers occur.[11]

Muller's work took time because his analyses for his dissertation and for the *Truncate* wings story took many generations of crosses, and Muller was meticulous in the detail he amassed to explore both problems. Of the four papers he published in 1914, two were based on his experimental work—the finding of a mutant, bent wings, that turned out to be an autosomal recessive that assorted independently of chromosomes II and III.[12] This completed the correlation of linkage groups to chromosomes in *D. melanogaster* and in Muller's mind, it clinched the case for the genetic basis of the chromosome theory of heredity. The other two papers involved interpretations of data and phenomena that he read in the literature. He worked out the Mendelian ratios to be expected for tetraploids when he read Gregory's evidence for a tetraploid composition of a new variety of *Primula*.[13] Those were fine for Morgan, but the fourth paper

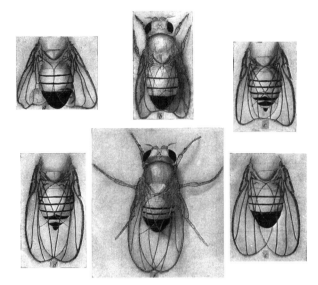

Figure 3. Truncate *wings and evolution. The range of expression of Truncate wings is determined by a chief gene (Truncate, now called dumpy, dp), modifier genes (diminishers or intensifiers), and temperature (higher temperature exaggerates the mutant phenotype; lower temperature renders it more normal in appearance). This continuous distribution, the presence of genetic modifiers, and the role of the environment in gene expression were all components of a Darwinian model of evolution by natural selection. (Reprinted from Altenburg E. and Muller H.J. 1920. The genetic basis of truncate wing—An inconstant and modifiable character in* Drosophila. Genetics **5:** 1–59.)

was a disappointment, and Morgan told him not to publish it. Muller submitted it anyway, using the title "The bearing of the selection experiments of Castle and Phillips on the variability of genes" (Figure 4).[14] Muller attacked the authors for proposing an instability of the gene through crossing as written "in a spirit of mysticism," a phrase bound to raise the

THE BEARING OF THE SELECTION EXPERIMENTS OF CASTLE AND PHILLIPS ON THE VARIABILITY OF GENES

CASTLE and Phillips have recently reviewed the results of six years' work in which they selected for and against "hoodedness" in rats.[1] In "hooded" or "piebald" rats only part of the coat is pigmented; the area of dark (versus white) coat varies greatly in different animals, but tends, in those of medium grade, to cover the head, shoulders and middle of the back, like a hood. Starting with a strain which was probably hybrid, although of unknown ancestry, and selecting during thirteen generations for a larger extent of colored coat ("plus" selection), they succeeded in obtaining animals with a greater and greater area of pigmentation. The average, the mode, and the extremes were raised. Conversely, selection for less pigmentation ("minus" selection) was accompanied by a gradual but decided and continual diminution in the dark area. "Return" selection also succeeded; that is, plus selection was effective even in a line which was already lighter than the average on account of a previous minus selection, and, *vice versa*, minus selection caused a lightening of a strain that had been made exceptionally dark by a prior plus selection.

Figure 4. *Muller disobeyed Morgan and published this attack on Castle's hooded rats analysis, causing a lasting enmity between the two on genetic issues. It also ruptured his relationship with Morgan. (Reprinted from Muller H.J. 1914. The bearing of the selection experiments of Castle and Phillips on the variability of genes. The American Naturalist 48: 567–576.)*

hackles of Castle, but not entirely without justification.

Morgan did not want to get involved in a polemic war with Castle who made the accusation that the *Drosophila* work of Morgan's school was going beyond the bounds of conventional science. According to Castle, Morgan's group took things on faith or added modifying factors that were not yet proven. Castle believed that the variations in *hooded* rats that he and Philips had selected represented fluctuations of the gene.[15] Like many early geneticists during the 1900–1910 period, Castle had been misled by Bateson's unit-character designation for what later became the gene, and they assumed that if a character trait varied, then it had to be the gene

that varied. Morgan agreed with Muller's interpretation and he certainly was miffed by Castle's attacks, but he felt it was detrimental to Muller's career to enter the fray and not a good idea to have one of his students do this because Castle would surely think that Morgan had put Muller up to it. Muller wrote the article because he had two years of unpublished data from his *Truncate* analysis to prove unequivocally that such modifier genes existed (see Muller Gives His Version of the Early Contributions below).

In 1915, Muller spent the entire year wrapping up his dissertation, hoping to finish it in time to get a 1915 Ph.D. Unfortunately, he missed the deadline to submit the thesis by several weeks so the actual award for the degree was dated 1916. Muller also had another reason to finish up in a hurry. Julian Huxley (Figure 5) had made two visits to

Figure 5. *When Rice University was founded, Julian Huxley was recruited as its first Biology Department Chair. He asked Morgan for a student and was given Muller. This led to a lifelong friendship between Huxley and Muller. (Reprinted from Huxley J. 1970. Memories. George Allen and Unwin, United Kingdom.)*

Morgan's laboratory, the first in 1912, en route to the dedication of the new Rice University in Houston, Texas, where Huxley was asked to set up a department.[16] In 1915, he asked Morgan if he could recommend a student. Morgan suggested Muller, who snapped at the offer. He owed no loyalty to Morgan, and he was fiercely independent and anxious to set on his own course; he could not think of a more sympathetic place to be than Rice, a new school with no traditions to limit him, and with Huxley, the Department Chair and the grandson of Darwin's most stalwart evolutionist. Thus, Muller left, leaving as formal evidence of his presence in the fly work two papers on bent wings, a polemical blast at Castle's erroneous interpretations of gene function, and a not yet written pair of articles on recombination. He also left behind the bittersweet memories of his contributions to the fly lab, and he set out to pursue his own Holy Grail—the nature of mutation and the gene.[17]

Muller Gives His Version of the Early Contributions

The following is an account from a single sheet of paper, undated, in which Muller wrote four columns of contributions of Morgan, Sturtevant, Bridges, and himself from 1910 to approximately 1930. The purpose of the sheet is unknown, but it may have been a rough guide to the topics he intended to cover in a work he called "The Baur Manuscript," which is a handwritten manuscript of several hundred pages, unpublished, in the Lilly archives at Indiana University. Baur had asked Muller to write a volume on *Drosophila* and Muller worked on that project on and off for approximately ten years before abandoning it.

Muller may also have set it aside for Mrs. Muller to give to me when he was in ill health and I had arranged to write his biography. He had also arranged for Altenburg to send me a copy of his correspondence with Altenburg (over 700 letters) and those letters are deposited in the Lilly Library. The sheet is slightly larger than the size of U.S. typing paper and more like the paper used in Europe, so this might have been written between 1932 and 1940 while he was abroad. It is too old to have been written in 1967 as a guide for my doing his biography.

I have organized the Muller sheet in the sequence he provides, but I have not used most of his abbreviations, preferring instead to make the document more intelligible. I cannot decipher some of his abbreviations and he did not date all of the entries. I included the original sheet at the beginning of this chapter as well as the following version for those interested in making a comparison.

MORGAN

- Usefulness of *Drosophila* for study of gene mutations and their inheritance, when intensively studied, 1909.
- Many sex-linked genes—with X chromosome, 1910–1911.
- Interchange of the genes and its explanation by Janssen's theory of chiasmatypie, 1911.
- Degree of linkage depends on distance in chromosomes (from theory of Janssen's), 1911.
- Second group of linked genes, which equal other chromosomes, 1912.

- No crossing-over in male, 1912.

- Abnormal abdomen—environmental influence, 1912.

- Various mutations for one character and their interaction (e.g., eye color), 1912.

- Lethal factors in X, 1912.

- Partial reverse mutation of white—against presence-and-absence theory, 1912.

STURTEVANT

- Mapping of chromosomes X and II, 1912 and 1913.

- Double crossing-over, 1912 and 1913.

- Third group of genes, 1912 and 1913.

- Inherited variation in linkage, 1912 and 1913.

- Explanation through inversions—inversions in nature in *D. melanogaster* and *D. simulans*, 1921 and 1926.

- Unequal crossing-over (Bar), 1925.

- Crossability and usability of *D. simulans*. Parallel mutations in *simulans*. Inversion in *simulans*; *melanogaster/simulans* crossover differences.

- Systematics of *Drosophila*.

- Intercellular influences in mosaics (Bar, vermilion, etc.), 1925.

- Suggested to and helped Dobzhansky with translocations and segregation in translocations, 1928.

- Position effect, 1932.

- Suggested explanation of secondary nondisjunction to Bridges, 1914.

BRIDGES

- Found and worked out position and interaction of most of the mutations; accumulated data and standard maps, 1911–1933.

- Linkage variation with age discovered, and at the suggestion of others, did the work on nondisjunction, 1913–1915 (a proof of chromosome theory).

- Deficiency, 1915.

- ("duplication") translocation, 1918.

- Haplo-IV; triplo-IV, 1920?

- Triploids—proving Muller's theory of sex determination by genic balance; tetraploids and haplo mosaics, 1921.

- Specific modifiers of eosin (1913); vortex (1918); giant (1928); etc.

MULLER

- Multiple factors; theoretical conception of gene (pleiotropy?) and interaction with others and environment and explanations of Castle's rats, 1911–1912.

- Dissection of "irreconcilable character" (*Truncate*) into constant but weak and interacting genes by linkage method, 1913–1914.

- Interference and its evidence for chromosomes, 1912; many genes studied together; interchange of groups of genes equals crossing-over, 1915.

- Consequence of bent wings; explanation of Y; parallel size and lengths of groups and chromosomes.

- Noncontamination, 1915.

- Nomenclature, 1913.

- Balanced lethals and explanation of *Oenothera*; degeneracy through mutation process, 1914–1918.

- Suggested explanation and test for deficiency, haplo-IV to Bridges, 1915; translocation (Whiting?); cytological proof of nondisjunction; suggested sex-determination mechanism to Morgan and Bridges; mechanism of gynanders and test (therefore?) and conclusion of indetermination.

- Linkage variation with X rays. Suggested Plough's disproof of Bateson.

- Nonquantitative alleles (truncate series, 1919; white series, 1918); collected principles concerning mutation, 1921; mutations in relation to alleles; gene attributes, 1914, 1918, and 1921.

- Calculate size and number of genes, 1913; genes as mutable autocatalysts and basis of life, 1921; gene interactions.

- Initiated quantitative studies on mutation (and techniques) with Altenburg; recognized characteristics of mutation—spectrum, degeneration, recessive, specific direction, and rates, 1918–1919.

- Effect of temperature, 1919 and 1926.

- (Effect) of X rays on gene mutation, 1927, and chromosome abnormalities; cytological demonstration of chromosome abnormalities; map confirms linear order; inert regions; and translocation in evolution; spindle fiber effect on crossing-over; explanation of crowding (of genes on tips?); reverse mutation; classification of mutations—hypomorphs, neomorphs, and twinning.

- Suggested explanation for Weinstein, Redfield, Gowen, Payne, Ward's Curly, Plough on Bateson, Patterson, and Hansen.

JOINT STUDIES

- Morgan and Bridges— "with," 1920.

- Morgan and Bridges—data on gynanders.

- Altogether: Observation of specific and multiple effects of one gene and influence of many genes and environment on same character; explanation of multiple alleles.

- Mrs. Morgan—attached Xs and their breakage; deleted X; X with duplication; ring chromosomes.

End Notes and References

1   Carlson E.A. 1981. *Genes, Radiation, and Society: The Life and Work of H.J. Muller.* Cornell University Press, Ithaca, New York.

2   Lock R.H. 1906. *Heredity, Variation, and Evolution.* J. Murray, London.

Lock lived and wrote in Ceylon. He was a botanist and greatly admired Bateson's papers on Mendelism. He noted in his preface that he would have entitled his book *Genetics*, the term just introduced by Bateson for the new field, but no one would recognize it and so he chose the more specific title for his book. Unlike Bateson, Lock embraced the chromosome theory of heredity. Muller's enthusiasm for the book was lifelong. He attributed his initial introduction to genetics and his motivation for making genetics his career to Lock's book. Wilson had recommended it to him and Muller read it during that summer of 1909, while working as a bellhop, by ducking under a flight of stairs during his break time.

3   I relied on my notes from an interview with Sturtevant in 1967. He described the fate of several of these lesser-known first-generation fruit fly workers. See Chapter 23, Table 1.

4   Edgar Altenburg. Interview with E.A. Carlson, Houston, Texas. 1967.

5   There are different views on the roles of these factions in Morgan's laboratory. For the Sturtevant-Bridges-Morgan view, see Schultz J. 1967. Innovators and Controversies. *Science* **157:** 296–301.

Altenburg's views can be found in his correspondence with Muller at the Lilly Library, Indiana University, Bloomington. See the Library appendix for relevant details from letters of January 19, 1946; March 4, 1946; March 23, 1946 (Altenburg to Muller); May 9, 1941; and March 24, 1946.

6   I base this on two sources. One is Muller's 1954 course on "Mutation and the gene" and the second is a sheet of paper (undated) that Muller prepared (see Muller Gives His Version of the Early Contributions on page 209).

7   Muller had two breakdowns. The first he cited in a biographical sketch for the National Academy of Sciences (Lilly Library, Indiana University, Bloomington). The second occurred in 1932 shortly before he departed for the Sixth International Congress of Genetics at Ithaca and the Third International Congress of Eugenics in New York City. In that second breakdown, while teaching at the University of Texas, Muller attempted suicide with an overdose of barbiturates (see Carlson E. 1981. op. cit. pp. 174–175).

8   I discovered one way to assess Muller's view when I was an elevator operator during the summer of 1952. Another operator expressed his frustration over our superintendent's indifference to the short (15 minute) lunch break we were assigned. He told me, "Bosses are all alike. They say, 'How can you be hungry when I just ate?'" I consider that remark one of the more illuminating I have encountered about human behavior.

9   In most cases, the disputed priorities went to Morgan's credit by default. He was careless about putting co-authors on his papers, and he would certainly favor a co-author who participated in the experiments rather than one who supplied the ideas.

10 Dexter J.S. 1914. The analysis of a case of continuous variation in *Drosophila* by a study of its linkage relations. *American Naturalist* **48:** 712–758.

11 Muller H.J. 1916. The mechanism of crossing over. I–IV. *American Naturalist* **50:** 193–221; see 284–305, 350–366, and 421–434.

12 Muller H.J. 1914. A gene for the fourth chromosome of *Drosophila. Journal of Experimental Zoology* **17:** 325–336; and Muller H.J. 1914. A factor for the fourth chromosome of *Drosophila. Science* **39:** 906.

13 Muller H.J. 1914. A new mode of segregation in Gregory's tetraploid primulas. *American Naturalist* **48:** 508–512.

14 Muller H.J. 1914. The bearing of the selection experi-

ments of Castle and Phillips on the variability of genes. *American Naturalist* **48:** 567–576.

In his "Mutation and the gene" course in 1954, Muller cited his defiance of Morgan's advice as the source of their rupture and Morgan's attitude that Muller was a troublemaker.

[15] Castle W.E. 1912. The inconstancy of unit-characters. *American Naturalist* **46:** 352–362; and Castle W.E. and Phillips J.C. 1914. *Piebald Rats and Selection.* Carnegie Institution of Washington Publication No. 195.

[16] Julian Huxley and Muller were friends for life. The year after Muller arrived, Huxley told him to take over the department so that he could enlist to fight in the war. While Huxley did most of his work in evolutionary studies, he immediately realized the value of the work done by the Morgan and Wilson school at Columbia in shaping the new science of genetics.

[17] Muller's graduate course, "Mutation and the gene," which he offered yearly, reflected his belief that in Morgan's laboratory, when he left, these were the two foremost problems of genetics and he would devote his career to making contributions to both of them; he did.

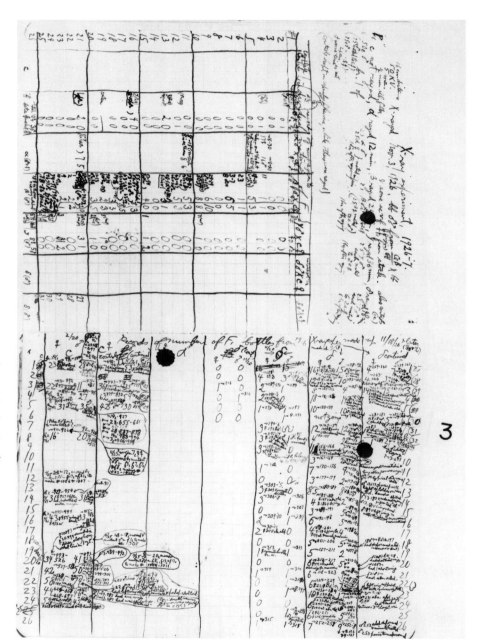

I found this page of Muller's notes for his 1926 data in a bound notebook in Muller's laboratory when I was sorting out items for shipment to the Lilly Library at Indiana University. It appears in a notebook titled "1921" and the date of November 3, 1926 is found in a later part of that notebook. Muller's crowded data and inkblots give a more accurate depiction of the way in which science is done, rather than what appears in the cleaned-up published versions. (Courtesy of the Lilly Library, Indiana University, Bloomington; for its provenance, see Carlson diary, volume 19, p. 97, entry date August 2, 1967.)

3

Drosophila *Genetics after 1915*

...There remains no reason to doubt the application of the dictum "all life from pre-existing life" and "every cell from a pre-existing cell" to the gene: "every gene from a pre-existing gene." We need at present make an exception here only of those very special conditions under which life itself, as a naked gene, originates.

H.J. MULLER[*]

Sturtevant was responsible even more than Morgan for that great development of genetics in terms of chromosome theory which was mainly based in the studies on the fruit fly Drosophila *that started at Columbia University in the decade 1910–1920.*

H.J. MULLER[†]

ALTHOUGH MORGAN'S MOST RENOWNED fly lab broke up with Muller's departure in 1915 to Rice University in Texas, the work of Morgan and his remaining students increased, and new members quickly lined up to be among Morgan's students. The fame of fruit fly genetics also spread to laboratories around the world. Geneticists in every major country worked with *Drosophila*. In 1939, the first volume (compiled by Muller) of the *Bibliography on the Genetics of* Drosophila included 130 pages of entries.[1] The initial laboratory at Columbia University found additional summer homes at Woods Hole and especially at Cold Spring Harbor. Dozens of fly labs in the United States began publishing papers throughout the period from 1915 to 1939. Morgan's move to California in 1924 and his formation of the new biology department at the California Institute of Technology initiated a West Coast rivalry with laboratories stimulated by Muller in his move to Texas and with laboratories along the East Coast and the Midwest stimulated by Morgan's original stay in New York. Around the world, a similar growth of fruit fly genetics took place, especially in Germany, Scandinavian countries, France, Great Britain, and the USSR.

---

[*]1936. Muller H.J. Bar duplication. *Science* **83:** 528–530; see p. 530.
[†] 1960. H.J. Muller to Carroll G. Bowen, October 5, 1960. Lilly Library, Indiana University.

Morgan Wins a Nobel Prize

With increasing administrative responsibilities and the fame he encountered for his founding of classical genetics, Morgan spent less time doing experiments and more time writing about genetics. The 1915 classic text that he wrote with his students was followed by *A Critique of the Theory of Evolution* (1916), *The Physical Basis of Heredity* (1919), *The Theory of the Gene* (1926), and *Embryology and Genetics* (1934) (Figures 1 and 2).[2] He wrote many other monographs and popularizations as full-length books. He also wrote dozens of reviews and appraisals of the work of his school. In 1933, he was awarded the Nobel Prize, but he delayed a year before picking up his prize and delivering his speech. Morgan's laboratory at the California Institute of Technology flourished. Bridges remained active there until his death in 1938, and Sturtevant ran the laboratory when Morgan died in 1945.

Figure 1. *T.H. Morgan in the 1930s at the California Institute of Technology. He had shifted from his fruit fly and developmental research to full-time administration, building a department of biology that reflected new trends in biology and genetics. (Reprinted from Thomas Hunt Morgan 1866–1945. 1945 Obituary Notices of Fellows of the Royal Society **5**: 451–466.)*

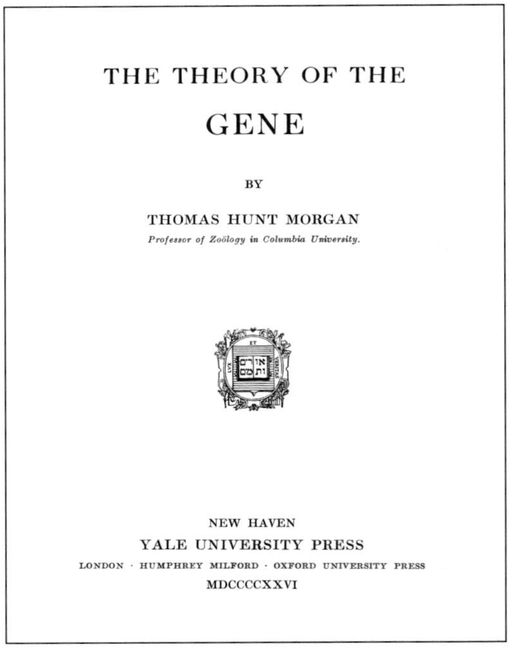

THE THEORY OF THE

GENE

BY

THOMAS HUNT MORGAN

Professor of Zoölogy in Columbia University.

NEW HAVEN
YALE UNIVERSITY PRESS
LONDON · HUMPHREY MILFORD · OXFORD UNIVERSITY PRESS
MDCCCCXXVI

Figure 2. *The title page of Morgan's* The Theory of the Gene. *Although he was initially skeptical of the stability of the gene, its association with chromosomes, and its role in evolution, Morgan gave way to the evidence of his own findings and abandoned his earlier preference for models that did not work out. The book canonizes the gene as the key to understanding genetics and evolution, a view that his students had embraced earlier with less evidence. (Reprinted from Morgan T.H. 1926. The Theory of the Gene. Yale University Press, New Haven, Connecticut.)*

Morgan Changes the Way Genetics is Done

As his work, especially that of his students, built a strong case for Darwinian natural selection, Morgan gave way and eventually abandoned the saltation models that he had favored. Sturtevant provided the evolutionary evidence and Bridges the cytological evidence for this evolution. Morgan's gift was in synthesizing vast amounts of knowledge and conveying ideas effectively. The classical genetics that we conjure in our minds—the fusion of Mendelism, meiosis, crossing-over, and X-linked inheritance; the correspondence between linkage groups and chromosome number; the various anomalies with genetic consequences such as nondisjunction, triploidy, mosaicism, gynandromorphism, modifier genes, lethal factors, and chromosome rearrangements; and the comparison of genetic maps among cognate species—arose through studies in Morgan's fly lab or were greatly reinforced with dramatic evidence from his fly work. Whether referred to as Morgan's school, the fly lab, the *Drosophila* group, or the Columbia genetics group, these researchers gained immediate recognition worldwide as having changed the way in which genetics was done. Genetics had shifted from German (the three rediscovers stressed breeding analysis) to English (the Bateson school stressed presence and absence, coupling and repulsion, and modified ratios) to American (the fly lab stressed an interdisciplinary cytogenetic and evolutionary approach) as the dominant way in which genetics would be perceived throughout the world. Along the way, battles were fought in which

Morgan and his students took on and defeated (often with support from other geneticists who were not working with fruit flies) many competitive models. This included de Vries' mutation theory,[3] Bateson's erroneous models of all mutations as losses of genes and his incorrect model of reduplication to explain the coupling and repulsion phenomena he encountered,[4] Castle's erroneous theory that genes underwent minor fluctuating variations through hybridization,[5] Castle's incorrect model of the genetic map as a three-dimensional interconnected network,[6] Goldschmidt's erroneous model of genetic exchange between chromosomes without crossing-over,[7] and many other attempts by other laboratories to cast doubt on the broad validity of Morgan's and his students' findings. When reading a review article or a textbook, it is easy to forget that what is presented is often a result of a contest of competing hypotheses, in which facts and experiments, not nationalism nor power groups imposing a consensus, laid low the models that failed to hold up to new evidence. I say this not in the philosophic sense as some sort of Hegelian "thesis and antithesis" with some kind of vector pointing the right way, but in an entirely different sense that operates among scientists—a respect for experiments and facts that are consistent with a theory and its predictions. Whatever Morgan's faults may have been as a skeptic who had to eat crow on many occasions, he stamped experimentation as the hallmark of what is considered among scientists throughout the world to be good science.

Sturtevant Finds That Genes Have Two Functions

Sturtevant stressed his evolutionary work and constructed maps for the various species of the genus *Drosophila* that were similar to *D. melanogaster*.[8] Each

map required finding mutants, a difficult job until 1927, when Muller's proof that X rays induce mutations made such mutants available in large numbers.

Figure 3. *George Beadle started as a maize geneticist but switched to fruit flies while studying with Sturtevant and Morgan at the California Institute of Technology. He teamed up with Boris Ephrussi (left) in France and worked out a method of transplanting embryonic eye rudiments to determine which gene mutations were autonomous and which were capable of being corrected by the surrounding larval environment. From these studies, Beadle and Ephrussi recognized that they could establish biochemical pathways for the synthesis of two eye pigments, one for brown eye color and the other for orange eye color. The red-eyed flies have both. Later, Beadle teamed with Edward Tatum and pursued biochemical pathways in the fungus, Neurospora. (Courtesy of the California Institute of Technology Archives.)*

Sturtevant also took an interest in developmental problems. He found that vermilion did not register if a stock such as yellow vermilion miniature was used in a cross where a somatic nondisjunction occurred, resulting in a gynandromorph. One part of the fly (the male) would show yellow body color, red eye color, and a miniature yellow wing, whereas the other (female) part would show the normal gray amber body color and wing color, normal wing shape, and red eye color. Why was there not a vermilion eye on the male side?

Sturtevant was the first to find that genes had two functions. Most were autonomous, and they functioned at the cell level, whether in a mosaic, gynandromorph, or whole-body environment. But some mutations, such as vermilion, were nonautonomous and something diffused from the normal tissue to the mutant tissue, correcting the defect in the mosaic condition or in the gynandromorph.[9] It was this finding that later led George Wells Beadle (1903–1989), originally in Sturtevant's laboratory, to seek nonautonomous eye color mutations in a celebrated developmental and biochemical genetic analysis of eye color synthesis in fruit flies. Beadle worked largely with Boris Ephrussi in Paris

after he left Morgan's laboratory (Figure 3).[10] They used transplantation of embryonic rudiments for the eyes from one mutant stock into another at the larval stage of development. The differentiation of the implanted eye bud (usually appearing in the abdomen of the emerging adult) demonstrated either its original color (signifying that it was autonomous) or a wild-type color (signifying that it was transformed by a biochemical product in its surroundings). From this, Beadle and Ephrussi worked out a sequence of eye color precursors for the two pigments produced by the normal eye. One was an orange pigment and the other, a brown pigment. When they were together in the eye, the red eye color was present; if the brown pigment was absent, the eye was bright orange; and if the orange pigment was absent, the eye was brown. They found that Morgan's original white-eyed fly had both pigments but no place to put them in the eye: The white-eyed fly lacked a transporting protein to reach the site at which the pigments are normally deposited. Eventually, Beadle's work led to collaboration with E.M. Tatum on the bread mold *Neurospora*, the beginnings of biochemical genetics, and the analysis of biochemical pathways.[11]

Bridges Makes Many Contributions

Bridges exploited the cytogenetics of fruit flies. He obtained triploids and used them for the analysis of the balance theory of sexual development, adding those results to the work on nondisjunction.[12] He was the first to recognize the severe developmental consequences of autosomal nondisjunction, and he predicted that such aneuploids for chromosome II or III would be lethal. Flies bearing only one of the small dot-like chromosomes—haplo-IV flies—were sad creatures of minute size with numerous anomalies.[13] Bridges was happy keeping the stocks going, constructing new stocks, and making sure that every gene was precisely mapped through numerous crosses with its nearest neighboring genes. He reported these in a periodical, Drosophila *Information Service*, that he founded and edited (with the help of M. Demerec) until his death.[14] In addition, he co-authored, with Morgan, several technical monographs of the mutants obtained in *D. melanogaster*.[15] When Bridges died in 1938, Katherine Brehme completed his 1944 classic, *The Mutants of* Drosophila melanogaster.

Bridges quickly adopted Painter's discovery of salivary chromosome maps and worked out a band-by-band study of them (Figure 4).[16] These proved invaluable (like an atlas) for those needing to check their salivary chromosome analysis of structural rearrangements. His theory of repeats was influential for both evolutionary studies and theories of "gene nests" when pseudoallelism later emerged during the era just before molecular biology arose from studies of bacterial and viral genes. A major reason that Bridges was able to dominate the field that Painter introduced was Bridges' prior knowledge of the mutants and structural rearrangements in fruit flies. Painter was not as immersed in the day-to-day mapping and stock construction that permitted Bridges a cytogenetic grasp of salivary chromosomes.[17] Bridges' premature death deprived him of the opportunity to exploit his own theory of repeats and a thorough cytogenetic study of evolution in *Drosophila*. Had he lived as long as Sturtevant, he would have made about 25 more years of scientific contributions.

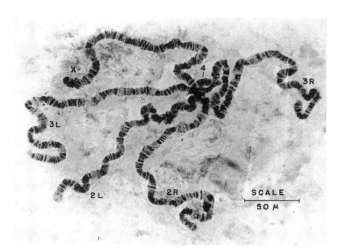

Figure 4. *Salivary chromosomes in fruit flies. The arms of the X, 2, and 3 chromosomes are indicated as joined in a chromocenter along with the shorter chromosome 4. These giant chromosomes are cables of 1000 or more replications of partially coiled chromosomes found in the salivary glands of fruit fly larvae. Berwind Kaufman's beautiful extended spread is a classic and an aesthetic rarity (none of the arms are overlapping). Theophilus Painter, a cytologist, discovered that salivary gland chromosomes are giant chromosomes with banding. He recognized their importance, but Bridges had stocks of hundreds of chromosomal rearrangements and quickly led the field of correlating such mutant defects with salivary chromosomes. (Reprinted, with permission, from Demerec M. and Kaufman B.P. 1986.* Drosophila Guide: Introduction to the Genetics and Cytology of Drosophila melanogaster, *9th edition. Carnegie Institution of Washington.)*

Muller Shows That Radiation Induces Mutations

Muller wasted no time in establishing a rival school, first at Rice University, then back at Columbia University while Morgan was on sabbatical leave in 1921, and again at Texas (this time at the University of Texas at Austin). In fact, Muller's career took him out of the United States for seven years as he went from Germany to the USSR to Spain and to Scotland before returning to the United States at Amherst and finally settling down at Indiana University for the last two decades of his career. Muller was as energetic as Morgan. He may not have had Morgan's breadth of interests in biology, but he possessed an unrivaled focus for fruit fly research. His work was his life, and his first marriage failed over that incompatibility. He was fortunate that his second wife accepted his work habits, which allowed him to work seven days a week until his retirement at the age of 72.

Muller was driven by his curiosity and his goals were big. At Rice, he began studies on mutation frequencies and formulated his conceptions of mutation as changes in the individual gene, a view he presented when he briefly returned to Columbia.[18] He balanced his experimental papers with his theoretical papers. He argued that the gene was the basis of life, a view that was not widely held when he proposed it in 1926 but that is largely accepted today.[19] He designed the stocks needed to prove that X rays induce mutations and his paper created a sensation when it appeared in *Science* in 1927 (with no data) and even more so that same summer in Berlin at the International Congress of Genetics, when he provided an avalanche of data supporting his findings.[20] Until that proof was available, Morgan, Sturtevant, and Bridges doubted that Muller had induced muta-

tions and they perceived him to be headed for self-destruction as a result of that claim. After all, Morgan had found no evidence from his radium treatments in 1909 or early 1910 that radiation induces mutations and many laboratories produced similar negative or ambiguous results. How could Muller claim a 150-fold increase in mutations when Morgan found next to nothing for his efforts? But Muller did, and the elegance of his ClB stock (Figure 5) for the analysis made him and the fly work even more celebrated for initiating a new generation of fruit fly genetics.[21]

Muller's proof involved a quantitative rather than qualitative approach to detect mutations. He chose a class called X-linked lethals. Morgan had found the first X-linked lethal in fruit flies in 1912. By the early 1920s, Muller knew that lethals occurred more frequently than visible mutations. The ClB stock he designed in 1921 had the following components. The C stood for a category of crossover suppressors (shown by Sturtevant five years later to be chromosome inversions), the l for the recessive lethal it contained, and the B for the dominant gene Bar eyes. This designation marked the chromosome when it was present in females (the ClB was lethal in males). Using heterozygous ClB females, he and Altenburg used the stock to measure spontaneous mutation frequencies and later, Muller applied the stock to test temperature as a possible mutagen, using the chemist's rule of thumb that a ten-degree rise in temperature led to a doubling of chemical activity (this relation is sometimes called a Q10 by chemists). Muller's initial assumption was that if Morgan, Blakeslee, and others had failed to find mutations with X rays, the mutation process must be chemical.

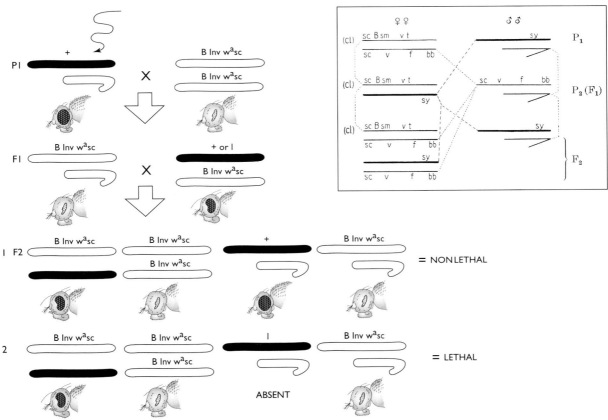

Figure 5. *The routine detection of X-linked lethals. (Above) Muller modified the ClB (pronounced "see ell bee") stock in the 1940s to facilitate keeping stocks of lethal or infertile mutations. He combined two viable inversions (one of them involving the scute* bristle region) with a dominant marker, Bar eyes, and a recessive eye color mutant, apricot eyes. He used Basc (pronounced "bask") as the acronym for this stock. Like the ClB test, each F₁ female represents one treated or testable sperm from the P₁ male. The individual female is mated with her brothers and in the F₂ a vial is scored as a nonlethal if both Bar apricot males and red-round-eyed males are present. The vial is scored as a lethal if the red-round-eyed male flies are missing. In the typical X-linked lethal test for mutagens (chemically or physically induced), the vials can also be assessed for X-linked visible mutations. A more efficient stock for obtaining such visible mutations is also available, along with dozens of stocks that Muller created for detecting the presence of rearrangements, semilethals, mosaicism, and other genetic events of interest. Muller reported most of these stock designs in the annual Drosophila Information Service (DIS), which Bridges and Demerec edited beginning in 1938. (Inset) The ClB stock designed by Muller to detect spontaneous or induced mutations. The lethal associated with the C stock (also a crossover inhibitor associated with an inversion) arose in a stock with the dominant marker, Bar eyes, and several recessive markers. Its absence, along with the absence of its homolog, when that homolog carries an induced recessive lethal mutation, leads to a vial harboring only females. This made a quantitative measure of radiation-induced damage simple to score and tabulate. Recessive lethals were about ten times more frequently induced than recessive visible mutations. ([Left] Figure drawn by Claudia Carlson. [Right] Reprinted from Muller H.J. 1928. The problem of genic modification. Zeitschrift für Induktive Abstammungs und Vererbungslehre (supplement) 1: 234–260.)*

In 1926, Muller (Figure 6) had changed his mind. He reread the literature on radiation and biology and reasoned that X rays might induce mutations. Since lethals were the most common form of mutations, he felt that he had a good chance of detecting X-ray-induced lethal mutations using his ClB stock. What shocked Muller in the fall of 1926 when he ran his first X-linked lethal tests was how frequently they arose at the dose he used (today estimated to be approximately 4000 roentgens or 40 Grays). Each X-bearing sperm exposed to X rays ended up in a different egg and hence, a different daughter. These heterozygous females were individually mated with a male whose X chromosome contained several recessive mutations as markers and as genes which could be used to map any induced mutation that arose. Females with an induced lethal mutation in the X chromosome produced no sons;

Figure 6. *Muller stopped to have his portrait taken en route to the International Congress of Genetics in 1927. (Photo taken in New York by Fabian Bachrach.)*

those vials were all female. This was because both the ClB X chromosome and the mutated X chromosome each contained an independent lethal. The C factor kept the genes from crossing over.

Radiation Genetics Becomes a Field of Its Own

The field of radiation genetics took off with remarkable consequences that netted Muller a Nobel Prize in 1946. Although Muller's work had priority of publication in 1927, the following year he accepted the independent confirmation by Lewis J. Stadler as a codiscovery because Stadler's corn and barley studies were set up at the same time that Muller was producing his first results from his radiation-treated flies. Compared to Stadler's work, Muller's was much like Darwin's work on natural selection compared to Wallace's, when Wallace independently arrived at the same theory of evolution.[22] Muller's experiments were far more extensive—his mutations carefully mapped and compared to those that had arisen spontaneously during the past 15 or so years. Muller showed that these mutations were induced in the irradiated sperm and that they usually resulted in whole-body visible mutations rather than mosaics. Of the few mosaics

he obtained, most did not transmit the mutant condition nor their mosaicism. Among the nonmutants tested, few showed evidence of continued sporting of delayed or new mutants, and he ruled out W.H. Eyster's genomere model of the gene.[23] At best, he concluded that the sperm chromatid is a doubled thread or molecule (or prematurely doubled) that occasionally produces mosaics from radiation exposure. Muller also showed that X rays induced mostly recessive lethal mutations as predicted from the spontaneous studies, but they also introduced something new—a category of mutations that he called dominant lethals, dead embryos that failed to develop (Figure 7).[24] Muller also reported many mutations that prevented crossing-over, and he suspected but did not yet have time to demonstrate that they were associated with the chromosome rearrangements recently described by Bridges and Sturtevant.

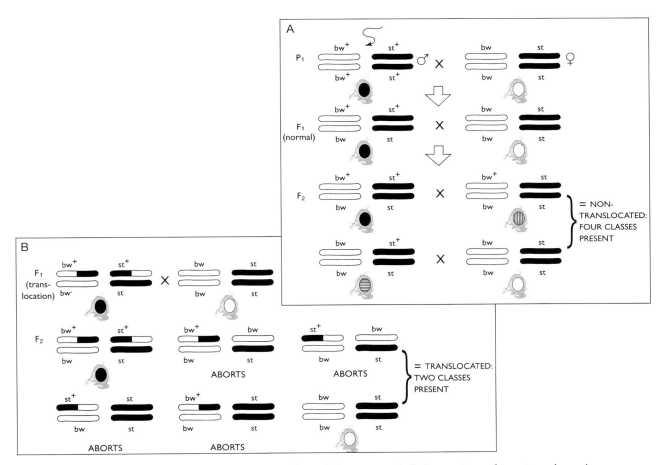

Figure 7. *Genetic scheme used to detect cytological aberrations. Muller's genetic test for reciprocal translo-cations eliminated the need to cytologically search for them. If a P₁ red-eyed male is treated with mutagens, some of the sperm will have broken chromosomes II and III. Those II-III translocations that are reciprocal, euploid, and eucentric will survive (the others that form aneucentric or aneuploid combinations will abort). To detect these, Muller used the eye color mutations brown (on chromosome II) and scarlet (on chromosome III). In the nontranslocated F₁ male mated to virgin (**bw;st**) females, independent assortment of the dihybrid produces four classes of sperm:* **bw⁺;st⁺**, **bw⁺;st**, **bw;st⁺**, *and* **bw;st**. *Each of these will fertilize a* **bw;st** *egg. Thus, four eye colors are present among the progeny in the F₂ in a given vial testing a P₁ male sperm. These are red, brown, scarlet, and white. Note that* **bw;st** *flies are white eyed due to the loss of two color factors (this is indistinguishable from the white eyes caused by a loss of a transport protein in the X-linked white that fails to bring pigment to the ommatidia of the flies eyes). If a translocation occurs, the pairing and distribution of the translocation heterozygote result in only two viable combinations: the* **t(bw⁺;st⁺)** *and the nontranslocated* **bw;st**. *All other combinations abort the embryo. Hence, only two classes of eye color are found in a vial with a translocation from the P₁ sperm: red-eyed flies and white-eyed flies. (Figure drawn by Claudia Carlson.)*

Muller Uses Radiation as a Tool for Genetic Analysis

Muller's radiation genetics led to the construction of new stocks, and he used these to produce "deleted X chromosomes," in which a small segment with a gene of interest could be used to study dose effects in otherwise diploid flies (and retain their fertility).[25] This allowed him to analyze gene number, gene size, and the relationship between gene mutations and position effects. Muller's work ushered in an era of complex sophisticated genetics, in which chromosomes could be engineered and genes intended as markers could be used to reveal structural rearrangements. He carried this with him from Texas to Berlin, Leningrad, Moscow, and then to Edinburgh, and at each of these places, he and his students left the influence of the Muller school just as surely as his mentor Morgan had left the imprint of the Morgan school. The main difference between Muller's work in those decades after 1915 and the Morgan school was that Sturtevant and Bridges remained planets in orbit around Morgan's greatness, whereas Muller did all he could to shed that image of reflected glory.

Drosophila Genetics Makes Advances Through 1950

Thousands of papers have been published on fruit fly genetics and this has continued as genetics shifted from classical to biochemical to molecular genetics. For many reasons, this trend will no doubt persist for decades to come. Mostly through his own crosses, Bridges put together a map of each of the thousand or more mutations known in his own time. In addition, the availability of stocks and the isolation of desired mutations, chromosome parts, and other tools for doing genetic research are well established. Fruit flies are also more complex than viruses and bacteria, and they permit studies of organ systems, behavior, sex determination, organogenesis, and many fundamental biological problems that could not be studied in yeast, bacteria, or viruses. The classical genetics story, however, did not end in 1915 with the publication of *The Mechanism of Mendelian Heredity*. For a good number of biologists, it *began* with that publication because the new approach allowed many investigators around the world to explore new problems.

In Germany, Curt Stern used fruit fly genetics to discover (independently of Muller) the phenomenon of dosage compensation.[26] This process permits females with two X chromosomes to produce the same gene activity as males with one dose of X-linked genes. Stern also applied the cytogenetic techniques of the Morgan school to recombinant genes attached to chromosomes that were morphologically altered through translocations, and the recombinant genotypes were associated, as predicted, with visibly recombined fragments of the chromosomes. This proved that Morgan's thesis of recombination involved a cytological exchange between homologous segments of chromosomes.[27] That work was also independently verified, but not in *Drosophila*; using maize, Barbara McClintock demonstrated this beautifully. Also in Germany, N.V. Timoféeff-Ressovsky carried out studies on gene mutation in the hopes of measuring gene size, bringing to this task physicists who helped launch what eventually became molecular genetics. Delbrück's participation was of particular importance in this new approach initiated by fruit fly studies.[28]

After Muller and Altenburg brought stocks of fruit flies to the Soviet Union, a flourishing school of genetics was established in the early 1920s. Some of the work proved false but innovative for future studies of gene structure and function. I.I. Agol and others developed a theory based on what today would be called complementation studies (use of phenotypes in heteroallelic combinations to predict the physical mapping of the gene); they called it "step allelomorphism."[29] One consequence of that effort was Muller's analysis of the boundaries of genes in nested sequences, such as the yellow-achaete-scute complex at the tip of the X chromosome. Through elaborate cytological tests (Muller called it the "left-right test"), Muller established that these individual genes had definite beginnings and ends, between which stretches of material could exist where breakage would not produce mutational damage.[30] By 1935, Muller considered the Soviet genetics program second only to the work done in the United States for laboratories committed to genetic research and the publications emanating from them. After 1936, that ended with the rise of Lysenkoism and the anti-genetic movement, which drove Muller out of the USSR and martyred dozens of geneticists.

In Great Britain, the fruit fly program begun by Charlotte Auerbach (Figure 8), while Muller was a guest investigator, moved in a new direction. After Muller had returned to the United States in 1940, Auerbach demonstrated that mustard gas and nitrogen mustard were potent chemical mutagens.[31] She could not tell Muller which compounds she was working with nor could she publish the results until the war ended because of the secrecy stamped on her work. Another Muller protégé, S.P. Ray-Chaudhuri, produced one of the key experiments for testing the

Figure 8. *Charlotte Auerbach was assigned to Muller as a postdoctoral student when he arrived in Edinburgh. He encouraged her to look for mutations with chemicals, and she found such agents working with the pharmacologist, J.M. Robson. Auerbach obtained abundant mutations with mustard gas and nitrogen mustard, regarding these as radiomimetic agents because their effects were similar to those of radiation. She worked out the differences between radiation mutagenesis and chemical mutagenesis. (Reprinted, with permission, from Kilbey B.J. 1995. Charlotte Auerbach [1899–1994]. Genetics* **141***: 1–5.)*

effects of low doses of radiation. He showed that an acute dose of 400 roentgens administered to the sperm of fruit fly males produced the same percentage of X-linked lethals as in males who were given a protracted dose of 400 roentgens of ionizing radiation administered slowly over 30 days.[32] To Muller, this demonstrated that chest X rays and other diagnostic X rays were mutagenic, although the individual risk to those exposed was individually small. However, dentists and radiologists who used a great deal of X raying (in an era that had not yet conceived of using lead aprons or leaving the room while the patient was being X rayed) were at risk. Muller made the first warnings of low-dose radiation hazards from the health uses of radiation, but these were not well received in Great Britain when he filed his report.

Drosophila Work Continues in Other Laboratories

In the United States, Demerec used *Drosophila virilis* to study variegated or ever-sporting patterns in the eyes of flies. Whatever Demerec's mutable genes were (he lost those cultures), something similar was found by a number of investigators (e.g., Jack Schultz), who showed that gene function was affected by the neighboring genes associated with a given gene.[33] When chromosomes were rearranged, these often led to variegated expression. For example, either crossing-over or uses of extra heterochromatin (usually additional Y chromosomes) would suppress the mutant expression. Thus, position effect was not limited to the Bar eye case studied by Morgan and Sturtevant. At the California Institute of Technology, Theodosius Dobzhansky worked with Sturtevant on chromosome rearrangements and then embarked on an extensive study of the speciation of the Drosphilidae family to work out evolutionary relations that he summed up in a highly influential book, *Genetics and the Origin of Species*.[34] This supplemented the mathematical approach of bringing breeding analysis and evolution into harmony, which involved the work of G.H. Hardy and W. Weinberg. Their famous algebraic expansion of gene frequencies, $(a + b)(a + b) = a^2 + 2ab + b^2 = 1$, provided the first insight into the way in which gene frequencies could be calculated in populations.[35] This early contribution to population genetics helped to end the split between Bateson's Mendelism and Pearson's biometricians. It also led to the much more extensive analysis of genes in populations and their evolutionary consequences, which was provided by R.A. Fisher, J.B.S. Haldane, and Sewall Wright in their numerous publications and in Fisher's influential book, *The Genetical Theory of Natural Selection*.[36]

In Austin, Texas, Painter's work on fruit fly cytology (presalivary and postsalivary gland chromosomes) was a major achievement that brought cytology and complex genetics together, especially the cytological verification of chromosome rearrangements that had been inferred from their genetic behavior.[37] His colleagues (and Muller's), J.T. Patterson and W.S. Stone, studied the evolution of chromosome number and interpreted the phylogenetic trees for dozens of species of *Drosophila* and related Diptera.[38] At Rice University, Altenburg demonstrated that ultraviolet radiation also induced mutations, permitting an opportunity to compare ionizing radiation with the peculiar form of nonionizing ultraviolet radiation that J. Schultz and T. Caspersson in Sweden associated with DNA damage as the basis of mutation production.[39]

At the end of this period, C.P. Oliver's and E.B. Lewis' discovery paved the way for the emergence of a new phenomenon. Although it superficially looked like another case of Bar eyes with duplicate genes, the alleles studied in the lozenge eye series proved that a fixed map order, and not an unequal crossing-over, was the basis for what they called pseudoallelism. It was pseudoallelism over the next 15 years that led to the "genetic fine structure" analysis of genes in viruses (bacteriophages by Seymour Benzer) and bacteria (*E. coli* by Demerec).[40] The history and significance of pseudoallelism are discussed more fully in Chapter 18.

These are just a few of the highlights among many hundreds of significant contributions that came out of the fruit fly explosion in the first half of the twentieth century. Classical genetics had emerged as the rival of atomic physics in capturing headlines and stimulating interest in basic science and its social applications. Although fruit fly research dominated the field, many organisms, especially mice, maize, fungi, algae, protozoa, bacteria, and viruses, contributed new ways of doing classical genetics that forced it into its long delayed marriage with biochemistry and molecular biology. The slowly growing real contributions to classical genetics from efforts to study human genetic medical disorders were also not to be ignored.

End Notes and References

[1] Muller H.J. 1939. *Bibliography on the Genetics of* Drosophila. Oliver and Boyd, Princeton.

[2] Morgan T.H. 1916. *A Critique of the Theory of Evolution.* Princeton University Press, New Jersey; Morgan T.H. 1919. *The Physical Basis of Heredity.* J. Lippincott Company, Philadelphia; Morgan T.H. 1926. *The Theory of the Gene.* Yale University Press, New Haven; and Morgan T.H. 1934. *Embryology and Genetics.* Columbia University Press, New York.

[3] de Vries H. 1901–1903. *Die Mutationstheorie.* Veit and Company, Leipzig.

[4] Bateson W. 1913. *Problems of Genetics.* Yale University Press, New Haven, Connecticut.

[5] Castle W.E. 1906. Yellow mice and gametic purity. *Science* **24:** 275–281.

[6] Castle W.E. 1919. Are genes linear or non-linear in arrangement? *Proceedings of the National Academy of Sciences* **5:** 500–506.

[7] Goldschmidt R.B. 1917. Crossing over ohne chiasmatypie? *Genetics* **2:** 82–95.

[8] Sturtevant A.H. 1921. *The North American Species of* Drosophila. Carnegie Institution of Washington Publication No. 301, 150 pages.

[9] Sturtevant A.H. 1920. The vermilion gene and gynandromorphism. *Proceedings of the Society for Experimental Biology and Medicine* **17:** 70–71.

[10] Beadle G.W. and Ephrussi B. 1937. Development of eye colors in *Drosophila:* Diffusible substances and their interrelations. *Genetics* **22:** 76–86.

[11] Beadle G.W. and Tatum E.L. 1941. Genetical control of biochemical reactions in Neurospora. *Proceedings of the National Academy of Sciences* **27:** 499–506.

[12] Bridges C.B. 1921. Triploid intersexes in *Drosophila melanogaster. Science* **54:** 252–254.

[13] Bridges C.B. 1921. Genetical and cytological proof of non-disjunction of the fourth chromosome of *Drosophila melanogaster. Proceedings of the National Academy of Sciences* **7:** 186–192.

[14] Bridges edited these with Demerec from 1934 until his death in 1938, after which Demerec continued them. They have long been a delight to fruit fly workers; the articles were short and not refereed. They were invaluable for obtaining information and techniques in different laboratories and were sent free of charge to any *Drosophila* laboratory worker listed in the issues.

[15] The major ones include Morgan T.H. and Bridges C.B. 1916. *Contributions to the genetics of* Drosophila melanogaster. I. Sex linked inheritance in *Drosophila.* Carnegie Institution of Washington Publication No. 277, pp. 1–88; Morgan T.H. and Bridges C.B. 1919. *Contributions to the genetics of* Drosophila melanogaster. I. The origin of gynandromorphs. Carnegie Institution of Washington Publication No. 278, pp. 1–122; Bridges C.B. and Morgan T.H. 1919. *Contributions to the genetics of* Drosophila melanogaster. II. The second chromosome group of mutant characters. Carnegie Institution of Washington Publication No. 278, pp. 123–304; and Bridges C.B. and Morgan T.H. 1923. *Contributions to the genetics of* Drosophila melanogaster. III. The third chromosome

group of mutant characters of *Drosophila melanogaster*. Carnegie Institution of Washington Publication No. 327, pp. 1–257.

[16] Bridges C.B. 1935. Salivary chromosome maps with a key to the banding of the chromosomes of *Drosophila melanogaster*. *Journal of Heredity* **26**: 60–64.

[17] Bridges C.B. 1936. Demonstrations (1) of the first translocation in *Drosophila melanogaster* and (2) of normal repeats in chromosomes. *American Naturalist* (abstract) **70**: 41.

[18] Muller H.J. 1922. Variation due to change in the individual gene. *American Naturalist* **56**: 32–50.

[19] Muller H.J. 1929. The gene as the basis of life. *Proceedings of the 4th International Congress of Plant Sciences, 1926, Ithaca* **1**: 897–921.

[20] Muller H.J. 1927. Artificial transmutation of the gene. *Science* **66**: 84–87; and Muller H.J. 1928. The problem of genic modification. *Verhandlung der V Internationale Kongress der Vererbunglehre, Berlin, 1927, ZiAV* (supplement) **1**: 234–260.

[21] Muller put the ClB stock to use in studying induced and spontaneous mutation rates. Created about five years earlier, it consisted of an **X** chromosome with the dominant marker Bar eyes and a large inversion of the **X** chromosome containing a recessive lethal. Hence, the Bar eye phenotype in females only existed in the heterozygous condition. If the homologous **X** carried an X-linked lethal, when mated it would produce only daughters and no sons. Later, Muller replaced the ClB with a viable **X** chromosome called Basc. It contained the dominant Bar eyes, the recessive apricot eyes, and a viable inversion. Inversion heterozygotes are used to balance lethal, multigenic, or infertile genetic stocks because odd-number recombinants lead to aneucentric chromosome formation that abort embryos or germ cell formation.

[22] Stadler L.J. 1928. Mutations in barley induced by X-rays and radium. *Science* **68**: 186–187.

[23] Eyster W.H. 1924. A genetic analysis of variegation. *Genetics* **9**: 372–404.

[24] Muller and his student, G. Pontecorvo, later showed that these were caused by radiation-induced chromosome breaks leading to aneucentric chromosome formation.

[25] Muller H.J. 1932. Further studies on the nature and causes of gene mutations. *Proceedings of the 6th International Congress of Genetics, Ithaca, 1932* **1**: 213–255.

[26] Stern C. 1931. Zytologisch-genetische untersuchungen als beweise für die Morgansche theorie des faktorenaustauchs. *Biologische Zentralblatt* **51**: 547–587.

[27] Stern C. 1932. Neuere ergebnisse über die genetik und zytologie des crossing over. *Proceedings of the 6th International Congress of Genetics, Ithaca, 1932* **1**: 295–303.

[28] Timoféeff-Ressovsky N.V., Zimmer K.G., and Delbrück M. 1935. Über die natur der genmutationen und der genstruktur. *Nachrichten an die Biologie der Gesellschaft der Wissenschaft Göttingen* **1**: 234–241.

[29] Agol I.J. 1930. Step allelomorphism in *Drosophila melanogaster*. *Genetics* **16**: 254–266.

[30] Muller H.J. 1956. On the relation between chromosome changes and gene mutations. *Brookhaven Symposia in Biology* **8**: 126–147.

Although most of this work was done with Daniel Raffel in the USSR, the circumstances there were so hectic that the work remained mostly unpublished until Muller got around to it in the 1950s, supplementing the work with many other rearrangements that had accumulated in the prior 20 years. Raffel, an American who went with Muller to the USSR, came back disillusioned. Because of his left-wing past, he could not get a job in genetics and spent the rest of his life in Maryland as a dairy farmer. Raffel's earlier work with Muller was Muller H.J. and Raffel D. 1938. The manifestation of position effect in three inversions at the scute locus. *Genetics* (abstract) **23**: 160.

[31] Auerbach C. and Robson J.M. 1946. Chemical production of mutation. *Nature* **157**: 302.

[32] Ray-Chaudhuri S.P. 1939. The validity of the Bunsen-Roscoe Law in the production of mutations by radiation of extremely low intensity. *Proceedings of the Seventh International Congress of Genetics, Edinburgh, 1939* (supplement), p. 146.

Sometimes a major piece of work gets buried because it was in an abstract and the work never got published in full due to negative circumstances—in fact, World War II broke out while the Congress was in session. This is a work of fundamental importance to studies of low-dose radiation (fallout, medical diagnostic radiation, disposal of nuclear waste, etc.) but it is almost never cited.

[33] Demerec M. 1929 Mutable genes in *Drosophila virilis*. *Proceedings of the 4th International Congress of Plant Science, Ithaca, 1926* **1:** 943–946.

[34] Dobzhansky T. 1937. *Genetics and the Origin of Species*. Columbia University Press, New York.

[35] Hardy G.H. 1908. Mendelian proportions in a mixed population. *Science* **28:** 49–50; and Weinberg W. 1908. Über den nachweis der vererbung beim menschen. *Jahresheftge des Vereins für Väterländische Naturkunde in Würtemburg* **64:** 368–382 (English translation in 1963. *Papers in Human Genetics,* Boyer S., ed. Prentice-Hall, Englewood Cliffs, New Jersey).

[36] Fisher R.A. 1930. *The Genetical Theory of Natural Selection*. Oxford University Press, United Kingdom.

[37] Painter T.S. 1934. Salivary chromosomes and the attack on the gene. *Journal of Heredity* **19:** 465–476.

[38] Patterson J.T. 1940. *Studies in the Genetics of Drosophila*. University of Texas Publication No. 4032, 250 pages. University of Texas, Austin.

[39] Caspersson T. and Schultz J. 1938. Nucleic acid metabolism of the chromosomes in relation to gene reproduction. *Nature* **142:** 294–295.

[40] Oliver C.P. 1940. A reversion to wild-type associated with crossing-over in *Drosophila melanogaster*. *Science* **26:** 452–454; and Lewis E.B. 1951. Pseudoallelism and gene evolution. *Cold Spring Harbor Symposia in Quantitative Biology* **16:** 159–174. See also Chapter 18.

Darwinism, Mendelism, and the New Synthesis

The procedure of science is to dismember natural phenomena into their constituent parts, and, after duly examining the parts and assaying their properties, to reconstruct the original order. It is hoped that by so doing a more communicable, if not more profound, insight into the nature of things may be gained than is possible to secure with the aid of other scientific methods. Some scientists obtain the greatest satisfaction from analysis and examination of parts; and others from synthesis of the whole from the parts; the former are attracted by the diversity and the latter by the unity of things. In the field of evolution, some investigators strive to discover the most widespread mechanisms of evolutionary changes, and others to detect the peculiarities of the evolutionary patterns in the separate lines of descent. It behooves us to recognize the legitimacy of both methods of approach.

THEODOSIUS DOBZHANSKY[*]

AS THE TWENTIETH CENTURY BEGAN, Darwinism was taking its lumps. Weismann's theory of the germ plasm made any Lamarckian modifications unlikely.[1] Saltationist theories abounded, and chief among those advocates were William Bateson in Great Britain, Thomas Hunt Morgan in the United States, and Hugo de Vries in Holland. The only stalwarts defending a pristine Darwinism were those of the biometric school of Karl Pearson and Walter Weldon and their supporters.[2] Straddling the fence was Sir Francis Galton, loyal to his cousin's theory of natural selection, but he supplemented it with generous inflows of discontinuous evolution based on chemical and physical models.[3] When Mendelism was confirmed in 1901, saltationists greeted it with enthusiasm, hoping to find justification for evolution through discontinuous events. The biometric school dismissed it as an irrelevancy or as a wrong-headed threat. For approximately the next 30 years, there would be personal and professional dislikes, disputes in published articles, and a belief that the contending views were irreconcilable.[4] The middle 30 years of the twentieth century brought almost universal agree-

---

*1937. In *Genetics and the Origin of Species*, p. 276. Third edition, revised, 1951. Columbia University Press, New York.

ment that the conflict was resolved and that a new synthesis had emerged.[5] In the last 30 years of the twentieth century, the reemergence of debates between neosaltationists and advocates of the new synthesis has made it likely that the larger picture of evolution still remains incomplete.[6] Unlike the chromosome theory of heredity, which was dominated by Morgan's school and which brought together breeding analysis and cytology largely in the United States, evolutionary theory in the first decades of the twentieth century was diffuse, lacking unity, and international in its participation. Both plant and animal studies were widely used, supplemented with theoreticians who relied heavily on mathematical models.

Cytological Approaches to Evolution Yield Results

Plant biology led the way in working out the cytological factors in plant evolution. The early insights of Davis and Gates were confirmed by *Oenothera* (evening primrose) work, especially through the second generation of *Oenothera* scientists including Otto Renner, Alfred Blakeslee, and Ralph Cleland (Figure 1).[7] The evolution of *Oenothera* was not a mainline event, but a curiosity in the history of these plants. In *Oenothera*, a series of interchanges among chromosomes led to the formation of rings or complexes (called Renner complexes) of translocated chromosomes which tended to segregate together as single units. *Oenothera* was also the first organism to show aneuploidy and polyploidy. Polyploidy (triploidy to octoploidy) was soon reported in many other plants, especially cereal grasses. Gates and Blakeslee showed

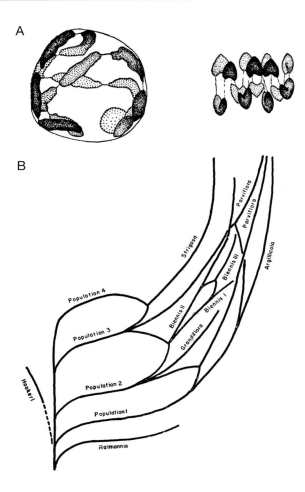

Figure 1. Oenothera *evolution. Ralph Cleland worked out the known and predicted variants of chromosome evolution among the* Oenothera. *All could be traced to the original form,* Oenothera lamarckiana. *([A] Reprinted from Cleland R.E. 1936. Some aspects of the cyto-genetics of Oenothera. Botanical Review 2: 316–348. [B] Reprinted, with permission, from Cleland R.E. 1956. Chromosome structure in Oenothera and its effect on the evolution of the genus. In Proceedings of the International Genetics Symposia, 1956. [supplemental volume of Cytologia, pp. 5–19].)*

that *Datura* was a far better plant to use for working out the trisomics for all 12 chromosomes in *Datura stramonium*.[8] Experimental evolution in plants was also successful. Hybrids could be made between two genera: the radish, *Raphanus*, and the cabbage, *Brassica*.[9] The resulting hybrid occasionally doubled its chromosomes to establish a fertile new species. Similar speciation by hybridization could also be accomplished in cereal grasses.[10] In the 1930s, O.J. Eigsti showed that colchicine acted as an agent that induced polyploidy (it caused a failure in the formation of the mitotic apparatus), and Blakeslee and others quickly used this method to induce artificial polyploids.[11]

Chromosome rearrangements were found in abundance when related species and genera were studied in Diptera. Flies in the Southwestern United States were studied carefully by Patterson and Stone, and their evolutionary history could be reconstructed by the chromosome rearrangements that had taken place (Figure 2).[12] Similar findings were noted in the Diptera of the Hawaiian Islands.[13] Such chromosome rearrangements contributed to an understanding of what geneticists began to call isolating mechanisms in evolution: A population with an inversion or translocation would produce fewer progeny with a cognate species lacking that rearrangement, and thus, the two species would tend to drift apart from each other in gene exchange and become more isolated as their genetic differences accumulated.[14] These chromosomal contributions to evolution had a significant part in Theodosius Dobzhansky's (Figure 3) influential book,

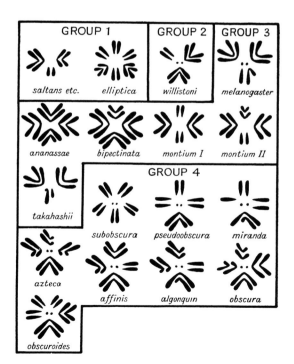

Figure 2. Drosophila *chromosome evolution. J.T. Patterson and W.S. Stone at Texas and T. Dobzhansky at the California Institute of Technology and Columbia worked out the evolution of the Drosophilidae. (Reprinted, with permission, from White M.J.D. 1948. Animal Cytology and Evolution. Cambridge University Press, United Kingdom.)*

Genetics and the Origin of Species (1937).[15] Barbara McClintock also helped to shape this cytological component of evolution. Her analysis of sterilizing or lethal effects on pollen or ovule production of chromosome rearrangements both supplemented and guided the parallel studies of chromosome rearrangements and their consequences in fruit flies.[16]

Figure 3. *In 1937, Theodosius Dobzhansky published* Genetics and the Origin of Species, *which stressed the cytogenetic component of the "new synthesis" that fused Darwinism and genetics. (Courtesy of the National Library of Medicine.)*

Assumptions for Evolution Change in the Twentieth Century

The debates at the start of the twentieth century were very different from those that would take place a generation later. In the first decade, Pearson and Weldon perceived evolution to be statistical, with small variations forming continuous traits on which selection acted.[17] The biometric school believed that the shape and changes of curves would reveal the evolutionary trends and allow predictions for when those changes would occur. Mathematics was applied to the process of evolution impinging on a largely unknown heredity. In addition, in that first decade, the Mendelians tried to focus on the hereditary material itself. Since Mendelian ratios were discontinuous and the extracted traits were unchanged by their residence in a hybrid state for one or more generations, the Mendelians believed that mutational changes in individual genes would lead to discontinuous evolutionary change, selection merely sifting what nature provided by producing new mutations. At the same time, Galton had convincingly demonstrated that many variable traits behaved in a statistical way; a regression to the mean for height could be objectively measured. To the Darwinists of the biometric school, the existence of dominance and recessiveness was unimportant because such Mendelian traits were considered trivial. They insisted that the traits that counted in evolution were quantitative, with hybrids showing blending or a reversion to the mean and not dominance.

These turn-of-the-century concerns were largely abandoned as the number of Mendelian traits in dozens of plants and animals reached into the hundreds. They included factors for disease resistance (in plant fungal infections), behavioral traits (waltzing mice), biochemical disturbances (Garrod's work on alkaptonuria and other disorders), fertility, viability, skeletal disturbances, and pathologies of all major organ systems. This forced a shift of the argument from the alleged triviality of Mendelian traits to the evolutionary role of monstrosities and pathologies or severe losses of normal traits. For the biometricians and orthodox Darwinians, such Mendelian traits were the culls of evolution, the stuff that failed to get passed on. No systematic search for beneficial mutations had been made, and it was questioned whether they even existed in Mendelian form.

By the 1920s, these arguments were giving way to a much more sophisticated study of gene interaction. Richard Goldschmidt, Sewall Wright, and H.J. Muller were among the numerous geneticists studying the complexity of character traits in the evolutionary process. Goldschmidt studied a series of alleles in the moth, *Lymantria dispar*, which showed how sexual selection could be analyzed to trace the genes involved in heterosexual and hermaphroditic strains (Figure 4).[18] Wright showed that coat color in guinea pigs was complex and that evolution acted not on individual genes, but on the relationships of genes participating in a common character.[19] Muller argued that he could use genetic analysis to dissect and isolate the components of variable traits—chief genes,

Figure 4. *Richard Goldschmidt had a gadfly role in genetics, frequently opposing established views and favoring physiological and complex models, such as denying the existence of discrete genes and favoring the sudden origins of higher taxonomic categories (so-called hopeful monsters). However, his work with the moth Lymantria was solid and established a continuum of sexual and intersexual forms. (Courtesy of the National Library of Medicine.)*

modifier genes, and environmental factors—in the expression of continuous or variable traits such as *Beaded* and *Truncate* wings.[20]

Quantitative Approaches to Mendelism Make an Impact

Pearson and Weldon were blinded by their distaste for Mendelism and even further driven to anger by Bateson's blunt and aggressive behavior regarding their views. But it was not that they were innocent victims of Bateson's pugnacious personality, because they heaped venom and vitriol on Bateson and anyone who championed Mendelism as significant for genetics or evolution. That bias extended to

mathematician G. Udney Yule, who tried to be helpful.[21] He saw no contradiction between Mendelism and Darwinian natural selection on small differences if the genes involved did produce small differences. Despite Yule's published articles and his efforts to convince Pearson that this would end the debate, Pearson rejected a shift away from his statistical approach to evolution. Pearson was only able to ease

Figure 5. *R.A. Fisher was the first to merge Darwinism with population genetics in his influential book,* The Genetical Theory of Natural Selection. *His contributions to population genetics helped to establish the new synthesis. (Courtesy of the American Philosophical Society, Bronson Price papers.)*

those views in a major work, *The Genetical Theory of Natural Selection* (Figure 6),[24] in which he emphasized the influence of natural selection on individual genes and supported the work of Morgan's school.

In Great Britain, J.B.S. Haldane (1892–1964) also displayed great gifts for mathematics and its applications to genetics and evolution (Figure 7).[25] Like Pearson, Haldane was a polymath with interests in many fields. He was largely self-taught in genetics, biochemistry, and the life sciences, his education

himself out of the debates when Weldon died.[22] He may have felt guilty that a younger man like Weldon was so loyal to him that it cost him his life, as a result of a premature heart attack. Despite his leaving the public arena and shifting to eugenics and other interests, Pearson was still unconvinced that a mathematical model using Mendelism would be relevant to evolution. He wrote discouraging comments to a young mathematician, R.A. Fisher (1890–1962) (Figure 5), who had originally attracted his notice for his contributions to statistics in biology. Fisher persisted, however, and soon established himself as a leading mathematical Mendelian.[23]

Fisher was severely myopic, conservative in his outlook, and an ardent supporter of the negative eugenics movement (state-enforced sterilization of the "unfit"). From 1918 to 1930, he produced a stream of significant papers that showed not only the compatibility of mathematics and Mendelism through population genetics, but also the applications of population genetics to evolution. In 1930, he assembled

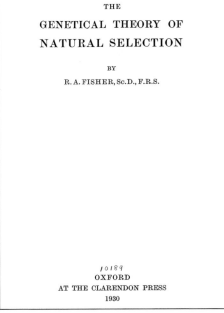

THE

GENETICAL THEORY OF NATURAL SELECTION

BY

R. A. FISHER, Sc.D., F.R.S.

10189
OXFORD
AT THE CLARENDON PRESS
1930

Figure 6. *R.A. Fisher's landmark publication,* The Genetical Theory of Natural Selection, *put together the mathematical basis for population genetics and its role in interpreting Darwinian evolution by natural selection. Fisher's work routed the last of the opposition from the biometric school of an earlier generation that rejected Mendelism as irrelevant to an understanding of evolution. (Reprinted from Fisher R.A. 1930.* The Genetical Theory of Natural Selection. *Clarendon Press, Oxford.)*

Figure 7. *J.B.S. Haldane, primarily a theoretical geneticist, helped to develop biochemical genetics and population genetics and mapped the first human genes on the* **X** *chromosome. (Reprinted, with permission, from Clark R.W. 1969. JBS: The Life and Work of J.B.S. Haldane. Coward-McCann, New York.)*

having focused on classics and mathematics (his bachelor's degree was in classics). He also had a strong interest in physiology. Haldane was the first to recognize linkage in mammals and to map human genes.[26] His papers on mathematical genetics also led him to formulate a general theory of population genetics that was fully compatible with Darwinian natural selection. In Haldane's versions, both individual genes and gene complexes were associated with selection. Politically, Haldane was a communist and for many years, he contributed articles to *The Daily Worker* and supported the Soviet Union. He was disillusioned after 1948 with the rise of the neo-Lamarckian Lysenkoism movement in the USSR that criticized genetics as bourgeois and false. Haldane eventually left the Communist party, but he was equally disillusioned with the politics of Great Britain. He went into exile in India, a country that had charmed him when as a young soldier he was

sent there to recuperate after a near-death encounter with gas warfare in France during the first World War.[27] Like Muller, Haldane was a supporter of the idealistic socialist application of eugenics. In 1921, he used his outlook to project the implications of eugenics into the distant future, predicting human cloning, human engineering of the oceans for algal-based food supplies, interstellar flight to populate new worlds, and atomic energy. The lecture forming the basis of his ideas became a small book, *Daedalus*, which inspired Aldous Huxley in a very different direction.[28] In 1932, Huxley showed the evil side of a society based on scientific control of behavior and reproductive biology, and Haldane's essay, *Daedalus*, was transformed into the novel, *Brave New World*. Haldane was also an eccentric and a favorite with the British press for his wit, biting humor, and capacity for insulting others.

Neither Fisher nor Haldane did much in experimental genetics. They preferred to interpret the data they found in the literature or predict the consequences of their assumptions about the fate of genes in populations; this was not the case for Sewall Wright (1889–1998) (Figure 8). Wright began his career as an experimentalist in mammalian genetics and his mathematics was largely self-taught. He was a student of Castle's, who guided his interests in coat color in guinea pigs.[29] Wright showed that a surprisingly large amount of gene interaction was involved in the development of coat color. This suggested to him that evolution might work by selection for such physiologically connected genes. He was also impressed by the genetic consequences of smaller populations and argued for an evolution in which geographical isolation by environmental factors broke up or maintained populations in small enough isolates to permit genetic drift from larger populations. Environmental effects on natural selection would thus have more varied genetic combinations present over contiguous

Figure 8. *Originally a mammalian geneticist from Castle's school, Sewall Wright shifted to population genetics and worked out the fate of gene distribution in small isolates (founder effects). (Courtesy of the American Philosophical Society, Wright papers.)*

areas than would natural selection acting on one very large population where such genetic differences would be largely hidden. Wright's contributions included the mathematics for such variable small populations, whose reproductive behavior would necessarily include a large amount of inbreeding.

All three of the founders of what is known as mathematical population genetics shared a belief that Mendelism was essential for an understanding of natural selection and evolution. All three incorporated the Hardy–Weinberg law and greatly extended the field of population genetics to incorporate effects of inbreeding, isolated populations, mutation frequencies, and the detrimental effects of genes in the homozygous or heterozygous condition. All of their equations apply to natural populations of plants and animals.

Human Population Genetics Is Problematic

The mathematical genetics of the evolution of human beings is somewhat different, because it may be argued that two of the features of being human are belonging to a culture and being aware of mating choice. Humans do not mate randomly (much of population genetics assumes a random mating for large populations). We practice what psychologists call assortative mating through choice, unconscious bias, or parental arranged marriages. Unfortunately, in human populations, the factors that favor assortative mating are largely unknown but they include looks, health, compatibility, and cultural factors, such as social status, education, religion, ethnicity, and connections to one's own family. Not all of these factors may have a genetic underpinning.

Two important contributions to the population genetics of humans derive from a set of equations first proposed by C.H. Danforth at the 1921 Second International Congress of Eugenics in New York.[30] Danforth showed that a gene would last in a population in proportion to the severity of its detrimental effect on the individual; more severe mutations are largely of recent origin, because they are among the most likely to be eliminated when homozygous. Moderate or milder genetic disorders are likely to persist for a longer history of generations before being eliminated. Muller followed up that set of equations in his 1950 paper, "Our load of mutations," and he also assumed that some of this elimination occurs through the partial dominance of the recessive mutant gene expressed in the heterozygous state.[31] Muller had a decent mathematical mind and could follow many of the key arguments about mathemati-

cal population genetics, but he was foremost an experimentalist and did not want to immerse himself in mathematics as had Fisher, Haldane, and Wright.

The New Synthesis Looks to the Future

The new synthesis brought scientists from many fields together. Paleontologists, population geneticists, cytologists, systematists, and Mendelian experimentalists joined together in the 1940s to shape a theory of Darwinian natural selection which argued that structural rearrangements, changes of chromosome number, mutation frequencies, gene physiology, modifier genes, detrimental effects in heterozygotes, inbreeding, drift, isolating factors, assortative mating, gene duplication, position effect, quantitative traits, polygenic inheritance, and many other factors had a role in evolution.[32] The net effect of all these components was consistent with an evolution of species largely through natural selection of the differences in populations over numerous generations. Factors left unresolved were involved in the tempo of evolution; some species seem to have had relatively rapid evolutionary change and others remained relatively constant for immensely long times.[33] Any teleological trends, ideas of progress in evolution, or noticeable saltations (sometimes called "hopeful monsters," Richard Goldschmidt's memorable phrase) were ruled out in this new synthesis. By 1950, the new synthesis had Darwinism fully restored and invigorated with the findings of a half-century of classical genetics.

The combined effect of cytogenetic studies of the evolution of plants and animals with the mathematical analysis of Mendelizing genes in populations greatly augmented classical genetics; the branches of breeding analysis, cytology, and evolution were confluent. The divisive issues at the start of the twentieth century were now resolved: Mendelian genes did participate in evolution, genes were associated with the chromosomes, chromosomal differences in number and rearrangements played a part in evolution, chromosome rearrangements had predictable genetic consequences, both continuous and discontinuous traits existed and had a Mendelian basis, and mathematical models of the biometric school were compatible with mathematical models of the new synthesis. Pockets of resistance still existed but from 1950 to 1975, classical genetics and the new synthesis were in harmony.

During the last 25 years of the twentieth century, once again a royal battle erupted over the tempo of evolution (the classic issue of saltationists versus Darwinian gradualists) and whether new species arose suddenly (or relatively rapidly) or fell within the model of evolution offered by the new synthesis. It was not, and still is not, easily resolved, as the twenty-first century begins, because the genetic events associated with speciation are not known in sufficient detail. Classical genetics could not, and cannot, solve that concern. Those details are likely to emerge from studies of comparative genomics, a twenty-first-century field that requires a major input from the molecular basis of gene and chromosome history.

End Notes and References

[1] Weismann A. 1885. The continuity of the germ-plasm as the foundation of a theory of heredity, pp. 163–256. In Poulton E.B, Schonland S., and Shipley A.E., eds. 1891. *Essays Upon Heredity and Kindred Biological Problems*, volume 1, Clarendon Press, Oxford.

[2] Provine W.B. 1971. *The Origins of Theoretical Population Genetics*. University of Chicago Press.

[3] Gillham N.W. 2001. *Sir Francis Galton: From African Exploration to the Birth of Eugenics*. Oxford University Press, United Kingdom. Also see Cowan R.S. 1969. *Sir Francis Galton and the Study of Heredity in the Nineteenth Century*. University Microfilms, Ann Arbor, Michigan.

[4] Provine W.B. 1971. op. cit. gives a blow-by-blow account of the lacerating feud. Also see Gillham N.W. 2001. op. cit. for Galton's role as fomenter and arbitrator of the disputes.

[5] Huxley J. 1942. *Evolution: The Modern Synthesis*. Harper and Row, New York.

[6] Gould S.J. and Eldridge N. 1977. Punctuated equilibria: The tempo and mode of evolution reconsidered. *Palaeobiology* **3**: 115–151.

[7] Cleland R.E. 1949. *Phylogenetic relationships in Oenothera. Hereditas* (supplemental volume): 173–188.

[8] Blakeslee A.F., Bergner A.D., and Avery A.G. 1937. Geographical distribution of chromosomal prime types in *Datura stramonium. Cytologia* (jubilee volume), pp. 1070–1093.

[9] Karpechenko G.D. 1928. Polyploid hybrids of *Raphanus sativus* L. x *Brassica oleracea* L. *Zeitschift für Induktive Abstammungs und Vererbungslehre* **48**: 1–85.

[10] Müntzing A. 1939. Studies on the properties and the ways of production of rye-wheat amphidiploids. *Hereditas* **32**: 521–549.

[11] Eigsti O.F. 1947. Colchicine bibliography. *Lloydia* **10**: 65–114; and Blakeslee A.F. and Avery A.G. 1937. Methods of inducing chromosome doubling in plants. *Journal of Heredity* **28**: 393–411.

Eigsti told me that he had worked out the colchicine method for inducing polyploidy, and he told Blakeslee of his discovery when he was a young summer investigator at Cold Spring Harbor Laboratory. Blakeslee then published a confirmation, but Eigsti had not yet published his finding. Blakeslee (or independently, in France, A.P. Dustin) is usually given the credit for this discovery.

[12] Patterson J.T. and Stone W.S. 1952. *Evolution in the Genus* Drosophila. Macmillan, New York.

[13] Carson H.L., Clayton F.E., and Stalker H.D. 1967. Karyotypic stability and speciation in Hawaiian *Drosophila. Proceedings of the National Academy of Sciences* **57**: 1280–1285.

[14] Muller H.J. 1940. Bearings of the *Drosophila* work on systematics. In Huxley J., ed. 1940. *The New Systematics*, pp. 165–268. Clarendon Press, Oxford.

[15] Dobzhansky T. 1937. *Genetics and the Origin of Species*. Columbia University Press, New York.

[16] Muller H.J. 1940. An analysis of the process of structural change in chromosomes in *Drosophila. Journal of Genetics* **40**: 1–66; see pp. 53–55.

Muller discusses the relation of his own work on structural rearrangements and that of McClintock on her ring chromosomes and the breakage-fusion-bridge cycle, including Muller's use of those ideas in interpreting what he called "radiation necrosis" (after 1945, this was renamed "radiation sickness"). This is a historically important article on the atomic age that presents the concerns at a genetic and cytological level in an era that had no inkling of how substantial the problem would become after Hiroshima.

[17] Provine W.B. 1971. op. cit.

[18] Goldschmidt R. 1934. Lymantria. *Bibliographia Genetica* **11**: 1–186.

[19] Wright S. 1916. *An Intensive Study of the Inheritance of Color and of Other Coat Characters in Guinea-Pigs, with Especial Reference to Graded Variation*. Carnegie Institution of Washington Publication No. 241.

[20] Altenburg E. and Muller H.J. 1920. The genetic basis of

truncate wing—An inconstant and modifiable character in *Drosophila. Genetics* **5**: 1–59.

21 Yule G.U. 1907. On the theory of inheritance of quantitative compound characters on the basis of Mendel's laws—A preliminary note. *Report of the Third International Conference 1906 on Genetics*, pp. 140–142. Royal Horticultural Society/Spottiswoode and Company, London.

22 Provine W.B. 1971. op. cit. p. 88.

23 Provine W.B. 1971. op. cit. pp. 140–153.

24 Fisher R.A. 1930. *The Genetical Theory of Natural Selection.* Clarendon Press, Oxford.

25 Dronamraju K., ed. 1968. *Haldane and Modern Biology.* Johns Hopkins University Press, Baltimore.

26 Bell J. and Haldane J.B.S. 1936. Linkage in man. *Nature* **138**: 759.

27 Clark R. 1969. *JBS: The Life and Work of JBS Haldane.* Coward-McCann, New York.

28 Dronamraju K., ed. 1995. *Haldane's Daedalus Revisited.* Oxford University Press, United Kingdom.

   Haldane's original essay and several comments on its influence are included in this source.

29 Provine W.B. 1986. *Sewall Wright and Evolutionary Biology.* University of Chicago Press.

30 Danforth C.H. 1923. The frequency of mutation and the incidence of hereditary traits in man. *Eugenics, Genetics, and the Family* **1**: 120–128.

31 Muller H.J. 1951. Our load of mutations. *American Journal of Human Genetics* **2**: 111–176.

32 Simpson G.G. 1944. *Tempo and Mode of Evolution.* Columbia University Press, New York; Stebbins G.L. 1966. *Processes of Organic Evolution.* Prentice-Hall, Englewood Cliffs, New Jersey; and Mayr E. 1942. *Systematics and the Origin of Species.* Columbia University Press, New York.

33 The issue involves stability of type. Certain fossils (especially mollusks) have shown relatively little change over long periods of time whereas other fossil histories show very rapid change (e.g., cichlid fish). Some argue that Darwinian natural selection accounts for both types of change. See Reference 32 for support of that Darwinian view. Opposed to this is the somewhat saltationist view of S.J. Gould and N. Eldridge (1977. op. cit.), which does not yet provide a genetic mechanism for these saltation-like rapid changes. Goldschmidt argued for some sort of "macromutation" of the entire genome, but no evidence exists for such massive reorganization other than through aneuploidy and polyploidy.

Throughout her life, Barbara McClintock was dedicated to maize research. She excelled in cytogenetics and worked out the genetic consequences of chromosome rearrangements. She was to maize what Bridges was to Drosophila cytogenetics. Her later work on movable genetic elements gained her a Nobel Prize. Shown here, McClintock works her field of corn. (Courtesy of the Cold Spring Harbor Laboratory Archives.)

Classical Genetics to the Mid-Twentieth Century

If doublet structures are repeats, as the evidence thus far indicates, then, judging from their widespread occurrence in the salivary gland chromosomes of Drosophila, it is likely that other multiple allelic series may be resolved into duplicate loci which act, by reason of a position effect, as a developmental unit.

E.B. LEWIS[*]

We are dealing with one gene and the ... alleles are due to mutation at ... different mutation sites of that gene, in every case the result of mutation being that of inactivating the gene. On this interpretation, in a heterozygote for two different mutant alleles, recombination between mutational sites could, and does in fact, occur.

G. PONTECORVO[†]

AT MID-CENTURY, CLASSICAL GENETICS was at its zenith and the Genetics Society of America held a golden jubilee at their meeting in Columbus, Ohio. This was intended as an honored remembrance of Mendel's rediscovery. Twenty-six invited speakers reflected on the origins of their fields or their ongoing research.[1] Six speakers who spoke on their work in biochemical or microbial genetics reflected the anticipation of an equally pro- ductive half-century to come. Classical genetics had moved beyond transmission genetics and was well penetrated into the thinking of developmental biology, cell biology, evolution, pathology, anthropology, and commercial plant and animal breeding. Noticeably absent from the panel of speakers and topics were the boundary issues of 1900 and 1950. In 1900, eugenics was very much a concern for the newly emerging field of genetics, and in 1950, all of the speakers knew that their colleagues in the USSR and its allies were in serious trouble from a political attack on the legitimacy of genetics itself. This political assault was known as Lysenkoism in the West and Michurinism in the USSR.

[*]1945. The relation of repeats to position effect in *Drosophila melanogaster*. *Genetics* **30**: 137–167.
[†]1952. Genetic formulation of gene structure and gene action. *Advances in Enzymology* **13**: 121–149.

Plant and Animal Cytogenetics Yield Useful Results

Studies of cytogenetics proved fruitful for investigators because chromosomes could be counted; their morphology in the hands of able cytologists often showed the presence of chromosome rearrangements, and most important, each of the chromosomal aberrations had its own genetic consequences and effects on reproduction. A good deal of the cytogenetics was worked out in maize, especially by the patient and carefully designed work of Barbara McClintock (Figure 1).[2] Both *Drosophila* and maize were useful for cytogenetic studies because many of their genes were mapped, and each system had a bonus of extra large

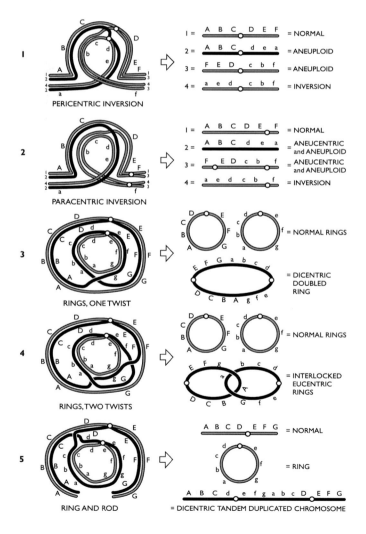

Figure 1. *Cytological rearrangements associated with McClintock's use of maize. Sturtevant, Muller, and especially Barbara McClintock worked out the consequences of crossing-over within inversion heterozygotes and other chromosome rearrangements. Maize turned out to be particularly well suited for demonstrating the predicted consequences of crossovers in these rearrangements. It also led McClintock to her interpretation of the breakage-fusion-bridge cycle. In (1), a pericentric inversion produces two aneuploid but eucentric chromosomes that abort the embryos. Fertility may thus be reduced, but the progeny that have the normal or the inverted chromosome are functionally healthy. In (2), a paracentric inversion prevents the formation of recombinant products (gametes or embryos) because the aneucentric chromosomes either fail to become gametes or prevent early cell divisions after fertilization. In (3), a single crossover in a pair of rings produces a double-size dicentric ring along with the two unaffected products of meiosis. In (4), a double crossover leads to interlocking rings and two unaffected products of meiosis. In the ring/rod heterozygote (5), a crossover produces a linear chromosome of double length (a tandem repeat) with two centromeres. The unaffected ring and rod products can form normal offspring. (Figure drawn by Claudia Carlson.)*

chromosomes differentiated along their length. In fruit flies, these giant chromosomes were found in salivary glands in which multiple rounds of replication led to cable-like extended chromosomes whose partial coiling resulted in regular banding patterns. T.S. Painter recognized their importance for cytogenetics but once he published his first accounts, a stampede of able cytologists around the world looked at genetic novelties and recognized their cytological basis. In maize, the prophase chromosomes had distinct knobs or chromomeres that permitted genetic identification, with small deletions or chromosome rearrangements expressing a known mutant phenotype. With patience, both Bridges (using salivary chromosomes) and McClintock (working on maize chromosomes) confirmed the sequence and relative distances of the known mutations from their recombination maps and showed that a parallel map existed in the chromosomes. McClintock explored the consequences of crossing-over in inversion heterozygotes and ring chro-

mosomes and soon students in genetics classes around the world were realizing that the Möbius strips of mathematics had a counterpart in the reality of interlocked rings and double-size dicentric rings in the products of crossing-over in ring chromosomes.[3]

Cytogenetics also had a major role in interpreting the evolution of plants. By 1950, the *Oenothera* story was a much-admired but unique side branch in evolutionary history. Cleland pursued each complex of translocations and rearranged chromosomes in every variety that he could locate (Figure 2).[4] The components were signatures of their past ancestry, and Cleland could use them to predict not-yet-discovered forms as well as (and then demonstrate by crossing) the parental forms that led to new species or varieties of *Oenothera*. The histories of polyploids and aneuploids in the divergence of plant species were much more frequent. It was not just a matter of doubling the diploid to the tetraploid form; many polyploids arose by the fertilization of cognate species that differed

Figure 2. *Ralph Cleland, at the microscope, and, to the right of his left shoulder, Salvador Luria, Tracy Sonneborn, and Hermann Muller at Indiana University (the other two are unidentified). The photo was taken shortly after Muller returned from Stockholm, where he received the Nobel Prize in Physiology or Medicine in 1946. (Courtesy of The Lilly Library, Indiana University, Bloomington, Indiana.)*

slightly in their chromosome composition. Such amphidiploids, as they were called, sometimes experienced a failure of meiotic separation, doubling each of the two different haploid sets of chromosomes and establishing an allotetraploid instead of the more familiar autotetraploid.[5] Commercial breeders, especially horticulturists and agronomists, made use of these techniques to establish polyploid strains that in plants usually resulted in larger fruits or larger yields of grain. These plant breeders quickly learned that many qualities were lost in such attempts, including texture, taste, or nutritional value, and these cytological forms had to be crossbred with kindred forms to introduce those genes that had commercial value. This was not always a successful effort. Eigsti applied his colchicine methods to produce triploid watermelons that were seedless and tasty, but they remained curios because for many people, part of the fun of eating watermelon is to spit out the seeds.

Overwhelmingly, the applications of classical genetics to agriculture were successful. The yields of cereal grains improved by 10–20% in the Swedish Agricultural Stations, where Arne Münzing and Åke Gustaffson used X-ray-induced mutations as well as isolated spontaneous mutations to assemble particularly productive strains for commercial use.[6] Similar studies in the United States and other industrialized countries with professionally trained geneticists found similar benefits. The promise of hybrid corn some 40 years earlier became fully realized, with virtually every farmer now using hybrid seed routinely.

Animal genetics showed similar outcomes from breeding experiments at agricultural stations. The yields of milk, butterfat, lean meat, marbled meat, and meat per feeding ratio all increased dramatically, to the benefit of millions of consumers for whom abundant food was never a concern.[7] Cytogenetics in mammals was minimal because triploid, tetraploid, and aneuploid forms either aborted or were rendered sterile. Cytogenetics shifted from the whole organism to tumor studies, in which chromosome rearrangements were debated as causes or consequences of tumor cell growth. It would not be until the 1960s that human cytogenetics became a major field in both basic science and clinical applications. When human instances of trisomic and monosomic conditions appeared in the early 1960s, counterpart conditions appeared in mice and other mammals. Calico traits, once limited to female cats, occasionally occurred in male cats who turned out to be **XXY**, the equivalent of Klinefelter syndrome human males.[8] Mice with the **XO** condition, however, turned out to be fertile instead of sterile.[9] Sometimes, animal cytogenetics preceded the discovery of similar conditions in humans. Murray Barr noted sex chromatin in his slides of cats and, after checking his records, determined that these were females (Figure 3).[10] He soon found the same phenomenon in humans. The phenomenon of **X** inactivation in mammals, interpreted for sex chromatin by Mary Lyon, became the mammalian method of dosage compensation.[11]

A Morphological Distinction between Neurones of the Male and Female, and the Behaviour of the Nucleolar Satellite during Accelerated Nucleoprotein Synthesis

GENETICISTS have long emphasized that 'maleness' and 'femaleness', so far as chromosome content is concerned, are projected from the fertilized ovum into the morphologically and functionally specialized somatic cells. It appears not to be generally known, however, that the sex of a somatic cell as highly differentiated as a neurone may be detected with no more elaborate equipment than a compound microscope following staining of the tissue by the routine Nissl method.

Figure 3. *Murray Barr first noted the presence of sex chromatin in the nuclei of female cat cells and extended it to other mammals and humans. This quickly led to a search for chromosome disorders in humans. (Reprinted, with permission, from Barr M.L. and Bertram E.G. 1949. A morphological distinction between neurons of the male and female, and the behaviour of the nucleolar satellite during accelerated nucleoprotein synthesis. Nature **163**: 676–677.)*

Model Genetic Systems Lead to Tumor Gene Studies

Mouse genetics, especially at Bar Harbor, Maine, became a major source of specialized genetic stocks for studying basic and applied genetic problems. There were mutations that affected the immune system and the capacity of adult organisms to accept or reject a transplant from nonkin. Numerous genes in mice were associated with tumor susceptibility, obesity, diabetes, and other human disorders. The genetics of allergies, memory, aging, neurological defects, and skeletal malformations set the stage for later mouse models of human biochemical disorders.[12] Immunogenetics made considerable progress using animals models, especially those of mice and cattle.

Fruit flies and mice both contributed to developmental genetics. In mice, the T series of alleles were remarkable for working out early stages of lethal action in the developing embryo. Similar studies in fruit flies had revealed the importance of what Hadorn called "lethal factors in development."[13] Tumor genes were common in fruit flies, with a history going back to the first decade of Morgan's fly lab. These genes are usually melanotic and appear as solitary or multiple black blobs in the larva or abdomen of the adult fly. Bridges and Stark analyzed a lethal X-linked mutant tumor strain[14] and some 25 years later, Leanne S. Russell (later in mouse genetics) identified its mode of death as a blockage of the intestines. The tumor was not malignant because transplanting it did not lead to metastatic growth. The susceptibility to tumor formation is dependent on genes in the X, II, and III chromosomes of fruit flies, and this was consistent with a multihit model of tumor formation in mammals, where tumors are more commonly found in aging adults.[15]

The Idea of Units Leads to the Problem of Gene Structure

Researchers in both plant and animal genetics (and later microbial genetics) found the idea of the gene extremely helpful. In its undefined state as a unit of inheritance, it was serviceable for all living things. But always in the mind of the geneticist was the hope of associating those genes with some material entity. The history of the unit that became the gene long predates classical genetics. The unit of inheritance that we today call a gene underwent numerous name changes between 1865 and 1909 when it received the name "gene" from Wilhelm Johannsen.[16] When Herbert Spencer speculated about the universe, he drew on a trend that had begun in 1810 with the publication of Dalton's atomic theory. Matter could be considered chemical combinations of atoms having specific valences and chemical formulae could be assigned to molecules. This inspired the search for additional units of composition of matter. The cell doctrine (1855) proposed by Virchow, which emerged out of Schleiden and Schwann's cell theory, raised a structural parallel for the composition of tissues, organs, and organisms. Ten years later, Spencer, Mendel, and Darwin independently conceived of units of inheritance. Spencer inferred these to be part of his theory of the universe. He believed everything was composed of hierarchies of units from atoms to molecules, to cells, to organisms, to societies. He correctly inferred that "physiological units" that served as hereditary units must exist between the chemist's small molecules and the cells.[17] These were far smaller than cells, but far larger than the chemist's typical molecules of that era. Unfortunately, imagination is always limited by reality, and Spencer had no way to flesh out the function of his physiological units nor locate them in the cell. His theory was lacking in predictive value and hence most scientists could not use it.

Mendel's units, which he called factors or "merkmale," were more specific because each such unit was associated with a specific character trait that could be followed through breeding analysis.[18] That union of hereditary units with a mechanism to follow the units signaled the beginning of genetics. Unfortunately for Mendel, his character traits were not of evolutionary significance and his rules of combination and ratios (3:1 or 9:3:3:1) were not universal when he tested them, on Carl Nägeli's suggestion, with *Hieracium*. These two weaknesses, from the perspective of nineteenth-century interests among biologists, made Mendel's units no more interesting than Spencer's. It took the rediscovery of Mendel's laws in 1900 to awaken new interest in these hereditary units and breeding analysis as a much-needed tool for understanding heredity and its relationship to evolution.

Darwin's concept of units was partially founded on Spencer's physiological units, but Darwin tried to make his "provisional theory of pangenesis" a much more comprehensive theory.[19] He liked the idea of a constant flow of hereditary units into the gonads, where they could be assembled in reproductive cells and transmit the collective state of the body and nervous system into the offspring. This explained why siblings do not resemble one another the way that identical twins do (siblings are conceived during different stages of their parents' life cycles). In addition, it explained why regenerated areas might show slightly different color or features other than those that developed when the original body part was present. This theory suffered from some of the same weaknesses as Spencer's physiological units—it did not predict outcomes when different strains were crossed and their descendants were observed. Even worse, and much to Darwin's disappointment, his own cousin, Francis Galton, failed to prove its basic premise that the gemmules (as Darwin called them) circulated. Galton showed that transfusions of blood from one color strain of rabbit into another had no effect on the progeny.[20] The progeny were always of the recipient's color and never flecked with a hereditary tainting by the donor blood's presence in the circulation.

When Mendelism was rediscovered, the units were known as factors or character units (or unit characters). The term "factor" was fairly neutral and implied nothing about structure or function. "Unit character" implied some one-to-one association of the unit character to the character it controlled. As multiple allelism, coupling, spurious allelomorphism, and epistatic models of modified 9:3:3:1 ratios began to mushroom, the term unit character became misleading. In 1909, the gene thus emerged as the truncated remnant of Darwin's pangenesis and Hugo de Vries' pangene. But this 1909 gene was deliberately created to represent the unit without implying anything about its composition, size, or structure.

The Idea of the Gene Develops from 1909 to 1922

The explosive growth of fruit fly genetics by Morgan's group led to many insights about the undefined gene. Sturtevant recognized that a normal gene could mutate to more than one state when he interpreted multiple allelomorphism (after the early 1920s, Shull introduced the shorter term, alleles, to replace allelomorphs).[21] Geneticists from Mendel on down to the present have demonstrated that the extracted recessive is indistinguishable from the original parental homozygous recessive used to generate a line of descent through hybridization. Thus, stability of the gene was one of the first characteristics to be established. This required a long battle between the schools of Castle and Morgan, which were at odds over the variegated appearance of certain characters that were assumed to be simple Mendelian factors yielding a 3:1 ratio.[22] The *hooded* rat was a prime example of a hybrid that produced a variation in hoodedness. Muller and the fly group argued that this was "residual inheritance," but Castle would have none of that because it seemed as though every time Morgan's flies had some new problem of unexpected variation, they invoked modifier genes. Castle needed evidence and the fruit fly work was unconvincing to him. But his own rats obliged; he had to back off and accept residual inheritance when he carried out the critical experiments in his own *hooded* rats.[23] Hence, gene stability was legitimate—genes did not contaminate each other in the heterozygous state.

Morgan produced a second finding from the eosin case, which arose as a partial reverse mutation from a white-eyed stock. Thus, white could not be a simple loss of the gene for the red eye color factor; it had to be more complex. Morgan and his students spent several years carrying out crosses, without success, looking for recombination of the alleged compound structure of the normal red-eyed factor or compound factors associated with eye color.[24] This weakened, if not destroyed, Bateson's model of presence and absence. Character traits (such as a shift from color to white) may be interpreted as losses, but this did not mean that the mutated gene was missing—it might just be rendered nonfunctional.

As dozens of mutations accumulated during the first decade of fruit fly research, Muller became interested in the properties of those mutations. He assembled evidence that the mutation process was punctiform: It occurred just to that one gene in a diploid cell, not to the gene and its homologous allele. Hence, whatever caused the mutation had to be of a similar highly focused nature.[25] Sturtevant, however, showed that functionally, genes could be autonomous or nonautonomous. He readily showed this autonomous property in mosaic individuals, such as a male fly with one eye white and the other eye red. Individuals could also be nonautonomous and, like vermilion eye color, never show such mosaics, with the normal tissue correcting the eye color expression, rendering it red.[26]

One of the other unusual properties of the eosin mutation was the property of X-linked genes, Bridges' finding that was extensively analyzed by Muller. Most of these genes showed the same character expression in the male (one **X** chromosome) as in the female (two **X** chromosomes). Thus, an apricot-eyed male looks exactly like an apricot-eyed female in intensity and color for this allele of the white-eyed series of multiple alleles. But a female that is homozygous for eosin is much darker in eye color than a male who is hemizygous for eosin; the effects are not dosage compensated.[27] Muller called this phenomenon "dosage compensation," and he proposed that modifier genes neutralized the dosage relationship between female and male for most of the X-linked genes (Figure 4). For some reason, eosin was not dosage compensated or something happened to it that prevented it from being dosage compensated.

Genes were also stable and rarely mutated, but mutation rates were inconceivable in the early days of fruit fly genetics because individual mutations were relatively rare. Morgan's laboratory tried "ball park" figures by estimating the number of mutations found to the number of flies raised (not everyone kept a record of every fly used in every experiment), but this method was flawed. Instead, while they were still at Rice University, Muller and Altenburg gene mutations should in the majority of cases involve more or less inactivation of the processes governed by the normal gene, and that these less active genes should more often act as recessives to the normal than as dominants. This implies that one dose of the normal gene usually has an effect more nearly like that of two doses than of no dose. Whether the latter principle is a primary one, however, or is due to the past selection of modifiers, is another question.

On the compensation of the effects of dosage differences between the sexes, and on dominance

In the above connection, it will be worth while to make somewhat of a digression, to consider a curious fact that has emerged from the results concerning hypomorphs. That is, it appears that in the great majority of the cases of hypomorphic sex linked genes, one dose in the male produces about as strong or at times even a slightly stronger effect in the direction of normality than do two doses in the female. This must of course be due to the interaction of other genes in the X chromosome, whose simultaneous change in dosage affects the reaction.[9] In some cases at least it has been possible to show, by studies of the effects of different chromosome pieces, (a) that genes other than the genes for sex are acting as the "modifiers" in question, (b) that the modifiers responsible for the dosage compensating effect on different loci are to some extent different from one another, and (c) that more than one modifier may be concerned for a specific locus.[10] I base these conclusions on various results obtained in work of OFFERMANN, who has been especially active in the study, of PATTERSON, and of myself.

We may for convenience call these genes "modifiers," but with the reser-

[9] We arrived at our main results and conclusions regarding this phenomenon of dosage compensation in the spring of 1930. Although we communicated our results to Doctor STERN at that time (prior to the remarks made by STERN and OGURA 1931, upon this topic), we withheld our preliminary report (MULLER, LEAGUE and OFFERMANN 1931) until after certain checks had been carried through.

[10] Judging by certain results recently reported by MORGAN, BRIDGES and SCHULTZ (1931), the second-chromosome mutation Pale (associated with BRIDGES' original translocation) has, in addition to a "diluting" effect, an effect on the different eye colors of the white series similar to that produced by lessening from two doses to one the gene or genes in the X chromosome that are responsible for the dosage-compensation of most members of this series (thus, those allelomorphs of white that are lighter in the male are lightened by Pale, but the others are darkened somewhat). This means that the chemical process affected by Pale is the same as, or in its effect similar to, that affected by the dosage compensator (s) of the X; but, since we have seen that there is no reason to identify the latter with the gene or genes in the X that decide sex, we have no reason to agree with the suggestion of the above authors that "the translocation (Pale) may be closely connected with the sex-determining reaction."

Figure 4. *Twenty years before Barr's discovery of sex chromatin, Muller and his students coined the term "dosage compensation" and worked out a noncytological mechanism for the process in fruit flies. Modifier genes (dosage compensators) were demonstrated to equalize the dose difference between the two X female and the one X male. (Reprinted from Jones D.F., ed. 1932. Proceedings of the Sixth International Congress of Genetics, Volume 1. Brooklyn Botanic Garden, New York.)*

attempted to make an estimate by using a class of mutations not subject to personal bias: They chose lethal factors, because they could tell by the 2 female:1 male ratio and the absence of one class of males that a lethal had occurred. These 1919 studies yielded the first mutation rates.[28] About one X chromosome per 1000 X chromosomes tested carried a new lethal mutation.

The one exception to this was the stock of Bar-eyed flies, which kept throwing off reversions to round eyes. Sturtevant and Morgan proved that this was associated with a recombinant event (the outside markers going in either direction for the round-eyed new mutant flies that emerged). In some strange way, Bar eyes was not a real mutation, but a new expression associated with a duplication of the genes involved. They called this a position effect.[29] Hence, some genes were subject to position effect if their neighboring genes were shifted away from them.

As early as 1915, Muller and Morgan also made estimates of the number of genes in fruit flies by taking the smallest known map distance and dividing it into the total of all map lengths for the three autosomes and X chromosome. This estimate came out to roughly 2000 genes (far less than the approximately 14,000 known in 2001 when the fly genome was fully sequenced for all its nucleotides).[30]

Gene Composition, Function, and Number Are Clarified After the Early 1920s

Some views of the gene were in error, as Castle's theory of allelic contamination indicates. W.H. Eyster's genomere theory, tested by Muller in his 1927 paper on genic modification, was also among the erroneous theories.[31] Eyster likened the gene to a bean bag of particles and assumed that these sorted out mutations in a variegated or mosaic pattern, depending on the generation of gene replications within the sac. Muller showed that after the F_1, the incidence of mosaics virtually disappeared (he used samples of F_2 and F_3 females to determine the percentage that carried new lethals sorting from a putative P_1 altered sperm). He found none: Like visible ones, the only mosaics he got were those present in the F_1 females. Muller guessed that chromatids in sperm were sometimes prematurely doubled. After 1953, double-helical complementary strands of DNA provided the interpretation for the occurrence of mosaics.[32]

A third erroneous idea was target theory. Some geneticists assumed that gene size and volume could be determined by using mutation frequencies induced by X rays and an assumed volume for the chromosomes being tested. Timoféeff-Ressovsky,

Zimmer, and Delbrück were early enthusiasts for the use of physics to estimate the size of objects in a contained space,[33] and this method is not without merit. John Preer used it to estimate the size of Sonneborn's killer factors in *Paramecia*, concluded that they were the size of small bacteria, and estimated their number in a *Paramecium*. Both predictions were confirmed cytologically when proper staining revealed these endosymbiotic bacteria.[34] What made target theory less effective for the gene was the unknown thread-like structure of the gene. Genes were not cocci-shaped objects strung like beads in chromosomes, a fallacy created for illustrative purposes in text books or popular articles but never taken seriously by Morgan and his students when they realized, after the origin of eosin from white eyes, that genes were in some ways complex.

In 1932, Muller proposed a theory of gene function that greatly extended the findings of Sturtevant for vermilion eye color.[35] He classified gene function as amorphic if the loss of function was total (e.g., a white-eyed mutation with no pigment production at all), hypomorphic if the loss of function was partial (what microbial geneticists a generation later called leaky mutants and what in the white-eyed series would be characteristic of apricot eyes), hypermorphic if an excess of gene product occurs (as in black body color), and neomorphic if the function is totally new. The best example of a neomorph was Bar eyes, a dominant eye-shape mutation that arose as a

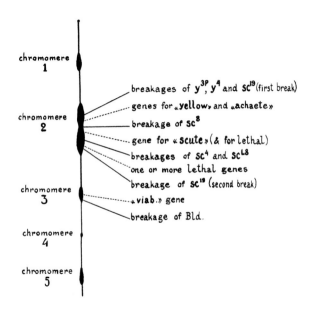

Figure 5. *Gene nests and the scute-19 insertion. Muller and his Russian colleagues identified a gene cluster from the X chromosome (yellow-achaete-scute) that was shifted from the X chromosome to the left arm of the second chromosome. They obtained cytological proof when salivary gland chromosomes were used for the analysis and proposed that duplications of genes, whether tandem or shifted, would be the basis of future gene evolution. (Reprinted from Muller H.J. and Prokofyeva A.A. 1934. The individual gene in relation to the chromomere and the chromosome. Proceedings of the National Academy of Sciences **21**: 16–26.)*

position effect after duplication of the gene involved. Muller considered neomorphs of particular importance to evolution because they could become the source of new functions for an organism.

It was not until salivary chromosome analysis was available that the problem of gene number and gene size could be estimated. One important rearrangement was the *scute-19* (sc^{19}) insertion (Figure 5).[36] A small piece of the chromosome from the tip of the X chromosomes was shifted into the left arm of the second chromosome, not too far proximal to the *dumpy* locus. Muller and Prokofyeva-Belgovskaya described this sc^{19} insertion, which contained the *yellow-achaete-scute* (*y-ac-sc*) region, as a thin band next to the much larger shoe-buckle band of the *dumpy* region. This allowed a more precise measurement of gene number and size, but even this was flawed because at the time, no one knew that much of eukaryotic DNA is inert or nongenic.

While still in the USSR, Muller and Daniel Raffel (also American) carried out another detailed study of the *yellow-achaete-scute* region (Figure 6) in 1936.[37] Their approach used heterozygous pairs of several inversions, each having in common a break near the *y-ac-sc-l* region (an X-linked lethal was the fourth of these adjacent genes). By combining the left end of one inversion and the right end of the other inversion, Muller and Raffel obtained deletions or duplications of one or more of these genes, and in a few instances, no changes at all in the four elements of this region. These latter recombinants demonstrated either precise borders of separation for these genes or inert material between adjacent genes, where losses of small amounts of material had no effect on their function. This operational proce-

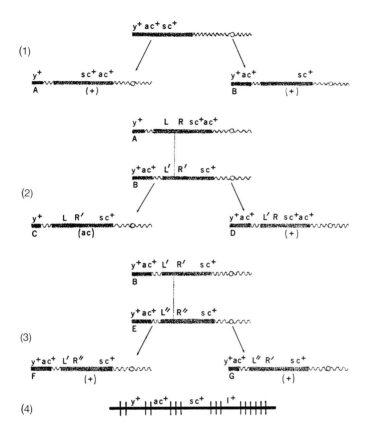

Figure 6. *The left-right test. Muller and his students worked out crosses between pairs of inversions whose left break was in the y-ac-sc-l region near the tip of the X chromosome. The second break was in the heterochromatic region at the other end of X. These recombinants demonstrated that breaks in the y-ac-sc-l region occurred in fixed boundaries and Muller assumed that some inert boundary material must exist between genes. Some 40 years later, these regions were shown to be associated with highly duplicated DNA, which is nongenic. (Reprinted, with permission, from Carlson E.A. 1966. The Gene: A Critical History. W.B. Saunders, Philadelphia.)*

dure led to the shorthand name for this study, the "left-right test," and ranks as one of the most subtle and informative genetic analyses of a gene nest done in premolecular times.

The most significant effort to analyze complex loci or gene nests resulted from the study of pseudoallelism. McClintock first named this relation in a 1945 study of a series of apparent alleles in maize.[38] She noted that yellow, pale green, and white genes are recessive. The yellow is allelic with white

and shows an intermediate color in the F_1 generation. Similarly, the pale green and white are intermediate in color in the F_1 but yellow when heterozygous with pale green as normal green leaf color. The normal phenotype suggests that a recombination should occur among them; this was in fact obtained. Their apparent allelism through white is the basis of their pseudoallelism. In McClintock's example, white was a deletion whose size encompassed the adjacent genes, yellow and pale green.

The Discovery of Pseudoallelism Has Future Implications

In 1940, C.P. Oliver found a case that at first resembled the Bar case—an apparent duplication for the *lozenge* eye mutations.[39] However, unlike the Bar case, there was no unequal crossing-over. The markers used (constituting what is called a four-point test) showed *lozenge-spectacle* to be to the left of *lozenge-1*. Additional work revealed a sequence of three sites with *lozenge-glossy* to the right of *lozenge-1*.[40] In the four-point test, two unrelated genes are used as markers. The generic formula is written as **a b/c d**, where **a** and **d** are the outside markers and **b** and **c** are the possible pseudoalleles. If recombination is present in the **b-c** region, a reading of ++++ would indicate that **b** is to the left of **c**. If the normal recombinant is **a ++ d**, then **b** is to the right of **c**. If both types of recombinant markers occur for the normal "reversion," then, like Bar, unequal crossing-over is obtained. Oliver and his student, Melvin Green, originally thought that they had found another Bar-like case, but the consistent placement of outside markers by the four-point test unequivocally showed that something new had been found. E.B. Lewis soon found that the allelic series of *Star* and *asteroid* yielded similar results, with *Star* to the left of *asteroid*. In the early 1950s, *stubble, vermilion, forked,* and *Notch* were added, as was an extensive analysis of white eyes. The work of M.M. Green, W. Welshons, W. Hexter, A. Chovnick, M.E. MacKendrick, and others confirmed what Oliver and Lewis had first found—multiple allelic series were complex and consisted of separable units.[41]

At the same time that Lewis was first demonstrating pseudoallelism in the *white* locus, G. Pontecorvo's laboratory was independently demonstrating that *white* was complex.[42] But Pontecorvo shifted his attention to a more favorable system for these studies, the fungus, *Aspergillus*. Instead of looking for a multiple allelic series, Pontecorvo identified many different mutant genes associated with vitamin synthesis. In every instance, whatever the phenotype or compound made by the gene, he found results identical to pseudoallelism when he applied the four-point test. To Pontecorvo, the results suggested that no pseudoalleles were present because all alleles were pseudoalleles. He proposed instead that a gene was divisible into smaller subunits or regions by intragenic crossing-over.[43] Lewis had argued that Bridges' repeats (or Muller's primary unequal crossing-over leading to tandem duplications) were the source of the pseudoallelism. Pontecorvo's was a structural interpretation of the genes (it can be analyzed by recombination into finer and finer units) and Lewis' was a functional or evolutionary interpretation of the genes (pseudoallelism is a form of position effect arising through an evolutionary history after an initial duplication) (Figure 7).[44]

I was a graduate student in 1953–1958, when these debates and findings first appeared, and in 1955, I was fortunate to find pseudoallelism in the *dumpy* locus. Unlike *white, lozenge, forked,* and the other cases emerging at the time in the fly work, the *dumpy* series had long been intriguing to Muller because decades earlier, he had studied *Truncate* to work out the gene-character model of chief genes and modifiers. Bridges had renamed the series *dumpy* because all of the factors involved could be tested and detected using *dumpy*. Muller had noted that, in addition to allelism with *dumpy*, some alleles in the series were not allelic in the mutually heterozygous state. Thus, the three factors were **o** = oblique wings, **v** = vortices with disturbed bristle patterns on the thorax, and **l** = a lethal factor. At the time, all except

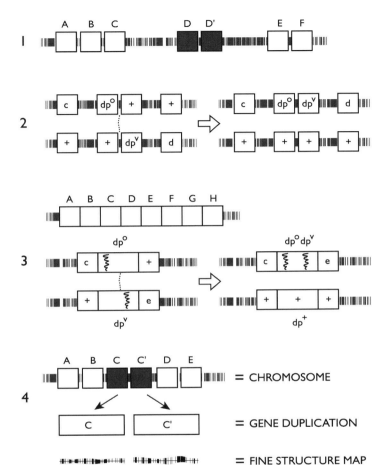

Figure 7. *Lewis' pseudoallelic analysis of genetic repeats. In the 1950s, two contending models of gene structure were debated. Bridges had argued for a past gene evolution by repeats, as shown in (1), where the duplicate genes D and D´ may undergo differential mutational histories. Lewis extended that model by proposing a pseudoallelic status for many of these repeat regions seen in salivary gland chromosomes. The four-point test for pseudoallelism, seen in (2) for the dumpy series of alleles, indicates that the alleles dp*⁰* and dp*ᵛ* are aligned in sequence dp*⁰*-dp*ᵛ* by the placement of outside markers. Crossing-over occurs between the duplicated genes. (3) Pontecorvo argued that in fungi, every gene he tested showed pseudoallelism using a four-point test; this suggested that crossing-over occurred as an intragenic event. Thus, dp*⁰* and dp*ᵛ* in Pontecorvo's model would be different lesions within the same functional gene. As it turned out, both were correct. The work of Benzer and Demerec established a genetic fine structure (4) within functional genes (called cistrons by Benzer) as well as recombination between cognate functional genes, such as the tryptophane series of genes studied by Demerec and his students. (Figure drawn by Claudia Carlson.)*

l alone were known. The rule for these was consistent: A combination of any two would only show what they had in common: **o/ov** was (**o**), **olv/v** was (**v**), and **ov/olv** was (**ov**). But **o/v**, **o/lv**, or **ol/v** were all normal in appearance. This normal appearance for two members of a multiple allelic series was called complementation. Hence, the *dumpy* series was the first pseudoallelic series involving alleles, pseudoalleles, and complementation. I mapped the region for my Ph.D. dissertation project, including the lethal alone which arose in the laboratory. The sequence was discontinuous: **l-ol-olv-o-lv-ov-v**.[45]

I like to think of those pseudoallelic maps as the last contribution of classical genetics to the problem of gene structure and function. From 1955 on, these attempts to interpret gene structure and function by breeding analysis and cytology yielded something new. In one sense, it was not molecular (although molecular biologists certainly thought of it as molecular genetics); it was microbial. Experiments with fungi, bacteria, and viruses were so much more efficient in revealing what came to be known as "genetic fine structure" that all of these fruit fly series, with one exception, were eclipsed by the hundreds of potential sites within a gene or sequence of contiguous genes that the microbial recombinations revealed. Those findings were chiefly the contributions of Milislav Demerec studying tryptophane genes in *Salmonella typhimurium* and Seymour Benzer studying a morphological plaque mutation called the *r*II region in bacteriophage.[46] In particular, Demerec's work made it likely that both Pontecorvo and Lewis were right. Some duplicated genes had an evolutionary history, but genetic fine structure could be found within each of these duplicated genes.[47] Benzer's analysis of the *r*II region, however, captured the imagination of every geneticist reading his papers because his work was seen as bringing about the union of classical genetics and molecular biology. It was the *r*II analysis that made

the ultimate small distances of fine structure merge into the nucleotides of the double-helix model of DNA.

The one exception in pseudoallelism that continued to flourish after the dazzling events of genetic fine structure in microbial systems was Lewis' analysis of the bithorax region (Figure 8).[48] When he began to study the bithorax region, Lewis quickly realized that this system must have played a part in the evolution of wing formation and body segments from the metamers that constituted more primitive insects. Lewis' mastery of dipteran embryology, his deep understanding of the evolution of insects, and his persistence in isolating and combining body segment alleles in this particularly favorable pseudoallelic system gave him the opportunity to merge classical genetics and developmental biology in a novel way. It led to the search for genes that regulated development.[49]

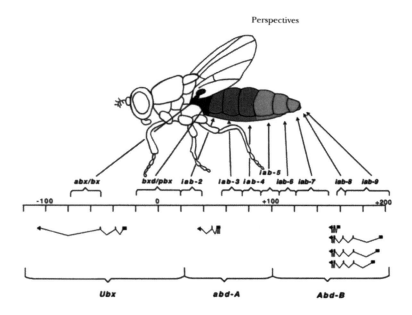

Perspectives

Figure 8. *Lewis' most significant gene nest was the homeotic series of mutations called bithorax in fruit flies. These genes demonstrated the evolution of thoracic components, including wings and halteres in the Diptera. Lewis' systematic analysis led to a search for developmental genes and a new field of developmental genetics. (Reprinted, with permission, from Duncan I. and Montgomery G. 2002. E.B. Lewis and the bithorax complex: Part II. From cis-trans test to the genetic control of development. Genetics* **161:** *1–10.)*

End Notes and References

1 Dunn L.C., ed. 1951. *Genetics in the 20th Century*. Macmillan, New York.

2 For a biography of her career and life, see Keller E.F. 1983. *A Feeling for the Organism: The Life and Work of Barbara McClintock*. W.H. Freeman and Company, San Francisco.

3 McClintock B. 1932. A correlation of ring-shaped chromosomes with variegation in *Zea mays. Proceedings of the National Academy of Sciences* **18:** 677–681.

4 Cleland R. 1936. Some aspects of the cytogenetics of *Oenothera. Botanical Revues* **2:** 316–348.

5 Clausen R.E. 1941 Polyploidy in *Nicotiana. American Naturalist* **75:** 291–306.

6 Müntzing A. 1951. Genetics and plant breeding, pp. 473–492. In Dunn L.C., ed. 1951. op. cit.

7 Lush J.L. 1951. Genetics and animal breeding, pp. 493–526. In Dunn L.C., ed. 1951. op. cit.

8 Cattanach B.M. 1962. XXY mice. *Genetical Research* **2:** 156–158.

9 Welshons W.J. and Russell L.B. 1959. The Y chromosome as the bearer of the male determining factors in the mouse. *Proceedings of the National Academy of Sciences* **45:** 560–566.

10 Barr M.L. and Bertram E.G. 1949. A morphological distinction between neurons of the male and female and the behavior of the nucleolar satellite during accelerated nucleoprotein synthesis. *Nature* **163:** 676–677.

11 Lyon M. 1988. The William Allen Memorial Award Address: X-chromosome inactivation and the location and expression of X-linked genes. *American Journal of Human Genetics* **42:** 8–16.

12 Little C.C. 1951. Genetics and the cancer problem, pp. 431–472. In Dunn L.C., ed. 1951. op. cit.

13 Hadorn E. 1961. *Developmental Genetics and Lethal Factors*. (Translated from German.) Methuen, London.

14 Stark M.B. 1918. An hereditary tumor in the fruit fly, *Drosophila. Journal of Cancer Research* **3:** 279–301; and

Stark M.B. and Bridges C.B. 1926. The linkage relations of a benign tumor in *Drosophila. Genetics* **11:** 249–266.

15 Russell E.S. 1942. The inheritance of tumors in *Drosophila melanogaster* with especial reference to an isogenic strain of St. Sr. tumor 36a. *Genetics* **27:** 612–618.

16 Johannsen W.J. 1909. *Elemente der Exakten Erblichkeitslehre*. G. Fischer, Jena.

17 Spencer H. 1866. *Principles of Biology*. D. Appleton and Company, New York.

18 Mendel G. 1865. Experiments in plant hybridization. *Verhandlung Naturforschung Verein in Brünn*, volume 4. For English translations of Mendel's two papers and other pertinent documents, see Stern C. and Sherwood R., eds. 1966. *The Origins of Genetics: A Mendel Source Book*. W.H. Freeman and Company, San Francisco.

19 Darwin C. 1868. Provisional hypothesis of pangenesis. *Animals and Plants Under Domestication*, volume II, pp. 428–483. Orange and Judd, New York.

20 Galton F. 1871. Experiments in pangenesis, by breeding from rabbits of a pure variety, into whose circulation blood taken from other varieties had previously been largely transfused. *Proceedings of the Royal Society (Biology)* **19:** 393–404.

21 Sturtevant A.H. 1913. The Himalayan rabbit case, with some considerations on multiple allelomorphs. *American Naturalist* **47:** 235–238.

22 Carlson E.A. 1966. *The Gene: A Critical History*, chapter 10, pp. 77–88. Saunders, Philadelphia; Castle W.E. 1914. Mr. Muller on the constancy of Mendelian factors. *American Naturalist* **49:** 37–42; and Muller H.J. 1914. The bearing of the selection experiments of Castle and Philips on the variability of genes. *American Naturalist* **48:** 567–576.

23 Castle W.E. 1919. Piebald rats and the theory of the gene. *Proceedings of the National Academy of Sciences* **5:** 126–130.

24 Morgan T.H. and Bridges C.B. 1913. Dilution effects and bicolorism in certain eye colors of *Drosophila. Journal of Experimental Zoology* **15:** 429–466.

[25] Muller H.J. 1923. Mutation. *Eugenics, Genetics, and the Family* **1**: 106–112.

[26] Sturtevant A.H. 1920. The vermilion gene and gynandromorphism. *Proceedings of the Society for Experimental Biology* **17**: 70–71.

[27] Morgan T.H. and Bridges C.B. 1913. op. cit.; and Muller H.J. 1932. Further studies on the nature and causes of gene mutations. *Proceedings of the Sixth International Congress of Genetics, Ithaca* **1**: 213–255.

[28] Muller H.J. and Altenburg E. 1919. The rate of change of hereditary factors in *Drosophila*. *Proceedings of the Society for Experimental Biology* **17**: 10–14.

[29] Sturtevant A.H. and Morgan T.H. 1923. Reverse mutation of the Bar gene correlated with crossing over. *Science* **57**: 746–747; and Sturtevant A.H. 1925. The effects of unequal crossing over at the Bar locus in *Drosophila melanogaster*. *Genetics* **10**: 117–147.

[30] Morgan T.H. 1926. *The Theory of the Gene*, pp. 309–310. Yale University Press, New Haven, Connecticut.

Morgan does not give the exact number in this commentary but Muller mentioned Morgan's estimate when I took Muller's class, "Mutation and the Gene," in 1954.

[31] Eyster W.H. 1924. A genetic analysis of variegation. *Genetics* **9**: 372–404; and Muller H.J. 1928. The problem of genic modification. *Proceedings of the Fifth International Congress of Genetics in Berlin, 1927* (*Zeitschrift für Induktive Abstammungs und Vererbungslehre* [supplementary volume] **1**: 234–260.)

[32] Muller H.J., Carlson E., and Schalet A. 1961. Mutation by alteration of the already existing gene. *Genetics* **46**: 213–226.

[33] Timoféeff-Ressovsky N.V., Zimmer K.G., and Delbrück M. 1935. Über die nature der genmutation und der genstrucktur. *Nachrichten auf der Gesellschaft. Der Wissenschaften Göttingen* **1**: 234–241.

[34] Preer J.R. 1948. The killer cytoplasmic factor, its rate of reproduction, the number of particles per cell, and its size. *American Naturalist* **82**: 35–42; and Chao P.K. 1953. Kappa concentration per cell in relation to the life cycle, genotype, and mating type in *P. aurelia*, variety 4. *Proceedings of the National Academy of Sciences* **39**: 103–112.

[35] Muller H.J. 1932. op. cit.

Later, microbial geneticists used the term "leaky" for hypomorph and "nonleaky" for amorph. The microbial system does not have a vocabulary for genes that are neomorphs and hypermorphs.

[36] Muller H.J. 1935 On the dimensions of chromosomes and genes in dipteran salivary glands. *American Naturalist* **69**: 405–411.

Muller's estimate of "at least 10,000 genes" is close to the 13,000 found from its molecular sequencing 65 years later.

[37] Muller H.J. and Raffel D. 1938. The manifestation and position effect in three inversions at the scute locus. *Genetics* (abstract) **23**: 160. See also a more detailed analysis of the data initiated by that abstract in Muller H.J. 1956. On the relation between chromosome changes and gene mutations. *Brookhaven Symposium in Biology* **8**: 126–147.

[38] McClintock B. 1944. The relation of homozygous deficiencies to mutations and allelic series in maize. *Genetics* **29**: 478–502.

[39] Oliver C.P. 1940. A reversion to wild type associated with crossing over in *Drosophila melanogaster*. *Proceedings of the National Academy of Sciences* **26**: 452–453.

[40] Green M.M. 1956. A cytogenetic analysis of the lozenge pseudoalleles. *Zeitschrift für Induktive Abstammungs und Vererbunglehre* **87**: 708–721.

[41] Carlson E.A. 1959. Comparative genetics of complex loci. *Quarterly Review of Biology* **34**: 33–67.

[42] MacKendrick M.E. and Pontecorvo G. 1952. Crossing over between alleles at the W locus in *Drosophila melanogaster*. *Experientia* **8**: 390; Pontecorvo G. 1952. Genetic formulation of gene structure and action. *Enzymology* **13**: 121–149; and Pontecorvo G. and Roper. A. 1956. Resolving power of genetic analysis. *Nature* **178**: 83–84.

Pontecorvo proposed a "milli-micro-molar" reaction process at the gene level to account for pseudoallelism in *Drosophila* and Lewis proposed a form of localized diffusion in *cis* and *trans* arrangements of genes that led to a position effect expression of heteroallelic genes. Both ideas were essentially abandoned when genetic fine structure and the operon models interpreted the way developmental or biochemical processes were regulated by genes.

[43] Pontecorvo G. 1958. *Trends in Genetic Analysis*. Columbia University Press, New York.

[44] Lewis E.B. 1950. The phenomenon of position effect. *Advances in Genetics* **3:** 73–115; and Lewis E.B. 1971. Pseudoallelism and gene evolution. *Cold Spring Harbor Symposia in Quantitative Biology* **16:** 159–174.

[45] Carlson E.A. 1959. Allelism, pseudoallelism, and complementation at the *dumpy* locus in *D. melanogaster*. *Genetics* **44:** 347–373.

[46] Benzer S. 1957. The elementary units of heredity. In *The Chemical Basis of Heredity* (ed. W.W. McElroy and B. Glass), pp. 70–93. Johns Hopkins University Press, Baltimore; and Demerec M. 1956. A comparative study of certain gene loci in Salmonella. *Cold Spring Harbor Symposia in Quantitative Biology* **21:** 113–122.

Benzer introduced some short-lived concepts for the gene based on Percy Bridgman's operational philosophy that had an attraction for theoretical physics. In these operational definitions, the gene as a unit of function was a cistron; as a unit of recombination, a recon; and as a unit of mutation, a muton. The operations involved defined their size and activity. For about ten years, papers referred to genes as cistrons, but that has largely faded. I suspect that the terms were too precise, and while they fitted microbial genes, they did not fare as well with eukaryotic genes that soon added split genes with introns and exons as well as assorted recognition or promoter sites. The more complex genes became, the more geneticists reverted back to the generic safety of Johanssen's unburdened term, gene.

[47] Demerec M. 1956. op. cit.

[48] Lewis E.B. 1957. Two wings or four? *Engineering and Science Monthly*, November. California Institute of Technology, Pasadena.

[49] Nüsslein-Volhard C. and Wieschaus E. 1980. Mutations affecting segment number and polarity in *Drosophila*. *Nature* **287:** 795–801. For an overview of this field of molecular developmental genetics, see Gerhart J. and Kirschner M. 1997. *Cells, Embryos and Evolution*. Blackwell, London.

Normal ($2n$)

$2n + 1 \cdot 2$ $2n + 3 \cdot 4$ $2n + 5 \cdot 6$ $2n + 7 \cdot 8$

$2n + 9 \cdot 10$ $2n + 11 \cdot 12$ $2n + 13 \cdot 14$ $2n + 15 \cdot 16$

$2n + 17 \cdot 18$ $2n + 19 \cdot 20$ $2n + 21 \cdot 22$ $2n + 23 \cdot 24$

Blakeslee's trisomics. (Reprinted, with permission, from Avery A.G., Satina S., and Rietsema J. 1959. Blakeslee: The Genus Datura. Ronald Press, New York.)

Classical Genetics and One-Celled Organisms

On the other hand, if these d'Hérelle bodies were really genes, fundamentally like our chromosome genes, they would give us an utterly new angle from which to attack the gene problem. They are filterable, to some extent isolable, can be handled in test tubes, and their properties, as shown by their effects on the bacteria, can then be studied after treatment. It would be very rash to call these bodies genes, and yet at present we must confess that there is no distinction known between the genes and them. Hence we cannot categorically deny that perhaps we may be able to grind genes in a mortar and cook them in a beaker after all. Must we geneticists become bacteriologists, physiological chemists and physicists, simultaneously with being zoologists and botanists? Let us hope so.

*H.J. Muller**

A TEST FOR A THEORY'S UNIVERSALITY was an early aim of those breeders who were inspired by the rediscovery papers of Mendel's findings. Many plants and animals were studied to make that confirmation in the first decade of the twentieth century. The chosen organisms were of horticultural, hobbyist, or agronomic importance. It might have occurred to some scientists then that an ideal system would be unicellular. Such organisms would breed rapidly, and experiments could be carried out in a matter of days, as indeed they were in the mid-twentieth century. But this was considered impractical for many reasons. Bacteria were largely of interest to physicians and bacterial health hazards were certainly of concern. The organisms were also described then as amitotic in their reproduction, although the actual mechanism of their replication took five more decades to work out. At that time, there was no effective way to use them for genetic purposes, and no one conceived then that they had or even could have had a sexual means of reproduction. Viruses were relatively new—the first isolated strain was tobacco mosaic virus found in the late 1890s—but since they were not yet visible under the best of microscopes, they were difficult to use for biological

*1922. Variation due to change in the individual gene. *American Naturalist* **56**: 32–50.

studies. This left three candidates: fungi, protozoa, and algae, which were eukaryotes. A relatively large number of them were of no threat to health and most did not require elaborate food for growth. Isolating them was only a minor difficulty (solvable with glass micropipettes) and an investigator would have had to use a dissecting microscope to study them. But technology was not the only reason that microbial systems were largely ignored. The first success in microbial genetics used a fungus. Despite that happy occasion, work in microbial genetics remained an oddity.

Blakeslee Discovers and Abandons Mating Types in *Mucor*

Shortly after the rediscovery of Mendelism in 1900, Blakeslee published an article on the genetics of sexuality in the bread mold, *Mucor* (Figure 1).[1] He showed that mating occurred between two strains that he called "plus and minus." Blakeslee did not study *Mucor* and one-celled organisms for long. He switched to plant genetics and made a substantial contribution by working out all 12 of the trisomics for *Datura* (jimsonweed).[2] Why did this microbial study, showing a clear Mendelian trait, fail to result in an early rush to study one-celled organisms? One explanation is the cost involved. Although molds could be grown on agar plates or slants in test tubes and did not require expensive equipment, this was not the case for organisms such as bacteria that required autoclaves, expensive compound microscopes, incubators, and a fair amount of training and safeguards (culturing bacteria carelessly could spread unwanted disease). Another factor was the tradition that bacterial studies belonged to medical schools and only

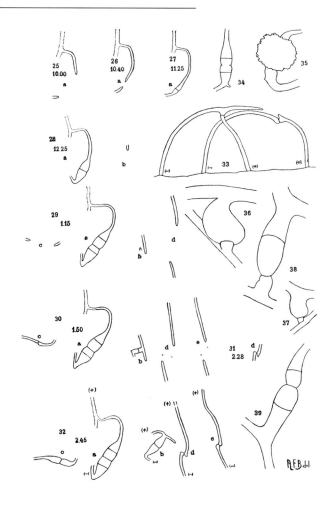

Figure 1. *Alfred Blakeslee contributed the first recognition of sex in microorganisms through his studies of the plus-and-minus strains of the bread mold,* Mucor. *(Reprinted from Blakeslee A.F. 1904. Sexual reproduction in the mucorinae.* Proceedings of the American Academy of Arts and Sciences **40**: 205–319.)

organisms of no interest to human health were appropriate for advancing pure biological science. The field of biochemistry, which could provide genetic studies of metabolic interest, did not yet exist. To identify

varieties, few morphological traits, such as color, were available. Their simplicity, advantageous for numbers of progeny and rapidity to study generations, was offset by the lack of useful phenotypes.

Jennings Assesses Microbial Genetics

More important, I believe, are the biological assumptions about the genetics of one-celled organisms in the first 20 years after the rediscovery of Mendelism. Herbert S. Jennings (1868–1947) (Figure 2) reviewed the status of that field at the 1921 Second International Congress of Eugenics held at the Museum of Natural History in New York City.[3] The major participants in the founding of classical genetics who presented their findings and assessments offered an opportunity for the emerging field of American eugenics to learn some basic genetics. Jennings, working at Johns Hopkins University, studied the ciliate, *Paramecium*, and an ameba-like organism, *Difflugia*. Both of these organisms could be readily grown in filtered media derived from hay infusions or desiccated

Figure 2. *Herbert S. Jennings used the ciliated protozoa,* Paramecium, *to work out pure lines. (Courtesy of the National Library of Medicine.)*

lettuce leaves. Jennings used *P. aurelia*. He started with a single cell and showed that all of the offspring were like identical twins and constituted what Johannsen, using the bean, *Phaseolus*, would call a pure line. Just as Johannsen found the existence of variation within a pure line, so did Jennings. He argued that "descent for thousands of generations is from a single parent; all the progeny have the same material or 'genes' as had the parent."[4] When he attempted to select within such a clonal line of descent, he found no change in the range of body size for the population of *P. aurelia*.

In *Difflugia*, the problem was even more instructive, because this protozoan contained a variable number of spines and the spines themselves varied in length and thickness. Despite his attempts to select the extremes in morphology, the range in a clonal line of descent remained constant. Once in a while, however, a variant did arise that established a new range, as was true with Johannsen's beans. Jennings claimed that "hereditary changes [are] due to alteration in the germplasm, quite uncomplicated by kaleidoscopic regrouping of the germinal substances."[5] However, Jennings did not believe that the type of changes he got were Mendelian. Instead, he argued, they were akin to Galton's model of ancestral correlations.

Jennings recognized that the world of microbial genetics differed in many respects from the reported findings, strange as they were, in higher plants and animals. He noted that V. Jollos found that *Paramecia* exposed to chemicals developed a long-lasting resistance or tolerance after exposure.[6] After several transfers in normal media, the descendant progeny would retain but eventually lose the tolerance to the chemical stimulus. Jollos called this a "dauer modification" (a temporary change that persisted for several generations). It was not quite Lamarckian because no opposing environment was required for the modified trait to gradually peter out. Jennings claimed that bacteria were different from typical plants or animals or even his protozoan species. Changes in virulence or host range seemed to be responsive or adaptive to the agent introduced. Some sort of direct environmental modification was present and "here in the lowest organisms 'acquired characters are inherited' on a rather large scale and must be an effective factor in evolutionary change."[7] He acknowledged, however, that this Lamarckian interpretation was a tentative conclusion because the techniques for isolating single bacteria were difficult and it was hard to follow their cell lineages, unlike in studies using *P. aurelia* or *Difflugia*.

In addition to studying pure lines in protozoa, Jennings also studied a problem that intrigued nineteenth-century microbiologists such as E. Maupas. Maupas claimed that if a line of descent from a single cell were followed for hundreds of generations, it would eventually die out. But occasionally, some

Figure 3. *Tracy Morton Sonneborn was Jennings' student at Johns Hopkins and spent his academic career at Indiana University. He taught courses on the genetics of one-celled organisms and devoted his research to Paramecium aurelia, whose mating types he worked out. He discovered a killer trait that he believed involved plasmagenes, but this turned out to be a consequence of an endosymbiont bacterial toxin instead. (Courtesy of The Lilly Library, Indiana University, Bloomington, Indiana.)*

event took place that led to rejuvenescence. This event was associated with a self-fertilization process called autogamy.[8] Jennings's student, Tracy Morton Sonneborn (1905–1981) (Figure 3), pursued this in the early 1930s and found some remarkable genetic properties associated with *P. aurelia*.

Sonneborn Promotes Cytoplasmic Inheritance

Sonneborn reported the discovery of multiple mating types in this species, any two of which would lead to conjugation.[9] During this process, an exchange of duplicated small nuclei (micronuclei), and a mutual fertilization took place. Each of the separated conjugants would reconstitute a new macronucleus and this would serve as the somatic nucleus. The genes from the macronucleus did not play a part in reproduction

during the conjugation process. This permitted a search for Mendelian traits in *P. aurelia* and Sonneborn found that these did indeed occur. He also found a second phenomenon involving some mating strains (Figure 4). One strain seemed to possess some toxic factor that led to the death of its mate.[10] He called the survivor the killer strain and he called the vulnerable cells the sensitive strain. What intrigued

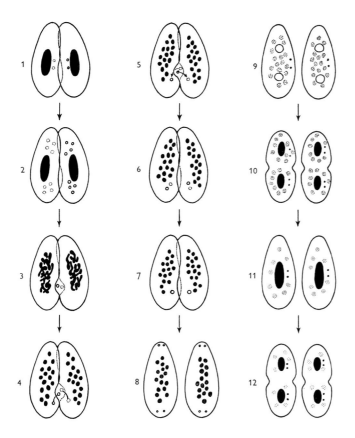

Figure 4. *Conjugation in Paramecium aurelia. (1) One macronucleus and two micronuclei per cell. In (2), the micronuclei divide and in (3), the macronucleus dissociates, accompanying one surviving micronucleus in each cell. (4–8) The micronuclei switch cells, divide, and fuse. The complex events continue with regeneration of the macronuclei (9), fission of the two cells, and finally the normal state returns (10–12) with one macronucleus and two micronuclei per cell. (Reprinted, with permission, from Beale G.H. 1954. The Genetics of Paramecium Aurelia. Cambridge University Press, United Kingdom.)*

Sonneborn was the observation that this killer trait was cytoplasmic but it was maintained by a nuclear gene. This gave him a chance to study nucleo-cytoplasmic inheritance, which became the leitmotif for the rest of his academic life. Sonneborn believed that classical genetics was dominated by the genetics of genes in nuclear chromosomes. To those geneticists, especially Morgan's school, the cytoplasm was "the playground of the genes" and to Sonneborn, the cytoplasm was the slighted co-partner of heredity,

with hereditary determinants of its own having fundamental roles at the level of cell organelles, metabolism, and the organized dynamics of the whole cell. Such genes were eventually found in mitochondria and chloroplasts, but not in other organelles. One needed the tools of molecular biology to demonstrate that plastid genes were derived from an evolutionary history of endosymbiosis (Figures 5 and 6). These genes did not support the plasmagene theory sought by Sonneborn.

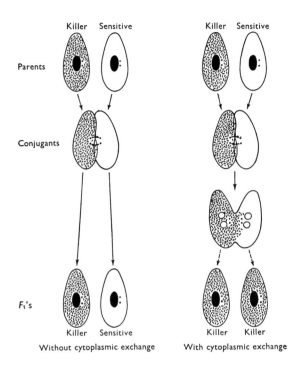

Figure 5. *Sonneborn discovered a strain of Paramecium aurelia that killed its conjugated mate by the transfer of a cytoplasmic factor (he called this kappa). He believed that the kappa were plasmagenes and devoted much of his career to the study of nucleocytoplasmic relationships. (Reprinted, with permission, from Beale G.H. 1954. The Genetics of Paramecium Aurelia. Cambridge University Press, United Kingdom.)*

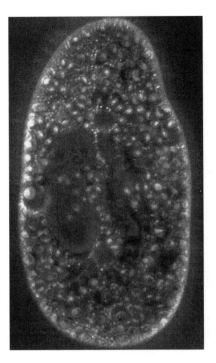

Figure 6. *Some animal cells harbor plant cells, a relationship of mutual benefit called endosymbiosis. The green algae can be isolated and cultivated in Paramecium bursaria. This suggested to Edgar Altenburg that a smaller particle (he called this a viroid) might be the source of the killer trait in P. aurelia. (Reprinted, with permission, from Wichterman R. 1953. The Biology of Paramecium. Blakiston Company, New York.)*

Muller and Sonneborn Become Estranged over Altenburg's Viroid Theory

Sonneborn did most of his work at Indiana University after leaving Johns Hopkins University, where he had studied with Jennings. He was also a great admirer of the work of H.J. Muller and became enthusiastic when Payne suggested that Muller might be a candidate for a senior job in Indiana

University's Department of Zoology. Payne asked Sonneborn to interview Muller at a meeting in St. Louis and then report back to him. Sonneborn gave Payne a glowing account of his discussions with Muller. In 1945, Muller was aware of Sonneborn's contributions to *Paramecium* genetics and he wanted to maintain a good relationship with Sonneborn. There was an understanding between the two that they could enjoy each other's comments as colleagues, but each would respect the other's domain. This was by no means easy because Muller, like Boveri and Wilson, believed that nuclear genes provided the essential genic basis of the cell and its contents. Sonneborn, of course, considered the cytoplasm to be more than just a passive partner of nuclear genes. To him, a vast unknown field of cytoplasmic inheritance determined cell structure, organelle organization, and a *terra incognita* of a pulsating cell physiology. Had Muller not gone to Indiana University, he would likely have ignored Sonneborn's work or rejected it.

Muller was grateful to Payne and Sonneborn and felt rescued since he despaired of a jobless future in genetics with a sick wife to support (Thea Muller had tuberculosis) and a toddler to raise (he was 51 when his daughter, Helen, was born).[11] But soon after Muller settled in, Altenburg came for a visit and toured Sonneborn's laboratory while he was a guest of the Mullers. He did not mention to Muller or Sonneborn that he was thinking about Sonneborn's work.[12] Since the days of Abraham Trembley some two centuries earlier, it had long been known that some animals harbored plant-like bodies called zoochlorellae. The relationship of the zoochlorellae to the host cell or organism is called endosymbiosis. Trembley worked with a coelenterate, a strain of hydra called *Hydra viridissima*, with zoochlorellae. He showed that if he placed these animals in a jar of water covered with a thick paper sleeve in which was cut the shape of a chevron, then exposure to sunlight would lead to a gathering of the hydras in the cut-out area. This demonstrated that they had a plant-like attraction to, or dependence on, sunlight to live.[13] A similar endosymbiosis existed in *Paramecium bursaria*. These cells harbored numerous photosynthetic zoochlorellae. Altenburg believed

that the killer strain might be a consequence of some invisible "viroid" that lived as an endosymbiont in strains of *P. aurelia*. He published his viroid theory shortly after he returned from his visit to Bloomington, Indiana.[14] Sonneborn was very familiar with *P. bursaria* and its visible zoochlorellae, but dismissed them as insignificant for his killer traits because no visible particles were present in his *P. aurelia* killer strains. Sonneborn was furious and Muller was embarrassed, feeling much like Morgan did when Muller published his attack on Castle's interpretation of the *hooded* rat analysis. This led to a strain between Sonneborn and Muller, with Sonneborn believing that Muller directly or indirectly supported publication of this attack on Sonneborn's work.

When I arrived at Indiana University in the fall of 1953, I was unaware of this tension between Muller and Sonneborn. Muller was on sabbatical leave in Hawaii for that year and Sonneborn was assigned as my temporary advisor. I began attending his seminars with his graduate students, but since my interest in fruit fly genetics sounded solidly in the nuclear camp, I was asked to leave the seminar group. I then read about a new finding of cytoplasmic inheritance in fruit flies. French geneticists had discovered a strain of flies that was sensitive to carbon dioxide and died in its presence, unlike strains that lacked this sensitivity, which was inherited maternally through the eggs.[15] I immediately asked Irwin Herskowitz, Muller's laboratory manager, if I could send away for a stock of these flies with which to work. Muller advised Herskowitz to tell me that if I did so, I should work in Sonneborn's laboratory because he did not want Sonneborn to think that Muller was horning in on Sonneborn's work.

The relationship between the two became more estranged during my years as a graduate student. Sonneborn scheduled one of his "genetics of microorganisms" graduate courses at the same time as

Muller's weekly laboratory meeting for his students. It was probably a coincidence, but Muller did not think so when we asked Muller to change his meeting time. We did not know that it was a Hopkins model of graduate education that we liked, but we sure knew that we wanted a multiple exposure of Indiana University's talented geneticists. To our delight, Sonneborn's disloyal associate, Wilhelm Van Wagtendonk, gave us blow-by-blow descriptions of the letters flying back and forth between Muller and Sonneborn and from both of them to President Wells, accusing each other of attacking the principle of academic freedom. It stopped being funny, however, when I went to Sonneborn's laboratory for conversations with his graduate students and found on the door a 3 × 5-inch index card with the penciled statement, "Keep door locked to keep out Drosophilae and other pests."

Like Castle and *hooded* rats, Sonneborn lost the war on his nucleo-cytoplasmic inheritance model for the killer trait. His own student, John Preer, noted that the killer trait died out when the *Paramecia* were well fed and encouraged to multiply rapidly. He used target theory and calculated the number of killer factors present in a cell of that strain. From this, he estimated that they were large enough to be visible in a compound microscope if properly stained. A few years later, Sonneborn's postdoctoral student, P.K. Chao, did the staining and found the killer particles, which resembled small bacteria (Figure 7). They were endosymbionts after all.[16]

Sonneborn's choice of organism was a major reason that the field failed to fulfill its initial start: It had a large chromosome number (hence the unlikelihood of linkage and chromosome maps), and the Mendelian phenotypes that Sonneborn and his students could detect were limited. When he started in

earnest to make *P. aurelia* the *Drosophila* of the microbial world, Sonneborn hoped that he could do what Beadle had done and he hired a biochemist, Wilhelm Van Wagtendonk. Unlike *Neurospora*, whose nutritional needs were simple for Beadle and Tatum to work out, the attempt to prepare a defined medium turned out to be a hellishly difficult problem. This frustrated his attempts to use a biochemical approach to *P. aurelia* genetics. Despite years of trying, Van Wagtendonk could not produce a totally defined medium (the lettuce distillate was normally supplemented with *Acetobacter* bacteria).

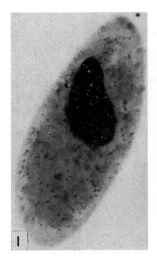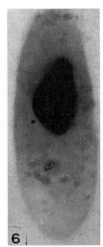

Figure 7. *The endosymbiont associated with the killer trait. John Preer was the first to calculate that the number of particles in* P. aurelia *associated with the killer trait made them likely to be of a size visible with an optical microscope, if properly stained. P.K. Chao's work rendered the particles stainable. (Reprinted, with permission, from Preer J. 1950. Microscopically visible bodies in the cytoplasm of the "killer" strains of* Paramecium aurelia. Genetics **35**: 344–362.)

Classical Genetics Leads to Molecular Genetics

For these reasons, I believe, the early hopes of a one-celled genetics gradually waned; it was classical genetics that pointed the way to molecular genetics.[17] I consider classical genetics to have worked its way to molecular genetics through five stages. The first stage came about as a result of the fruit fly studies. In 1927, Muller had induced mutations with X rays, which suggested to physicists that the gene was an accessible material unit, especially because of the punctiform way in which mutations arose. This also suggested the use of sperm volume as a suitable system for target theory, and three authors combined their talents in this effort. Max Delbrück was a physicist inspired by Niels Bohr to look for the relationship of quanta and life (photosynthesis and X-ray-induced mutations were possible examples) (Figure 8). He worked in Berlin with K.G. Zimmer, a

Figure 9. *After being forced into exile by the rise of Hitler, Erwin Schrödinger reflected on Delbrück's target theory and mutation work; he wrote* What is Life? *A popularization of genetics as seen through the perspective of a physicist, it was influential in attracting physical scientists to the new molecular biology. (©The Nobel Foundation.)*

Figure 8. *Max Delbrück, one of the founders of the phage school of molecular genetics, was a physicist turned biologist. He worked out the life cycle of the virus after having chosen it as a way to approach the gene by physical and chemical means. (Courtesy of the California Institute of Technology Archives.)*

biophysicist well versed in target theory, and N.V. Timoféeff-Ressovsky, a fruit fly geneticist in whose laboratory Muller had recently spent a year on a Guggenheim Fellowship. The resulting paper attempted to measure gene size by target theory, using carefully measured induced mutation frequencies for different doses of X rays.[18] Delbrück also tried to develop a quantum model of mutation and later told me that "it was a silly piece of work." Despite Delbrück's assessment of what was then called the "dreimännerwerk" (co-authors were still uncommon in the 1930s) or the green paper (the reprint had green paper covers), it was avidly read by fellow physicist Erwin Schrödinger (Figure 9), who derived his knowledge of genetics from it. In fact, it became the centerpiece of his highly influential book, *What is Life?*[19]

T. Caspersson and J.S. Schultz made the third of these first-stage contributions.[20] They studied ultraviolet-induced mutations in fruit flies and showed that the highest frequency of induced mutations occurred with wavelengths that were maximally absorbed by DNA. This suggested that DNA had a major role in heredity, a view that was proposed by Wilson in 1896 but abandoned after 1913, when Phoebus Levene's "tetranucleotide theory" made DNA seem like a dull crystal with no better function than to provide scaffolding for proteins.[21]

Beadle and Ephrussi contributed the second stage by working out the biochemical pathway for eye color synthesis in fruit flies. They used implanted embryonic anlagen of eyes stuck in larval hosts[22] as well as Sturtevant's recognition that some mutations were nonautonomous. After trying reciprocal implants on dozens of eye color mutations, Beadle and Ephrussi had a good idea of how genes functioned in carrying out a biochemical synthesis. Beadle teamed up with E.W. Tatum to use the orange bread mold, *Neurospora*, as a more suitable organism.[23] Tatum's biochemical knowledge was handy in selecting vitamins for synthesis, and the biochemical pathways that they revealed launched biochemical genetics as a field and gave new incentives for studying one-celled organisms (Figure 10).

Stage three is more appropriately molecular rather than biochemical in approach because it focused on DNA itself. In 1944, O.T. Avery, C.M. MacLeod, and M. McCarty showed that DNA was essential to induce bacterial transformation, a genetic transfer of information that Muller had interpreted on hearing of their results (Figure 11). He believed that fragments of DNA bearing one or more genes were inserted into host cells and then underwent crossing-over with the host DNA.[24] A.D. Hershey and M. Chase followed this with a more systematic study of radioactively labeled DNA and protein in bacteriophage to demonstrate that the bacteriophage proteins and plans for assembling mature viruses were provided by its DNA and not by its protein components. They depended on Delbrück's and Luria's earlier studies demonstrating that bacteriophage had a life cycle, occasionally mutated, and that mutant differences could be

Figure 10. *George Beadle (left) switched from fruit flies to the mold, Neurospora crassa, to work out biochemical pathways for the production of vitamins and other metabolic components of the cell. With Edward Tatum (right), he proposed a one-gene–one-enzyme model that was essentially correct for microbial genes. (Courtesy of the California Institute of Technology.)*

Figure 11. *Oswald T. Avery (left), Colin MacLeod (middle), and Maclyn McCarty (right) worked as a team at The Rockefeller University to isolate, purify, and demonstrate that DNA is the agent associated with the phenomenon of bacterial transformation. A biochemical triumph, it was a founding event in the transformation of classical genetics to molecular genetics. (Reprinted, with permission, from Dubos R.J. 1976. The Professor, the Institute, and DNA. Rockefeller University Press, New York.)*

mapped like genes in classical organisms by multiple infections of a host cell.[25] The pneumococcal transformation studies revived an interest in bacterial genetics that was ably developed by Joshua Lederberg (Figure 12).[26] The contributions of the phage group (founded by Delbrück, Luria, and Hershey, starting in 1939) revived the long-neglected bacteriophage studies of F. Twort and F. d'Hérelle.[27] In the early 1920s, those studies stressed the potential of bacteriophage for the medical treatment of bacterial infections. The technology (as Muller's prophetic quote at the head of this chapter suggests) for analyzing viruses as "naked genes" was too primitive in the 1920s. Although the phage life cycle preceded a working out of its biochemical components, the reverse was true for bacteria, whose sexuality and genetics were worked out after its DNA was identified as its genic component.

Figure 12. *Joshua Lederberg discovered sexuality among bacterial strains and worked out methods to isolate mutations and map them. He and Tatum worked jointly on much of the associated research. (Courtesy of the National Library of Medicine.)*

The fourth stage involved the fruit fly pseudoallelic studies of E.B. Lewis and G. Pontecorvo and their refinements using bacteriophage (the rII region of coliphage T4 studied by Seymour Benzer) and *Salmonella typhimurium* (studied by Milislav Demerec). A key feature of these studies was the recognition of genetic fine structure as having dimensions consistent with nucleotide sequences of DNA.[28]

The fifth stage clinching the molecular approaches to genetics involved the operon model of F. Jacob and J. Monod and the solution of adaptive enzyme formation, a cognate problem of the dauer modifications at the start of the twentieth century.[29]

All five of these stages required knowledge of classical genetics and used many of the conceptual tools of that era, which had been developed in the first half of the twentieth century. Essential for these studies were inducing mutations, mapping genes, crossing strains, determining dominance relations, and obtaining desired recombinants. The structure of DNA as a double helix by Watson and Crick did *not* require these tools, but it did require Muller's central thesis of classical genetics—the gene as the basis of life (Figure 13)—and it is no coincidence that the second *Nature* paper to appear in 1953 stressed the molecularization of classical genetics that was now possible: the nature of mutation, the potential discovery of a genetic code, and the mechanisms of gene replication.[30] All aspects of the physical struc-

---

THE GENE AS THE BASIS OF LIFE[1]

H. J. MULLER

University of Texas, Austin, Texas

1. THE LOCALIZATION OF GENES

What is meant in this paper by the term "gene" material is any substance which, in given surroundings— protoplasmic or otherwise—is capable of causing the reproduction of its own specific composition, but which can nevertheless change repeatedly—"mutate"—and yet retain the property of reproducing itself in its various new forms.   There is clear evidence that such material is to be found in the chromatin (where it is linearly arranged), and to some extent in the chloroplastid primordia and their derivatives, and there is no reason to believe that it exists anywhere else within the cell.   In this connection it may also be noted that in the most primitive organisms which contain a chlorophyll-like substance the chromosomes do not seem yet to have become distinctly grouped apart from their cytoplasmic surroundings, into a walled-off nucleus; hence the genes more directly associated with the chlorophyll, on the one hand, and the nuclear genes, on the other hand, may well have had a common origin, and may only in later phylogeny have become separated.

Figure 13. *The root belief of molecular biology. The belief that the gene was the basis of life, shown here in the title page of Muller's 1926 address, was the unquestioned tenet of those entering molecular biology and seeking to find a biochemical or structural composition of the gene as DNA. (Reprinted from Muller H.J. 1929. The gene as the basis of life. In* Proceedings of the International Congress of Plant Science *[ed. B.M. Duggar]. George Banta Publishing, Menasha, Wisconsin.)*

---

ture of DNA as a double helix were realized through molecular genetics with little, if any, reliance on the tools of classical genetics. One could argue, however, that the permissible aperiodic sequence of base pairs was the only aspect motivated by classical genetics. It was fully compatible with the gene as an "aperiodic crystal," a view so designated by Schrödinger in his profound work, *What is Life?*

End Notes and References

[1] Blakeslee, A.F. 1906. Differentiation of sex in thallus gametophyte and sporophyte. *Botanical Gazette* **42:** 161.

[2] Blakeslee A.F. 1929. Cryptic types in *Datura. Journal of Heredity* **20:** 177–190.

[3] Jennings H.S. 1923. Inheritance in unicellular organisms. *Eugenics, Genetics, and the Family* **I:** 59–64. Second International Congress of Eugenics, New York, 1921.

[4] Jennings H.S. 1923. op. cit. p. 59.

[5] Jennings H.S. 1923. op. cit. p. 59.

[6] Jollos V. 1921. Untersuchungen über variabilitat und vererbung bei infusorien. *Archiv für Protistenkunden* **43**: 1–222.

[7] Jennings H.S. 1923. op. cit. p. 63.

[8] Maupas E. 1889. La réjeunissement karyogamique chez les ciliés. *Archive de Zoologique Experimentale and Generale* **7**: 149–517.

[9] Sonneborn T.M. 1937. Sex, sex inheritance, and sex determination in *P. aurelia*. *Proceedings of the National Academy of Sciences* **23**: 378–395. For an overview of *Paramecium* genetics, see Beale G.H. 1954. *The Genetics of* Paramecium aurelia. Cambridge University Press, United Kingdom.

A sympathetic interpretation of Sonneborn's school and the harsh treatment that cytoplasmic inheritance received from Wilson's school (Morgan and Muller) and Emerson's school (especially Demerec and Beadle) is offered by Sapp J. 1987. *Beyond the Gene: Cytoplasmic Inheritance and the Struggle for Authority in Genetics*. Oxford University Press, New York. Sapp's view was that the nuclear chromosome camp engaged in a fair amount of bullying that prevented the cytoplasmic school from receiving its fair share of attention and respect. I believe, instead, that the real cause of this slighting was the overwhelming evidence for the presence of a chromosomal and not a cytoplasmic inheritance in virtually all organisms studied, from viruses to eukaryotes. The Jacob–Monod operon model and the Watson–Crick double helix were nails in the coffin of cytoplasmic inheritance. Unless the plasmagene adherents find more compelling evidence for a new category of hereditary functions, chromosomal geneticists will continue to ignore claims for a substantial role of plasmagenes in cellular heredity.

[10] Sonneborn T.M. 1943. Genes and cytoplasm. I. The determination and inheritance of the killer character in variety 4 of *P. aurelia*. II. The bearing of determination and inheritance of characters in *P. aurelia* on problems of cytoplasmic inheritance, pneumococcus transformations, mutations, and development. *Proceedings of the National Academy of Sciences* **29**: 329–343.

[11] Carlson E. 1981. *Genes Radiation, and Society: The Life and Work of H.J. Muller, pp. 284–285*. Cornell University Press, Ithaca.

[12] During one of our meetings in the 1970s, Dorothea (Thea) Muller gave me the background on the strain between Altenburg and Sonneborn.

[13] Baker J.A. 1952. *Trembley: Scientist and Philosopher 1710–1784*. Edward Arnold and Company, London. See also Lenhoff S. and Lenhoff H. 1986. *Hydra and the Birth of Experimental Biology—1744. Abraham Trembley's Memoirs Concerning the Natural History of a Freshwater Polyp with Arms Shaped Like Horns.* Boxwood, Pacific Grove, California.

The Lenhoffs' book includes an English translation of Trembley's major works on hydra.

[14] Altenburg E. 1946. The symbiont theory in explanation of the apparent cytoplasmic inheritance in *Paramecium. American Naturalist* **80**: 661–662.

Sonneborn, of course, knew that *P. bursaria* had endosymbionts. He discounted that possibility for the killer factor because no such bodies were present in *P. aurelia*. What Sonneborn did not know was that they were much smaller than zoochlorellae. Altenburg had no experimental evidence to show such endosymbionts; it was just speculation (at the time) and that annoyed Sonneborn. Sonneborn could not imagine that after finding mating types and elevating cytoplasmic inheritance to legitimacy that the killer trait would turn out to be just another case of endosymbiosis. This is a classic case of a scientist wedded to a belief and going down with it, as was true for Bateson, Castle, and other very fine scientists. Altenburg was never as gifted an experimentalist as Sonneborn, nor as gifted a thinker. In science, it is not how famous, talented, or bright you are that determines who is right; it is the determination of the facts.

[15] l'Héritier P. 1948. Sensitivity to CO_2 in *Drosophila*— Review. *Heredity* **2**: 325–348.

Ironically, carbon dioxide sensitivity also turned out to be endosymbiotic or parasitic. The sensitive strains harbor an intracellular microbe.

[16] Preer J. 1948. The killer cytoplasmic factor, its rate of reproduction, the number of particles per cell, and its size. *American Naturalist* **82**: 35–42. Sonneborn's acceptance of kappa as endosymbiotic bacteria is in Sonneborn T.M. 1959. Kappa and related particles in *Paramecium. Advances in Virus Research* **6**: 229–356.

[17] Carlson E.A. 1974. An unacknowledged founding of molecular biology: H.J. Muller's contributions to gene theory, 1910–1936. *Journal of the History of Biology* **4:** 149–170.

[18] Timoféeff-Ressovsky N.V., Zimmer K.G., and Delbrück M. 1935. Über die natur der genmutationen und der genstruktur. *Nachrichten die Gesellschaft Wissenschaften Göttingen* **1:** 234–241.

[19] Schrödinger E. 1945. *What is Life?* Cambridge University Press, United Kingdom.

[20] Caspersson T. and Schultz J. 1950. Cytochemical measurements in the study of the gene, pp. 155–172. In Dunn L.C., ed. *Genetics in the 20th Century.* Macmillan, New York.

[21] Levene P. 1919. The structure of yeast nucleic acid. *Journal of Biological Chemistry* **40:** 415–424; and Levene P. 1921. On the structure of thymus nucleic acid and on its possible bearing on the structure of plant nucleic acid. *Journal of Biochemistry* **48:** 119–125.

   In the 1920s, DNA was known as thymus nucleic acid and RNA was known as yeast nucleic acid. Because yeast was a plant and the thymus was derived from cattle, it was mistakenly believed that plants had RNA and animals had DNA. That was a minor confusion compared to the pessimistic interpretation of Levene's papers, which demonstrated that the base composition of the nucleic acids was roughly a 1A:1G:1C:1T ratio (the "tetranucleotide theory," as it was named) and thus, too simple a crystal to be of genetic worth.

[22] Beadle G.W. and Ephrussi B. 1936. The differentiation of eye pigments in *Drosophila* studied by transplantations. *Genetics* **21:** 225–247.

[23] Beadle G.W. and Tatum E. 1941. Genetic control of biochemical reactions in *Neurospora. Proceedings of the National Academy of Sciences* **27:** 499–506.

[24] Muller H.J. 1947. The Gene. Pilgrim Trust Lecture, 1945. *Proceedings of the Royal Society, B* **134:** 1–37. Muller had read the Rockefeller group's important paper on bacterial transformation: Avery O.T., MacLeod C.M., and McCarty M. 1944. Studies on the chemical transformation of pneumococcal types. *Journal of Experimental Medicine* **79:** 137–158.

[25] Cairns J., Stent G.S., and Watson J.D. 1966. *Phage and the Origins of Molecular Biology.* Cold Spring Harbor Laboratory of Quantitative Biology, Cold Spring Harbor, New York. See also the isotope labeling study of Hershey A.D. and Chase M. 1952. Independent functions of viral protein and nucleic acid in growth of bacteriophage. *Journal of General Physiology* **36:** 39–56.

[26] Lederberg J. 1949. Problems in microbial genetics. *Heredity* **2:** 145–198.

[27] Twort F. 1915. An investigation on the nature of the ultramicroscopic viruses. *Lancet* **II:** 1241; and d'Hérelle F. 1917. Sur un microbe invisible antagoniste des bacilles dysentériques. *Comptes Rendus de l'Academie des Sciences* **165:** 373.

[28] Benzer S. 1955. Fine structure of a genetic region in bacteriophage. *Proceedings of the National Academy of Sciences* **41:** 344–354.

[29] Jacob F. and Monod J. 1961. Genetic regulatory mechanisms in the synthesis of proteins. *Journal of Molecular Biology* **3:** 318–356.

[30] Watson J.D. and Crick F.H.C. 1953. Genetical implications of the structure of deoxyribonucleic acid. *Nature* **171:** 964.

Classical Genetics in the Service of Politics

Do male eugenists suffer from the illusion that intelligent women love to be pregnant, and to endure not only the physical disabilities but also the shame and humiliation, and the difficulties of maintaining a job, that pregnancy involves in our society? That they love the frightful ordeal of childbirth, so seldom relieved by competent medical treatment? That they love to spend forty or fifty thousand hours washing diapers, getting up in the night, tending colic, stewing soups and milks, meeting in a city flat their little savage's requirements for safe outdoor activity and companionship, acting as a household drudge, and either abstaining from the life of the outside world entirely or else staggering under the double burden of a very inferior position outside and work in the home as well?

H.J. MULLER*

MOST SCIENTIFIC DISPUTES in the life sciences involve one group of scientists at odds with another group, each interpreting the data in different ways. These disputes are usually settled by new experiments, use of new organisms, new technologies that offer more insights into older problems, and occasionally a new theory that makes older interpretations seem limited. The conflict between Castle's group and Morgan's group on several issues falls into this category. A few of the genetic disputes involve political or ideological issues; I discuss here three such episodes that were mired in politics. The first, which roughly spanned the years 1900–1945, attempted to apply a still nascent field of genetics to an idealistic improvement of humanity. That effort was called eugenics. The second dispute, called the Lysenko controversy involved an attack on the credibility of classical genetics. It arose in the Soviet Union and lasted from the mid-1930s to the mid-1960s. The third controversy was not so much ideological in its premises, but showed how various groups of sincere and accomplished scientists differed in assessing the genetic effects of radiation on humans. We call this the radiation controversy and it was most polemic during the years 1946–1970. There is good reason to believe that the scientific integrity of both groups was not at issue.

I include all three of these controversies because classical genetics can be (and was) applied to agriculture, medicine, social problems, and issues of national concern. Frequently, the most prominent

---

*1933. The dominance of economics over eugenics. *Scientific Monthly* **37**: 40–47.

scientists are drawn into these discussions. Classical genetics was heavily engaged not only in these controversies, but in additional ones as well. It is worth noting that the criteria used for resolving disputes in controversies of scientists who disagree on purely scientific issues are very different from that used to resolve scientific issues enmeshed in political disputes.

The Eugenics Movement Has Three Phases

Although classical genetics has brought immense benefit to humanity through its hybrid corn, green revolution, disease-resistant crops, and the production of antibiotics, classical genetics has also brought misery to people victimized by individuals and their policies who depicted some classes or categories of human beings as unworthy of reproduction. The movement that we properly condemn is called the eugenics movement but there were actually three largely independent phases. I categorize the first eugenics movement as the *attempt to define unfit people and limit their presence in society.* This movement goes back to antiquity but in its more biological form, it dates to the 1700s, occurring long before the Darwinian and Mendelian revolutions.[1] I categorize the second eugenics movement as the *desire to improve humanity by breeding the most talented and healthiest people.* This movement originated with Francis Galton and his studies of hereditary genius in the 1860s (Figure 1) and was inspired by Darwinism and a faith in progress.[2] The third eugenics movement began with the introduction of Mendelism, and it was quickly taken over in the United States by advocates of the first eugenics movement. I classify this as the *American eugenics movement.* It attempted to apply Mendelism to human social problems.[3]

Figure 1. *Francis Galton initiated several fields including twin studies, population genetics, eugenics, and correlation analysis. He disproved Darwin's theory of pangenesis and encouraged both the biometric school and Bateson's Mendelian school regarding the future of genetics.*

Defining Unfit People

The history of unfit people is a long one. In classical Greek mythology, the unfit included condemned families such as those in the House of Atreas, whose members (Antigone, Oedipus, and Creon) were believed to be doomed to tragic lives because of a fixed curse that could not be reversed. This resembles biblical transgressions of a similar era in which groups such as the Amelekites were condemned in perpetuity for sins against God. Biological theories of degeneracy (the root cause of being unfit) date to the early 1700s, when masturbation was the first environmental activity that was associated with physical and mental deterioration of the onanists (an earlier name for those who masturbate) and with puny or defective offspring. In the 1870s such degenerate individuals were grouped into classes, the most common of which included paupers, vagrants, the psychotic, and the mentally slow (then called feebleminded). Alleged families of degenerates, such as the Jukes in the Hudson River valley of New York and the Tribe of Ishmael in the Midwest, were held as prototypes of the tendency of a few unfit individuals to spread like weeds through a population.[4]

Before Mendelism, there was no pedigree analysis and the genetic evidence for unfit people was indirect. The guiding principles were not those of classical genetics. They involved the archaic folk belief that "like breeds like" and hence, degenerates produced degenerates, paupers produced paupers, and feebleminded people produced feebleminded offspring. This simplistic model permitted both identification and a series of attempted solutions throughout the nineteenth century. Those who accepted the innate nature of degeneracy (a transformation associated with Weismann's theory of the germ plasm in the 1880s) argued that isolation in mental institutions, exile to less populated territories, or long prison terms were the only protections society had. A few believed that simply passing restrictive marriage laws would prevent these groups from attempting to raise their families among decent people. Others believed that a shunning of the unfit through education would keep them from contaminating the otherwise decent stock of Americans. This branch of eugenics eventually came to be called negative eugenics.

Breeding the Best

Galton's model was quite different. He named the field of eugenics in 1883 and became interested in it because he believed humans had the intelligence and capacity to take evolution into their own hands. He advocated that the successful in life, especially those achieving eminence, have a moral duty to produce more than an average number of children. He also advocated that such successful families should include eugenic considerations in marriage; he most admired the German professors who married the daughters of other German professors.

Galton's books and articles tried to put this eugenics movement on a scientific basis, that it was not strictly based on a like-for-like inheritance. He studied dictionaries of biography and obituary notices of the eminent. He concluded that eminence occurs about once in every 4000 births. But among the offspring or parents of an eminent person, there was about a 20% probability that another eminent member would be found. To provide a control, Galton used the offspring of nepotically raised children in the households of bishops and cardinals.[5] He found that individuals raised in stimulating environments with opportunities for education did not produce eminent children more often than the 1 in 4000 frequency for ordinary people. Hence, Galton concluded, eminence is moderately inherited. But 1 in 5 odds are immensely better than 1 in 4000, and it was this alleged scientific advantage that motivated Galton to propose his eugenics movement to improve humanity. His thesis also rested on what can be called the heroic model of human civilization. In this model, culture is a product of a minute number of unusually gifted people whose names become household words—Pasteur, Napoleon, Lincoln, Newton, Darwin, and Mendel. This movement eventually came to be called positive eugenics.

American Eugenics Movement

Just as Americans led the way in fusing breeding analysis with cytology to establish the chromosome theory of heredity, they led the way in fusing breeding analysis with eugenics. Very shortly after Mendelism came to America, the American Breeder's Association (ABA) came into being in 1903 and it soon overwhelmingly adopted a third group that desired to apply Mendelism to human affairs, just as they had done to plant and animal breeding.[6] The key figures in its establishment were not Mendelians, but they supported the effort to introduce that new approach to the negative eugenics movement with which they were familiar. David Starr Jordan (Figure 2), a prominent educator and evolutionist, had studied and promoted the Tribe of Ishmael, a miscegenated fugitive tribe in the Midwest, as an example of "hereditary parasitism" in humans.[7] Alexander Graham Bell, best known for his contributions leading to the invention of the telephone, was long interested in deaf people and feared that their isolation through sign language would result in a "race of deaf people."[8] He also carried out breeding experiments on supernumerary breasts in sheep. The ABA picked a younger person as the group's leader. He had recently shifted to natural history from engineering and was asked to organize and seek funding for a formal American eugenics program. That young man was Charles Davenport.[9] Davenport was an excellent fund-raiser and manager. He established three institutions on the grounds of Cold Spring Harbor Laboratory on Long Island in New York. One activity, funded by private benefactors, many of them living in the Cold Spring Harbor area, was the teaching of science to teachers, especially experimental methods and the new genetics coming into being. The second, funded by the Carnegie Institution of Washington, was a center of experimental evolution, a basic science approach to using Mendelism, cytology, and evolu-

Figure 2. *David Starr Jordan's career bridged the late nineteenth and early twentieth centuries. He was an ichthyologist and evolutionist but also took an interest in eugenics and promoted it as a science through numerous popular books. (Courtesy of the National Library of Medicine.)*

Figure 3. *One of the three roles of the Cold Spring Harbor complex was the study of eugenics. The Eugenics Record Office had its own building and staff and combined research, field studies, lobbying, and education in a somewhat unsophisticated approach laden with hidden biases. Davenport is shown in the front row, fourth from the left. Laughlin is first on the left in the second row. (Courtesy of the Cold Spring Harbor Laboratory Archives.)*

tionary studies to solve common biological issues. The third, funded with money donated by the Harriman family, was the Eugenics Record Office (Figures 3 and 4). This would apply Mendelism and new findings in genetics to human genetic traits, most of them social traits associated with success or failure in American society. Like the European model of the unfit, the American eugenics movement came to be known as negative eugenics. Both movements shared a common belief that the unfit needed to be contained and they owed their degenerate traits to innate factors. Davenport, Laughlin, and other enthusiasts for the American eugenics movement were convinced that most of the social failure in America was a result of bad germ plasm and that negative Mendelian traits could be weeded out by judicious breeding programs.

Figure 4. *Charles Davenport ran the Cold Spring Harbor complex and assigned Harry Laughlin to the management of the Eugenics Record Office. They were the chief architects of the American eugenics movement, which stressed compulsory sterilization laws and restrictive immigration laws as responses to the alleged increases in genetically unfit people. (Courtesy of the Cold Spring Harbor Laboratory Archives.)*

Compulsory Sterilization and Restrictive Immigration
Become Missions of American Eugenics

One such program was initiated just before Mendelism was rediscovered. In 1899, a highly respected Chicago physician, Albert Ochsner, proposed sterilization of degenerates by vasectomy as a humane procedure that would only cut off defective germ plasm and not interfere with sexual enjoyment or the physiology of reproduction.[10] In that same year, his suggestion was put into practice at the Jeffersonville Reformatory in Indiana by Harry Clay Sharp, an idealistic prison physician who had learned of the importance of public health in human affairs from his medical school training in Kentucky. It was Sharp who campaigned for a compulsory sterilization law and got it passed and signed, making Indiana in 1907 the first place in the world to legalize compulsory sterilization of the unfit.[11] Well over 200 prisoners were sterilized by Sharp before a new governor ordered a stop to the practice. At medical meetings, Sharp continued to advocate for sterilization and he had a major hand in getting some 30 states to adopt such laws. These laws were upheld in a test case engineered by the Eugenics Record Office and the State of Virginia in 1927: In the Buck v Bell decision, the Supreme Court upheld the right of the State of Virginia to sterilize its unfit.[12]

A second objective of the Eugenics Record Office and its chief administrator, Harry Laughlin, was to provide "expert witness" support for restrictive immigration laws.[13] The American eugenics movement saw the waves of immigrants from southern and eastern Europe as a threat to the genetic worth of American stock. To limit such contamination, Congress was urged to limit immigration to mostly northern and western Europeans. This was achieved by using the 1896 census as the basis for immigrant quotas. In 1921 and 1925, Congress adopted such laws.

From about 1914 to 1940, the Eugenics Record Office attempted to collect data on the genetic worth of Americans. A staff of trained field workers conducted interviews of families who had socially desired traits such as musical and leadership abilities, engineering skills, or inventing and writing talents. They contrasted such pedigrees with those of feebleminded, psychotic, and pauper families.[14] The basic assumption used by Laughlin and Davenport was one of a Batesonian unit-character model. In almost all instances, they claimed a single-gene basis, and their major disagreements were over dominant or recessive modes of inheritance for these traits. Despite the Mendelian ratios that would allow production of nondegenerate offspring, they could not let go of the strong belief that sterilization of the defective was the best way to remedy bad germ plasm. Laughlin and Davenport did not understand this or they ignored the reproductive implications of the Hardy–Weinberg law or the models of chief genes and modifiers or multiple factor inheritance for quantitative traits in plants and animals. They continued to rely on pedigrees from their field office workers and mental institutions as clear evidence of a single-gene basis for a variety of traits that many sociologists, psychologists, and geneticists regarded as environmental rather than genetic or as genetic tendencies with enormous modification by appropriate environments. It is troubling that Davenport made this error. He had an engineering background and was an early enthusiast for applying statistics to biological problems and research. He was one of the first to demonstrate a quantitative trait in humans (human skin color) and to work out a Mendelian multiple factor model for this trait.[15] His initial objection to sterilization of the unfit was not that it was a bad genetic model, but that it might make men released to society promiscuous, because they would not have to support any children.[16]

Socialists Endorse Eugenics

The eugenics movement was filled with ironies. While positive eugenics owed its origin to the elitist sentiments of a class-based society (Galton belonged to its top leadership), the generation of Muller and his contemporaries took it over as a program appropriate for a socialist society. In this coalition were some major contributors to classical genetics including Muller, Huxley, and Haldane. They argued that genetic worth would be revealed in a classless socialist society where equal opportunities abound. In such a society, the most talented and healthy individuals would be encouraged to have more children, a proper and needed obligation to society. This also fit into their model of a socialist planned society, in which decision making would be based on reason, especially scientific principles. Muller's idealism or naiveté even led him to send a letter to Stalin (with a Russian translation of his book) urging that the USSR adopt his positive eugenics program outlined in his popular nonfiction work *Out of the Night* (1935).[17] Stalin, a former seminarian, was hardly interested in a genetic approach to social change and Muller was lucky to escape from the USSR before being arrested and executed, as were his students Solomon Agol and Isador Levitt. Muller used the ruse of enlisting to fight in the Spanish Civil War for his "safe exit." Although Muller's faith in the socialist revolution was severely shaken, his faith in positive eugenics remained intact. He was never a fan of negative eugenics, and in 1932, at the third and last International Congress of Eugenics, Muller denounced that movement in his talk, "The dominance of economics over eugenics," whose title revealed his socialist slant (Figure 5).[18] Muller defended virtually every intended victim of the negative eugenics movement with genetic, economic, and psychological analysis, showing that the movement produced no evidence of an innate basis for the plight of neglected and abused populations.

> ### THE DOMINANCE OF ECONOMICS OVER EUGENICS
>
> #### H. J. MULLER
> *University of Texas*
>
> It is now about fifty years since Francis Galton promulgated the doctrine of Eugenics. It has become a highly popular subject for parlor talk and best sellers. Yet, aside from the sterilization of imbeciles, we are today further than ever from putting eugenic principles into actual operation.
>
> That imbeciles should be sterilized is of course unquestionable, but we should not delude ourselves concerning the importance of the benefits thereof. Following Haldane, we may recall the fact that if (as is commonly claimed) most imbecility is due to the same recessive gene, then the sterilization of all imbeciles in every generation would not reduce their number to half until about ten generations had elapsed, and subsequent elimination would be even slower. And after all, actual imbecility represents only a very small part of the hereditary weaknesses which should be eliminated, and this particular defect is not as onerous as many others, firstly because imbeciles do not suffer from the consciousness of their own defect, and secondly because it is not inhumane to segregate them into institutions. Here they constitute much less of an economic and psychological burden on their fellow men than they would in the community at large, where most lesser defectives must remain.

Figure 5. *Muller's leftist ideology. In his speech to the Third International Congress of Eugenics, Muller denounced the American eugenics movement for its stereotypes and bias. Shown here is the title page of that speech, which Davenport and his colleagues tried to bar. (Reprinted from Muller H.J. 1934. The dominance of economics over eugenics. In A Decade of Progress in Eugenics. Williams and Wilkins, Baltimore.)*

Nazi Ideology Makes Eugenics a Repulsive Concept

The American eugenics movement was met with hardy approval from Nazi Germany. It fit their racist ideology in which a mythical Aryan ancestry was accorded the status of positive eugenics; an equally mythic belief in the innate degeneracy of Jews was part of their negative eugenics program. Nazi Germany greatly extended and applied compulsory sterilization to its defective Aryan stock and, as soon as World War II began, it used that opportunity to gradually deport Jews and kill them in increasingly larger numbers, especially after the infamous Wannsee Conference in early 1942, which revealed the "final solution" as approved by Hitler, Goering, and Himmler.[19] The high regard by Nazi supporters for the contributions of the Eugenics Record Office

to the formation of the Nuremberg laws was acknowledged with an award of an honorary doctorate degree to Harry Laughlin in 1937 by the University of Heidelberg. An important distinction between the American eugenics movement (led by Davenport and Laughlin) and the Nazi eugenics movement (run by authority of the Nazi party) was the attempt, inadequate and distorted as it was, to rationalize eugenics through classical genetics in the American movement (Figure 6). No such presentation of alleged Mendelian traits or other classical genetic tools was used in the race hygiene movement of Nazi eugenics to identify the components of alleged racial inferiority, hereditary degeneracy, parasitism, or antisocial behavior in the victims of Nazi ideology.

a

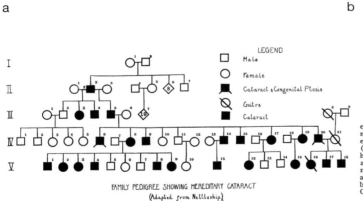

FAMILY PEDIGREE SHOWING HEREDITARY CATARACT
(Adapted from Nettleship)

b

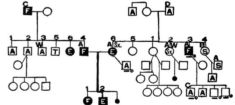

Fig. 6. This chart shows the product of a feeble-minded man and an epileptic, opium-eating, unchaste woman. The father's father was feeble-minded and a "criminal" and, besides the man in question, he had an epileptic son, and three alcoholics, of whom one had the vagrant tendency (W). The mother's germ-plasm does not show up much better, for she has a feeble-minded and alcoholic brother, who lives at the almshouse, an alcoholic sister who is a prostitute and a vagrant, and three alcoholic nephews of whom one (C) has been in jail (4). Two children were born alive to this pair 35 odd years ago. The first was feeble-minded and died before she was fourteen, the second is at the State Village for Epileptics. Case 2857.

Figure 6. *"Good" and "bad" pedigrees. The Eugenics Record Office was criticized for its uneven interpretation of human behavioral traits as being Mendelian. Here are examples of legitimate and questionable pedigrees, the first involving hereditary cataracts as a dominant mutation in a family (a) and the second involving the alleged inheritance of feeblemindedness and epilepsy. Note Davenport's innocent use of defamatory language to describe the family (b). (Reprinted from [a] Howe L. 1921. A bibliography of hereditary eye defects. Eugenics Record Office Bulletin No. 21. Cold Spring Harbor, New York. [b] Davenport C.B. and Weeks D.F. 1911. A first study of inheritance in epilepsy. Eugenics Record Office Bulletin No. 4. Cold Spring Harbor, New York. Reprinted from 1911. Journal of Nervous and Mental Disease **38**: 641–670.)*

Eugenics Is Purged after World War II

World War II was the nail in the coffin of the American eugenics movement. After the war, Muller urged purging eugenics from the new field of human genetics that was emerging as the most recent spin-off of classical genetics. Muller served as first president of the American Society for Human Genetics and he advised the editors of its new journal to shun the pedigree charts of social traits and the eugenics propaganda of the American eugenics movement.[20] Instead, he advocated a human genetics based on a search for cytological analysis of human defects, a serious application of population genetics to study gene frequencies, a biochemical attempt to identify genetic disorders, and a mission to search for and reveal the facts and principles of human genetics rather than apply poorly understood findings for eugenic aims.

Lamarckism Inspires
One Wing of Soviet Genetics

The roots of Lysenkoism extend to Russian science before the Bolshevik Revolution. Like biologists in almost all industrial countries, Russians in the early 1900s were split on seeing inheritance as basically Lamarckian or basically Weismannian. Chief among the early Lamarckists was I.V. Michurin (Figure 7), Russia's equivalent kin spirit to Luther Burbank.[21] Like Burbank, Michurin developed new strains of edible fruits and vegetables by careful selection and grafting favorable lines as a means of their propagation if they were infertile. After the Bolshevik victory, he cast his lot with the new Soviet government and became a folk hero for his dedication to improving Soviet agriculture. N.I. Vavilov was on the Weismannian side and was thus an avid Mendelian when the new field of genetics first reached him (Figure 8).[22] An energetic plant geneticist, his main claim to fame in the early 1920s was his theory of centers of origins for domesticated plants. He collected seeds of cereal grain varieties and other crops from around the world, enjoying tagging along on anthropological surveys. Like Michurin, Vavilov was a strong supporter of the Bolshevik cause, and he

Figure 7. *I.V. Michurin was a pomologist who used grafting and selection as his chief tools to develop many new varieties of fruit trees. He embraced Lamarckism and, like Burbank, had a good eye for new recombinants or mutants that arose in his orchards. (Reprinted from Michurin I.V. 1949. Selected works. Foreign Language Publishing House, Moscow.)*

Figure 8. *N.I. Vavilov was an enthusiastic Mendelian who had studied with Bateson. He established field stations modeled after the U.S. agricultural program and amassed a huge collection of seeds from around the world. Ousted by Lysenko, he died in a Siberian prison camp. (Reprinted, with permission, from Dobzhansky T. 1947. N.I. Vavilov, A martyr of genetics.* Journal of Heredity **38:** *227–231.)*

embracing Weismann's theory of the germ plasm, which doomed improvements by assigning fixed defective hereditary to the germ plasm. In its place, Lysenko recommended studying Michurin's works and methods based on the theory that heredity could be trained by the environment. To justify that this was not based on theory alone, Lysenko claimed that he and his students had "shattered" the heredity of cereal grains by cold shock (freezing and thawing seeds and seedlings) in a process called vernalization. From these shattered strains, the proper environment could transform spring wheat into winter wheat or rye into oats. Lysenko argued that Michurinism was progressive and that "Mendelism-Morganism-Weismannism" was reactionary. The debates extended from newspapers and journal articles and editorials to public debates at the agricultural societies. Muller, then a guest investigator at a

rose in power to become the equivalent of Secretary of Agriculture (he was head of the Lenin All-Union Academy of Agricultural Sciences—the LAAAS). Vavilov greatly admired the U.S. agricultural stations and he used this model to establish agricultural research stations in the USSR. He hoped that his collection of thousands of seed varieties could be tested out on a pilot scale with considerable crossing and selection for strains that would withstand the short growing season, harsh weather conditions, and relatively arid climate characteristic of most Soviet agriculture.

In 1935, an agronomist from Odessa, Trofim D. Lysenko, began a series of attacks on classical genetics (Figure 9).[23] He argued that classical genetics was bad science: It was constructed by a monk, required a very lengthy time to produce beneficial changes or combinations, lent its support to Fascist and racist doctrines of human inferiority directed at oppressed classes, and sent a pall of pessimism by

Figure 9. *In a bizarre episode of science history, Trofim Denisovitch Lysenko used politics to launch a career in plant breeding based on Lamarckism and a strange concept of "shattering" and retraining heredity in plants. The Lysenko controversy began in the mid-1930s and culminated during the Cold War in the mid-1960s, when Lysenko's methods were revealed to be fraudulent or based on self-deception. (Reprinted from Medvedev Z.A. 1969.* The Rise and Fall of T.D. Lysenko. *Columbia University Press, New York.)*

genetic institute in Moscow, defended Vavilov and classical genetics at the December 1936 LAAAS meeting, where he denounced Lysenko as a shamanist and fraud. Muller had to leave soon after this.[24] World War II interrupted the debate, but it was too late for Vavilov, who was arrested in 1939 as a British spy after the signing of the Hitler–Stalin pact, which initiated the Nazi invasion of Poland (he had worked as a postdoctoral student in Bateson's John Innes Institute before World War I) .

Lysenko Climbs to Domination of Soviet Genetics, 1935–1960

After the war, Lysenko's supporters began gaining strength and held a showdown LAAAS meeting in 1948, at which Lysenko announced that his views of Michurinism had been endorsed by the Central Committee of the Communist party.[25] Those charged words had an electrifying effect on the audience of Lysenko's supporters and detractors. One by one, the major detractors came forward, confessed their errors, and promised to cleanse their lectures of references to Mendelism and to champion the new Soviet science of Michurinism. The debate spilled over to the West. French communists strongly supported Lysenkoism, and a homegrown Lamarckian series of papers began to appear on environmentally transformed ducks. Mendelism was clearly favored in Great Britain and the United States, but it forced many geneticists to stop their political support of the USSR. Muller denounced Lysenkoism in his presidential address to the 1948 International Congress of Genetics in Stockholm, and the Eastern bloc delegates walked out and left for home.[26]

Lysenko eventually defeated himself. He was self-deceived and used questionnaires to obtain evidence of successful yields from his vernalized strains. Those who reported their actual yields were punished as saboteurs, and glowing reports from collective farms supporting vernalized crops very quickly poured in. To do this, the farmers had to tighten their belts, take from their own share to give to the state, and make good on the falsified yields they reported. By the early 1960s, the system collapsed. Too many people reported to their party representatives what they had to do to survive. Lysenko was abruptly dropped from favor and classical genetics was ordered back, with leading newspapers being asked to contact once-condemned geneticists to write articles that brought the public up to date on advances in genetics since the 1930s.[27]

The Radiation Controversy Is a Product of the Cold War

Lysenkoism as a bona fide science was never a major threat to classical genetics in the western scientific world but outside the Soviet bloc, it was a handy tool of propaganda that could be used to describe the repression of intellectual freedom behind the Iron Curtain. The Cold War involved classical genetics in the West in a quite different way. At issue was the development of military (and to a lesser extent, industrial) uses of atomic energy. The hastening of the end of the war in Japan in 1945 with the detonation of two atomic bombs initiated an arms race that would last for more than 30 years.

Atomic bombs kill by three methods. The immediate consequence is a very powerful incinerating blast that can level cities and extinguish virtually all life within a radius of a mile, as had occurrred in the cases of Hiroshima and Nagasaki. Second, over the next few weeks or months, immense numbers of survivors (located up to two miles away in radius from where the bomb exploded) can come down with what is called radiation sickness. Depending on the dose, some will be at high risk of death and others will recover. The third cause of death was of particular concern to geneticists, moralists, and the general public. The mutations induced by radiation in this third category could lead to increased risk of earlier onset of cancers (using the multihit model of tumor cell formation) and to future generations receiving the induced recessive mutations in their reproductive tissue. These mutations would continue to be present for dozens of generations. The 1922 Danforth equations and Muller's later modification and use of them in his famous essay, "Our load of mutations," had a direct role in this assessment. From this genetic point of view, atomic weapons would punish innocent generations to come, until the induced detrimental mutations (i.e., most mutations induced by radiation) were milked out by natural selection.[28]

The interpretation of radiation sickness also depended on classical genetics. In 1938, Barbara McClintock worked out the breakage-fusion-bridge cycle in maize to interpret unusual sectoring and spotting patterns in maize kernels (Figure 10). In 1940, Muller and Pontecorvo worked out a comparable finding in fruit flies to account for dominant lethal formation induced by ionizing radiation.[29] Muller and Pontecorvo demonstrated that dominant lethals were aborted embryos caused by the induction of dicentric

chromosomes in irradiated sperm that fertilized normal eggs and prevented cell division or led to such profound distress of the damaged cells that they would die. In humans, most adult tissues contain cells that do not divide or do so very rarely. When applied to adult human dividing tissues such as skin, the lining of blood vessels or endothelial tissue, bone marrow, and the lining of the guts, the skin reddens and blisters and has difficulty healing; the capillaries rupture, causing internal and surface bleeding into intercellular spaces; the blood becomes anemic and loses the capacity to block infections; and the intestinal tract bleeds and perforates. Muller associated these symptoms of radiation sickness after the Hiroshima disaster with the

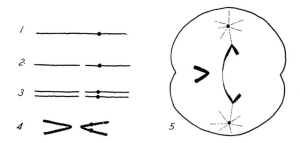

Figure 10. *The mechanism of radiation-induced cell death and Muller's rendition of the breakage-fusion-bridge cycle in relation to radiation necrosis (later called radiation sickness). (1) The intact chromosome has one centromere, (2) a chromosome break forms two fragments, (3) the two fragments replicate, and (4) each fragment is joined to its sister fragment. The left fragment, lacking a centromere, is called acentric; the right fragment contains two centromeres and is called dicentric. In (5), the acentric fragment cannot go to either pole. The dicentric fragment forms a chromosome bridge that will break and repeat the cycle. McClintock named this process, which led to spotting and streaking of corn kernels, but in animal cells, the effects are more deadly, leading to cell death. The finding that Muller and Pontecorvo associated with dominant lethals in fruit flies was extended by Muller to human radiation sickness. (Reprinted, with permission, from Muller H.J. 1950. Radiation damage to the genetic material. Part II. American Scientist **38:** 1–56.)*

induction of aneucentric chromosomes in dividing tissues.[30] The atomic bomb was not only a military weapon intended to cause great physical damage to its target area, but also a genetic weapon, the first devised by humanity, whether consciously or not.

Physicists tried to organize other scientists to stop the development of new atomic weapons and the proliferation of nuclear weapons, especially as the USSR developed its own atomic weapons program and was soon testing larger and more powerful bombs. Adding to the controversy was the unexpected consequence of testing the first U.S. hydrogen bombs in the far Pacific Ocean. Japanese fishermen about 75 or more miles away became coated with the fallout of the explosion, contracted radiation sickness, and several died. The United States was engaged in what it believed was a life-or-death struggle with the USSR. It tried to keep the effects of fallout secret with a deliberate policy of misinformation to the press. This included denial of fallout, denial of radiation sickness for the Japanese fishermen who died, and denial that low doses of radiation had any effects on the world's population.[31] The policy also included accusing critics of being communist sympathizers, dupes, or double agents, and it tried to discredit the findings of geneticists by alleging that the experiments were poorly done or not applicable to humans. One consequence of these tactics was that Muller was simultaneously attacked both by the USSR as a double agent sent by the United States to wreck Soviet agriculture and by the United States as a double agent sent back from the USSR to sabotage the development of the hydrogen bomb.

The reality was more complex than the propaganda provided by both sides. The U.S. government favored the view that low doses from worldwide fallout were considered negligible, unproven, or offset by the evolutionary benefits of occasional useful (or even heterozygous) mutations. Willard Libby and Edward Teller championed that interpretation. Although low doses produce very low risks to those receiving them,

these low rates must be multiplied on a world level for global fallout. This can amount to several thousands of leukemias and induced germ-line mutations per year of exposure.[32] This was a view championed by Linus Pauling (Figure 11), who led the worldwide protest against nuclear weapons testing in the atmosphere. Muller argued that this fallout-induced frequency was minuscule compared to the spontaneous incidence of these mutations and accounted for a tiny portion of all leukemias and germ-line mutations. Muller tried to generate a third position favoring testing based on political necessity, recognizing the small risks involved (rather than denying them), and seeking a political solution to weapons testing based on their acknowledged genetic harm if used in war. Geneticists who

Figure 11. *During the Cold War radioactive fallout became a political as well as genetic issue. To Linus Pauling (above), small but significant amounts of deaths and illness resulted from worldwide fallout. To Edward Teller, no evidence of such mortality or morbidity existed and he believed that any that might be invoked was offset by evolutionary benefits of new mutations. (Courtesy of the California Institute of Technology Archives.)*

raised these questions in classes or tried to respond to questions from their students had great difficulty assessing risk without specifying their own values. Most people would elect to have their (and their children's) teeth and chests X rayed for routine diagnostic examinations and take the individual risk because it is low. If somewhat knowledgeable about genetics, every parent of a child with leukemia or a sporadic mutation (such as retinoblastoma) that is life threatening or incapacitating will wonder if this particular case resulted from a medical or occupational risk or from the Cold War testing programs in the atmosphere.

Lessons Are Learned from the Abuses of Classical Genetics

What is called pure science often involves an attempt to wrest from nature an understanding of how things work. Scientists use observation and experimentation as their major approaches to such efforts. Conflicting interpretations of an emerging field are settled through amassing more observations and experiments that limit the number and extent of competing interpretations. Applied science has an additional objective of *utility* that may not be present in pure science. The value of utility is some public good, whether it is for health, commerce, or society. Very frequently in the debates of contending scientific groups involving applied science, neither group looks at its own values or the underlying utility that motivates the group. The scientists in the dispute are often convinced that their science is genuine and that their opponent's science is distorted by a lack of objectivity.

Eugenics, in hindsight, was a naïve belief that, even with an inadequate understanding of human genetics, society could nevertheless begin efforts to reduce alleged bad heredity and promote multiplication of alleged good genetics. The evidence for a genetic basis for alleged bad traits (mostly rooted in social failure) was then, as it is now, shallow and tainted with social bias. Similarly, the evidence for alleged good heredity was based less on genetic analysis than on admiration of successful accomplishments or social standing. At the same time, eugenics makes us uncomfortable because it reminds us (correctly) that humans also evolve and are subject to the same accumulation of mutations as plants and animals. Of real interest to future generations are issues such as the rate of such mutations, the fate of these mutations in the population under conditions of adequate medical attention, the effect of increasing amounts of such mutations rescued by medical attention, and the role of unconscious assortative mating (nonrandom mate choice) in addressing that problem. The hunt for major genetic components of traits that are powerfully influenced by religion, politics, upbringing, and social class may turn out to be spurious.

Both Nazi eugenics (as race hygiene) and Lysenkoism shared a common goal. Each movement sought the state's endorsement to apply its views and each succeeded. We have learned that governmental endorsement is dangerous. Science should remain controversial and it should vigorously debate its efforts to understand how things work and how they are applied. When the state becomes the umpire, the winning side owes its success less to scientific merit than to chance and politics. What is surprising in both cases is how effective totalitarian systems are in forcing science into conformity to ideology. What is hopeful in both cases is the rebirth of the science that was repressed within a few years after those movements collapsed. Classical genetics flourishes today in both Germany and the USSR without an official endorsement from the state.

End Notes and References

1. Carlson E.A. 2001. *The Unfit: A History of a Bad Idea*. Cold Spring Harbor Laboratory Press, New York. For a view of the eugenics movement in the twentieth century, see Kevles D. 1985. *In the Name of Eugenics: Genetics and the Uses of Human Heredity*. Knopf, New York.

2. Darwin was not a eugenicist and steered away from applications of natural selection to social problems. Curiously, Karl Marx, Herbert Spencer, and Francis Galton all drew inspiration from Darwin's ideas and imposed a notion of progress on it. Galton's books and articles on hereditary genius led to his own involvement with (and naming of) the eugenics movement. Galton's most cited work was 1869. *Hereditary Genius*. Macmillan, London.

3. The fallacy that vitiated all three eugenics movements was the belief that human behavioral traits had strong hereditary underpinnings. I do not argue that that is wrong; I argue that it has not been proven in a satisfactory way. The evidence is indirect and based on analogy, correlation studies, and frequently flawed twin studies. The genetics that is convincing in classical genetics has been experimental, with evidence for specific gene ties to specific traits. Even in animal genetics with mice, Mendelian behavioral traits are limited to pathologies such as waltzing. At best, learning traits in mice are quantitative. The eugenics movement too often attempted to identify alleged single-gene instances of retardation, psychosis, and various social conditions.

4. Carlson E.A. 2001. op. cit. Chapter 10, The Jukes and the Tribe of Ishmael.

5. Galton F. 1869. op. cit. p. 42.

6. Carlson E.A. 2001. op. cit. pp. 192–193.

7. Jordan D.S. 1898. *Footnotes to Evolution: A Series of Popular Addresses on the Evolution of Life*. D. Appleton and Company, New York.

8. Bell A.G. 1912. Sheep-breeding experiments on Beinn Breagh. *Science* **36:** 378–384.

9. MacDowell E.C. 1946. Charles Benedict Davenport 1866–1944: A study of conflicting influences. *Bios* **17:** 903–923.

10. Ochsner A.J. 1899. Surgical treatment of habitual criminals. *Journal of the American Medical Association* **32:** 867–868.

11. Sharp H.C. 1902. The severing of the vasa deferentia and its relation to the neuropsychopathic constitution. *New York Medical Journal* **75:** 411–414.

12. Lombardo P. 1985. Three generations, no imbeciles: New light on Buck v Bell. *New York Law Review* **60:** 31–52.

    This is an outstanding paper on the way in which law and science were misused in this landmark case, often cited as the Supreme Court's worst decision.

13. Hassencahl F.J. 1971. Harry H. Laughlin, expert eugenics agent for the House Committee on Immigration and Naturalization, 1921 to 1931. Ph.D. dissertation. Case Western Reserve University. Available from University Microfilms International, Ann Arbor, Michigan.

14. The field office amassed thousands of pedigrees mostly from middle-class professionals. They also had a smaller number of pedigrees from field interviews with circus freaks and from records provided by orphanages.

15. Davenport C.B. 1913. *Heredity of Skin Color in Negro–White Crosses*. Carnegie Institution of Washington Publication No. 188, 106 pages.

16. Although males were more frequently sterilized than females from 1909 to 1929, females in asylums were more frequently sterilized after the Great Depression because they could work without fear of getting pregnant. Davenport did not raise the question of promiscuity or prostitution for such women.

17. Muller H.J. 1936. *Out of the Night*. Vanguard, New York.

18. Muller H.J. 1932. The dominance of economics over eugenics. *Proceedings of the Third International Congress of Eugenics, New York 1932*. (*A Decade of Progress in Eugenics*.) Williams and Wilkins, Baltimore.

19. Arendt H. 1976. *Eichmann in Jerusalem: A Study of the Banality of Evil*. Penguin, New York.

20. Muller H.J. 1949. Progress and prospects in human genetics (preface). *American Journal of Human Genetics* **1:** 1–18.

[21] Michurin I.V. 1949. *Selected Works*. Foreign Language Publishing House, Moscow. For a scholarly appraisal of the Lysenko affair, see Joravsky D. 1961. *Soviet Marxism and Natural Science, 1917–1932*. Routledge and Kegan Paul, London.

   Joravsky stresses the agricultural debate and tends to avoid unpublished accounts of the alleged fraud and application of Lysenkoism to human biology.

[22] Vavilov was a friend of Muller's and had invited Muller to the USSR in 1933. Vavilov's career is described in Joravsky D. 1961. op. cit.

[23] Medvedev Z. 1969. *The Rise and Fall of T.D. Lysenko*. Columbia University, New York.

   Medvedev provides an insider's account of crimes, bullying, self-deception, and the bad science involved in Lysenko's confrontation with Vavilov and the Soviet classical geneticists.

[24] Carlson E. 1981. *Genes, Radiation, and Society: The Life and Work of H.J. Muller*. Chapter 17, A safe exit through the International Brigade. Cornell University, Ithaca.

[25] Anonymous. 1948. *Proceedings of the Lenin All-Union Academy of Agricultural Sciences*. Foreign Language Publishing House, Moscow.

[26] Muller H.J. 1949. Genetics in the scheme of things. *Proceedings of the Eighth International Congress of Genetics, 1948, Stockholm. Hereditas* (supplement): 96–127.

[27] Medvedev Z. 1969. op. cit. Lysenko's Agrobiology, Chapter 8; see p. 155.

[28] Muller H.J. 1950 Our loads of mutations. *American Journal of Human Genetics* **2**: 111–176.

[29] McClintock B. 1938. The fusion of broken ends of sister half-chromatids following chromatid breakage at meiotic anaphase. *Research Bulletin of the Missouri Agricultural Research Station No. 290*, 488 pages; and Pontecorvo G. and Muller H.J. 1941. The lethality of dicentric chromosomes in *Drosophila. Genetics* (abstract) **26**: 165.

[30] Muller H.J. 1950. Radiation damage to the genetic material. Part I. Effects manifested mainly in the descendants. Part II. Effects mainly in the exposed individuals. Reprinted with changes in pagination from *American Scientist* **38** (January and July), 58 pages.

[31] At the time (1950s), Muller suspected that the U.S. government had a policy of misrepresentation and he wrote many letters of corrections (mostly unpublished). Most of these abuses were admitted after the Freedom of Information Act in the 1970s and 1980s. When I wrote the Muller biography, I tried to get the FBI file on Muller (over 600 pages), but I was turned down after an initial approval.

[32] For the public, this was very confusing. Linus Pauling would argue for a cessation of atmospheric testing because he calculated the world population share of newly induced mutations from fallout; opposing him, Edward Teller would tell college audiences that a little bit of radiation is good for humanity because it speeds up evolution.

Classical Genetics and Human Genetics

...With chromosome mapping there also goes the possibility of studying such genetically observable chromosome behavior as crossing over, non-disjunction, and structural change. These too must someday become objects of genetic research, even in human beings.

H.J. MULLER*

WHAT WE CALL MEDICAL GENETICS today is largely an outgrowth of genetics since the 1950s. Prior to the end of the second World War, human genetics was enmeshed with eugenics. The relationship was poisonous and the revelations of the Holocaust and the abuses of Nazi genetics in promoting its ideology of race hygiene forced a divorce of the two fields. Out of that dung heap of racism and genocide arose the lotus of human genetics. Medical genetics came later, in the 1960s, originating from human genetics, when human cytogenetics and human biochemical genetics made treatment and diagnosis both a hope and a reality.

Human Genetics Is of Little Medical Interest in the Nineteenth Century

Before the twentieth century, human genetics was largely left to the field of pathology. Babies born with birth defects rarely survived infancy and usually succumbed to infections because they were frail. Diagnosis was inadequate because there was no biochemistry, let alone one that could be applied to medicine. Mendelism was still waiting for its rediscovery to awaken interest in pedigrees and their analysis. Human cytology was considered largely irrelevant to medicine. The little that was recognized as inherited involved trivial traits that survived into adult life such as short and stubby fingers, occasional instances of dwarfs producing dwarfed children, or isolated families with Huntington's disease. If one were to ask about heredity and health in the late nineteenth century, the concerns would overwhelmingly be directed to the incidences of tuberculosis, diabetes, and cancer, which were considered "constitutional" problems because they seemed to run in

---

*1949. Progress and prospects in human genetics. *American Journal of Human Genetics* **1**: 1–18.

families. Many physicians lumped such constitutional defects together with degeneracy theory or they interpreted such families as eugenically disadvantaged. The rare birth defects or sports were of little medical interest because they had no known treatment.

Eugenics Spurs Interest in Medical Genetics

Most of the eugenics movement before 1900 concentrated on behavioral traits of social failures and this was not considered an issue of prime medical interest.[1] Isolated physicians such as A.J. Ochsner and H.C. Sharp may have taken up the eugenics cause, especially after 1900, but most American physicians kept their distance. After Mendelism was introduced, Davenport (and his staff in the United States) and Karl Pearson (and the Galton laboratory in Great Britain) led the way in trying to find medical connections to genetic transmission. Ruggles Gates became an enthusiast for this effort and published a major compendium of known Mendelian traits in humans.[2] Despite Davenport's intentions, he never succeeded in bringing medicine into eugenics in a major way.

There were many reasons for the slow growth of human genetics. The obvious reasons were ethical and practical: Unlike mice, fruit flies, or maize, humans could not be used as experimental subjects for breeding. This meant that human testing had to rely on accidents of marriage and these were abundant enough for alert physicians to detect pedigrees for various human traits. Such observable pedigrees included oculocutaneous albinism and a few exotic disorders of the nervous system afflicting children. But medical schools in this era before antibiotics taught that the major concerns of physicians should be infectious diseases, malnutrition, and accidents. Few physicians thought it worthwhile to spend time studying rare untreatable disorders.[3] Until antibiotics

came into everyday life after World War II, so many normal infants and children died that the majority of physicians believed it silly to look into the causes of those who were abnormal in appearance and who also died. On the practical side, physicians doubted that anything could be learned in three-generation human histories because records were inadequate, people moved, they had relatively few children, and only the rare physician had the opportunity to know three generations of one family when mean life expectancy was much lower than it is today.

Mendelism did have an impact, however, on the first decade of the twentieth century in identifying disorders as following a Mendelian course. Brachydactyly (short stubby fingers and toes), a relatively benign condition, did run in families and followed a Mendelian dominant distribution; the parent with the trait passed it on to half the offspring.[4] Sir Archibald Garrod was impressed by Bateson's papers and applied the findings to some metabolic disorders he was studying (Figure 1). One of these was alkaptonuria, a condition that led to red-colored urine and the discomforts of arthritic pain. In 1902, he suggested that the condition was Mendelian recessive and commonly associated with cousin marriages. Bateson confirmed this and stated that "The facts published by Garrod make it likely that the disease follows recessive lines, for of 17 families in which cases have been seen, 8 were offspring of first cousins."[5] Garrod called these "chemical sports."

Figure 1. *The recognition that human metabolic disorders involve a recessive mutation, one of the foundations of human genetics, originated from studies of Archibald Garrod. He inferred the genetic relationship from an excess of cousin marriages among the parents of his patients with these disorders. (Reprinted, with permission, from 1967.* Genetics: A Periodical Record of Investigations Bearing on Heredity and Variation, *volume 56, frontispiece.)*

Independently, Weinberg demonstrated that consanguinity (inbreeding) was a major factor in recessive disorders. In humans, albinism was different in humans from mice and other animals because it was associated with defects in vision (nystagmus or quivering of the eyes) and numerous types of albinism were already known in human families. Bateson also confirmed the dominant nature of Huntington's disease (then known as "hereditary chorea"): "Adding all the families apparently from the heterozygous

(DR) parents recorded as mating with unaffected persons I get the total 117 affected, 99 unaffected, as nearly approaching the normal equality as we can expect when the nature of the evidence is remembered."[6]

Although Garrod's work, like Mendel's, would fall into a state of neglect for more than three decades, it did generate an interest in the use of pedigrees for studying human traits. Unfortunately, the appearance of a pedigree had an intimidating effect on those who were not geneticists and Davenport's field staff used poor judgment in constructing family histories that allegedly showed dominant inheritance for many social traits. This was also abused in mental facilities for the retarded (feebleminded) who were seen as victims of a single-gene dominant disorder or the unhappy victims of a double dose of the recessive from both parents. There is no doubt that some of these mental retardation cases, especially instances of phenylketonuria, might have been Mendelian, but the cause of such a defect was not worked out until the 1930s. Until then, it was held as a prototype for the way in which a mental defect was inherited. Since children with phenylketonuria look normal but are profoundly retarded, they fit into the model of the time of the low-end grade of feeblemindedness.[7] This was misapplied in the infamous Buck v Bell case that went to the Supreme Court in 1927, and the pedigree convinced the judges, 8-1, that Carrie Buck was part of a three-generation family of imbeciles.

Population Genetics Is Applied to Human Heredity

Mathematician G.H. Hardy in England and the physician W. Weinberg in Germany demonstrated that gene frequency tended to remain constant for Mendelian traits if they were panmictic (randomly breeding) and did not involve a strong selective disadvantage for one of the alleles, and the population

Figure 2. *Ronald Punnett asked the mathematician, G.H. Hardy (right), to estimate the spread of a gene in a population if it arose as a new mutation. He prepared the binomial expansion* $A^2 + 2AB + B^2 = 1$ *that today is called the Hardy–Weinberg law. Hardy considered the effort so trivial (and removed from pure mathematics) that he did not even mention it in his autobiography,* A Mathematician's Apology. *In applying the Hardy–Weinberg expansion, the incidence of* $A + B = 1$ *represents the number of sperm or eggs containing the alleles A and B. If B is the cystic fibrosis gene and A its normal allele, and if we know (or have calculated) that the incidence of the cystic fibrosis allele, B, in the reproductive cells is 1 in 50* $(= 0.02)$*, then A is present in 0.98 of the sperm or eggs. Calculating* A^2 *yields 0.95 of the population as homozygous normal. The number of births that are heterozygous will be or* $2AB$ *or* $(2)(0.2)(0.98) = 0.039$ *or 1 in 25 births, and the number of children born with cystic fibrosis will be* B^2 *or 0.0004 or 1 in 2500 births. The formula allows genetic counselors and population geneticists to calculate all the percentages of homozygotes and heterozygotes in a population given the incidence of a hereditary disorder at birth. (Courtesy of the Masters and Fellows of Trinity College, Cambridge, United Kingdom.)*

was sufficiently large.[8] The familiar equation, $A^2 + 2AB + B^2 = 1$, is the expansion of $(A + B)^2 = 1$, where A is the frequency of one gene in a population and B is the frequency of its allele (see Figure 2). Hardy's was an assessment for the way in which gene frequencies could be calculated in populations. His model laid to rest the eugenic fears that a new mutation would quickly swamp out the normal gene in a population, a fear raised in the 1870s when the Jukes family was first seen as a threat to the hereditary health of their neighbors. Weinberg had worked out the same formula for a very different purpose: to see how constant or novel human medical traits were and how many people in a population were carriers for the disorder in contrast to those who were homozygous for them. Neither Hardy nor Weinberg saw the evolutionary implications of their model because, at the time, the feud between the biometric school and the Mendelian school was based on an

argument about the quality of traits involved in evolution. To the biometric school the traits were continuous and to the Mendelian school they were discontinuous. Neither Bateson nor Pearson and Weldon made the evolutionary connection at that time. To make that connection required the work of R.A. Fisher, J.B.S. Haldane, and S. Wright.

Weinberg showed that mathematics could be applied to human populations to resolve many issues of genetic interest. He applied his reasoning to the incidence of twins and was able to prove that the trait for dizygotic twins was inherited but monozygotic twins did not vary in frequency from population to population. He also showed that programs based on like-for-like models of heredity would be thwarted if the real mechanism were Mendelian and the incidence of carriers was more than 100 times the incidence of individuals showing recessive genetic disorders.

Human Genetics Appears in the 1940s

The field of human genetics was largely of interest to Ph.D.s who looked with interest on the biology of the human species. The field of medical genetics was largely a medical field in which specific medical disorders were of interest to those physicians, especially specialists, who encountered them. Both fields were not quite as well defined as we think of them today. Even in 1941, Laurence H. Snyder, author of one of the first texts in medical genetics, took the view that "Much of human heredity is still to be investigated, and much of the future investigation must be done by those trained in medicine."[9] Snyder was not an M.D.; he held a D.Sc. and he argued that medical genetics would help in diagnosis; provide medicolegal advice in court; help prospective couples make reproductive decisions, including marriage; and "may provide the necessary information for setting up eugenic and euthenic programs for the protection of society, in which every physician should be able to take an intelligent part, based upon experimental data, not opinion, prejudice, and over-exaggerating the uncertainties." Snyder had taught medical school genetics at Ohio State for eight years before being invited to give his lectures at Duke. He covered medicolegal applications (largely blood groups), mental disorders, disorders of the senses, systemic disorders (involving the skin, muscle, and skeleton), blood disorders, susceptibilities to disease, and cancer. He hoped that medical geneticists would someday provide maps of the human chromosomes with hereditary defects assigned to their proper chromosomes, so that effective advice could be given to family members at reproductive risk. This hope was not based on projections from plant and animal genetics. In 1937, J.B.S. Haldane and Julia Bell had begun assigning human X-linked disorders to the X chromosome. Five such X-linked disorders were

mapped at the time of Snyder's lectures as a visiting professor to Duke University medical students.[10]

That advice was provided by another new field, genetic counseling, first proposed and named in 1940 by Sheldon Reed (Figure 3) at the then Dight Institute of Eugenics at the University of Minnesota. Reed argued that the advice should be provided to clients free of charge and without eugenic advice. This would prevent abuse and respect the autonomy of clients to use the information as they saw fit.[11] For a long time, it was a free service, but as medical genetics became more extensive and tests became available to identify those at risk, the field became a profession rather than a sideline hobby of Ph.D.s or M.D.s interested in human and medical genetics. By the early 1980s, the field had national standards for certification and a formal training program (usually a two-year master's degree).

Figure 3. *Sheldon Reed coined the term "genetic counseling" and argued that it should be offered as a medical service for clients seeking information about hereditary illness for which they or members of their family are at risk. He also urged that it be nondirective, respecting the autonomy of the client to use that information as they see fit. (Reprinted, with permission, from Reed S.C. and Anderson V.E. 1972. Genetics of psychoses. Stadler Genetics Symposium 4: 107–124.)*

Human Genetics Becomes Purged of Eugenics in 1948

The abuses of the eugenics movement in the United States and in Europe led to major changes in human and medical genetics. The American Society for Human Genetics was established in 1948, with H.J. Muller as its first president. He urged that its new journal avoid articles on eugenics; emphasize research to determine the principles of human genetics; and begin a search for human cytogenetics, human biochemical genetics, and gene frequencies that might be of use in working out the evolutionary history of human populations.[12] At the time, Muller's own human genetics interests stressed mutation rates in humans; the fate of genes in populations based on their detrimental effects (especially in a heterozygous condition); and the consequences of radiation damage from medical, industrial, and military applications of atomic energy and ionizing radiation.[13]

The shift from articles primarily dealing with human genetics to articles that were largely clinical took place in the 1950s. A major reason for this was the discovery that the human chromosome number could be readily obtained with new techniques introduced by S.H. Tjio and A. Levan, designating the human chromosome number firmly at 46 (from the 1920s until 1955, Painter's estimate of 48 was widely cited in textbooks and articles).[14] In 1958, J. Lejeune and his school in France reported finding a chromo-

SUMMARY

An analysis of somatic-cell karotypes of boys and girls afflicted with mongolism provided a count of 47 chromosomes instead of the normal total of 46. The supernumary chromosome is a small telocentric one that is V-shaped and has small hererochromatic arms (V_h). This identification led to the linking of this chromosomic aberration with a non-disjunction of the elements of the V_h pair during meiosis.

The fertilisation of a diploid gamete for the V_h chromosome by a normal haploid gamete appears to involve the formation of a trisomic zygote for this V_h chromosome. This aberration — which is the first example of human trisomia compatible with life and even with reproduction — accounts for the characteristic peculiarities of mongolism. Since certain non-disjunction mechanisms are, in the case of Drosophila, influenced by the ageing of the female fly, the curious correlation between advanced age of the mother and frequency of mongolism might even be explained by this trisomia.

Figure 4. *In 1959, Lejeune, Turpin, and Gautier worked out the first human nondisjunctional karyotypes. Down syndrome (then called Mongoloid idiocy) was the clinical outcome of an extra chromosome 21 (trisomy 21) present in the cells of a newborn. At the time, chromosomes were not numbered; the number 21 was assigned later. (Reprinted, with permission, from Lejeune J., Turpin R., and Gautier M. 1959. Le mongolisme premier exemple d'aberration autosomique humaine. Annales de Genetique 35: 41–49 [©Elsevier].)*

some number of 47 for Down syndrome (Figure 4).[15] Ten years later, the first prenatal diagnosis of chromosome abnormalities was attempted by amniocentesis. Trisomies 13 (Patau syndrome) and 18 (Edward syndrome) joined trisomy 21 (Down syndrome) as routinely diagnosable.[16]

Human Genetics Contributes to Evolutionary Biology

The evolutionary connection to human genetics was stimulated by A. Allison's observation that a blood disorder found primarily among Africans and people of African descent was associated with equatorial regions of Africa, suggesting a tie to resistance to (or at least survival from) malarial infections. The malignant form of that disease, involving the falciparum form of sporozoite, caused a frequently fatal disease called blackwater fever (associated with the breakdown of blood and sometimes severe damage to the central nervous system). J.V. Neel demonstrated that the genetics of the resistance was associated with the heterozygotes and that sickle-cell anemia, **ss**, was fatal through its anemia and **SS** was fatal through death from malaria, but heterozygotes,

Ss, survived the infections and lived to reproduce (Figure 5). This produced a condition called balanced polymorphism. Selection for the heterozygote led to an abnormally high incidence of the **s** gene in the equatorial populations (up to 40% carry the **s** allele). In countries that had conquered malaria such as the United States, African Americans were

---

The Inheritance of Sickle Cell Anemia[1]

James V. Neel

Heredity Clinic, Laboratory of Vertebrate Biology, University of Michigan, Ann Arbor

IF A DROP OF BLOOD is collected from each member of a randomly assembled series of American Negroes and sealed under a cover slip with vaseline, to be observed at intervals up to 72 hours, in the case of about 8 percent of the individuals composing the series a high proportion of the erythrocytes will be observed to assume various bizarre oat, sickle, or holly leaf shapes. This ability of the erythrocytes to "sickle," as the phenomenon is commonly described, appears to be attended by no pathological consequences in the majority of these individuals, and they are spoken of as having sicklemia, or the sickle cell trait. However, a certain proportion of the individuals who sickle are the victims of a severe, chronic, hemolytic type of anemia known as sickle cell anemia. This proportion has been variously estimated at between 1:1.4 (8) and 1:40 (4). The essential difference between sicklemia and sickle cell anemia appears at present to depend at least in part upon the relative ease with which sickling takes place. In sickle cell anemia the erythrocytes may frequently sickle under the conditions encountered in the circulating blood, whereas in sicklemia sickling does not usually occur under these conditions (12). This difference has been attributed to a greater tendency of the erythrocytes of sickle cell anemia to sickle when the O_2-tension is reduced, although recently this viewpoint has been challenged (13). Perhaps because of this difference—although there may be other factors involved, such as the aniso- and poikilocytosis to be observed in some individuals with the disease, and a greater resistance to hemolysis of trait cells when sickled than sickle cell anemia cells when sickled—the erythrocytes of a patient with sickle cell anemia have a greatly short-

---

Figure 5. *Balanced polymorphism. Working out the genetics of sickle-cell anemia, J.V. Neel found that the gene is maintained in the population by natural selection. Those with normal members, AA, die more frequently of malarial infections. The sickle-cell anemic, aa, dies more frequently of their blood disorder. The heterozygotes, Aa, survive their malarial infections, which run a milder course. This leads to increased incidence of Aa in a population and hence, more marriages of Aa × Aa. (Reprinted, with permission, from Neel J.V. 1949. The inheritance if sickle cell anemia. Science 110: 64–66.)*

slowly depleting the **s** gene through deaths of **ss** homozygotes.[17]

Evolutionary studies were also provided by anthropologists who saw in the human blood groups an opportunity to work out the history of human migrations and proliferation as our species moved across the world from a presumed African origin. The findings were important in showing that no particular blood group was unique to any one racial type but that different populations had certain fixed alleles at higher or lower percentages. Studies of founder effects and recognition that large populations of humanity may arise from a small initial population of migrant founders provided powerful insights into human evolution.[18] Such founder effects could be demonstrated in island populations (Pitcairn Island and Tristan da Cunha) as well as in religious isolates (the Amish in Pennsylvania and the Midwest) or political isolates (the French Canadians).

Biochemical Genetics Brings Medicine Back to Human Genetics

Biochemical analysis of genetic traits had a long quiescence between Garrod's work in 1902 and the discovery of phenylketonuria as a metabolic cause of a severe mental retardation.[19] Even more hopeful was the finding that a synthetic diet could prevent the mental retardation if the infants were diagnosed within the first few weeks of birth and placed on a special diet low in phenylalanine. A. Følling's analysis and the work that stemmed from it began the shift of medical genetics from primarily a diagnostic tool to one that could provide treatment. Even earlier, retinoblastoma, recognized as an autosomal dominant disorder, was quickly placed in the medical treatment category because surgery (removal of the affected eye or eyes) or radiation therapy (especially by X rays in the early days after Roentgen's discovery) would allow such children to reach reproductive maturity.

After the introduction of genetic services in the 1960s, the field of medical genetics generated considerable public awareness and changes in the teaching of medical students. In the 1950s, few medical schools bothered to include a course in medical genetics whereas today virtually all medical schools offer a sequence or a course in this subject. Medical genetics is no longer a field set aside for pathologists; it is a major part of obstetric and pediatric medicine. In addition, physicians are taught medical genetics without an appended social philosophy of eugenics. The shift from paternalism in medicine to the autonomy of the patient to make decisions about personal and family health occurred during the 1980s as cytogenetics, biochemical genetics, and the rapid growth of single-gene defects in humans accumulated in Victor McKusick's much cited and studied catalogs (Figure 6).

The human genome project has once again changed the way a field evolves. The opportunity to look for virtually every gene associated with the thousands of monogenic traits reported in McKusick's catalogs and the ability to construct genetic chips for diagnosis of many potential disorders based on ethnicity or severity will have dramatic effects in decades to come.[20] Critics of human genetics in its various forms fear a return to the abuses of the past but among the thousands of health professionals in the field of genetic services, few are active in eugenics movements or see the role

Figure 6. *Victor McKusick studied genetic disorders in isolates (such as the Amish) and pioneered a useful compendium of single-gene disorders in humans that has had numerous editions and is now available online for medical geneticists. Each entry is accompanied by an updated bibliography and a short paragraph containing useful information about the specific disorder. (Courtesy of The Alan Mason Chesney Medical Archives of The Johns Hopkins Medical Institutions.)*

of this field as one of altering society for a social good. The one value that presently dominates the field is the provision of information and available options for those at risk or those who need to employ that knowledge as they see fit. Treating genetic disease is a major motivation of those in the field of medical genetics. The use of genetic services to prevent the birth of children with untreatable disorders is seen, by those who provide them, as a personal choice and not a eugenic good.

End Notes and References

[1] For a history of human genetics, see Dronamraju K. 1989. *The Foundations of Human Genetics.* Charles C. Thomas, Springfield, Illinois. The only published history of human genetics, this book tries to combine Kuhn's paradigm theory with different trends in human genetics, but readers should discount this if they are not persuaded by philosophic treatments of history. It contains much useful information on the branches of human genetics and how they converged.

Dronamraju K., ed. 1992. *The History and Development of Human Genetics: Progress in Different Countries.* World Scientific, Singapore. This book provides accounts from about a dozen countries on the history of human genetics in those places. Not all countries made a transition from eugenics to medical genetics, as did Great Britain, France, Germany, and the United States.

[2] Gates R.R. 1946. *Human Genetics*, two volumes. Macmillan, New York.

[3] The epidemiology is quite striking: In 1900, the top killers of humanity in developed countries were tuberculosis, pneumonia, and gastritis (the last two mostly in infants and children). Only after 1930 did the trends shift to push infectious diseases toward the bottom of the top ten killers. Today, heart disease, cancer, and stroke (mostly associated with aging) are the top killers. One consequence of the public health and antibiotic revolutions of the twentieth century is the dramatic rise in mean life expectancy, from about 50 years in 1900 to about 75 years as the twenty-first century enters its first decade, almost all because of the reduction of infant mortality from infectious disease.

[4] Farrabee W.C. 1903. Hereditary and sexual influence on meristic variation and study of digital malformations in man. Ph.D. thesis. Harvard University.

[5] Bateson W. 1909. *Mendel's Principles of Heredity*, p. 227. Cambridge University Press, United Kingdom.

[6] Bateson W. 1909. op. cit. p. 229.

[7] The story is more complicated because females who are homozygous for PKU, whether treated or not, produce a secondary PKU with microcephaly, cardiac anomalies,

and other birth defects as a result of toxicity of the high phenylalanine in the uterine environment. This may have convinced some eugenicists to believe that PKU was Mendelian recessive (it is), but others believed that it was a like-for-like inheritance (maternal PKU creates that illusion).

8   Hardy G.H. 1909. Mendelian proportions in a mixed population. *Science* **28:** 49–50; and Weinberg W. 1908. Über den nachweis der vererbung beim menschen. *Jahrschafte ver vaterlandische Naturkunde Würtemburg* **64:** 368–382.

9   Snyder L. 1941. *Medical Genetics*, p. 9. Duke University Press, Durham, North Carolina.

10   Bell J. and Haldane J.B.S. 1937. The linkage between the genes for colour-blindness and haemophilia in man. *Proceedings of the Royal Society, London, B* **123:** 119–150.

11   I got to know Sheldon Reed while I was a visiting professor at the University of Minnesota and he was still active in human genetics. (He and his wife [also in human genetics] had second careers as professional dancers.) He told me why he came up with genetic counseling as an important adjunct to medicine and at the time he was still providing that service free of charge, whereas the physicians and Ph.D.s in human genetics that he trained were charging a fee for that same service.

12   Muller H.J. 1949. Progress and prospects in human genetics (preface). *American Journal of Human Genetics* **1:** 1–18.

13   Muller H.J. 1950. Radiation damage to the genetic material. *American Scientist* **38:** 35–39; 399–425.

14   Tjio J.H. and Levan A. 1956. The chromosome number of man. *Hereditas* **42:** 1–6.

15   Lejeune J., Turpin R., and Gautier M. 1959. Étude des chromosomes somatique de neuf enfants mongoliens. *Comptes Rendus de l'Academie des Sciences* **248:** 1721–1722.

16   Very rapidly after Lejeune's team announced their Down syndrome results, trisomies 13 (Patau syndrome) and 18 (Edward syndrome) and several sex-linked nondisjunctional conditions (Klinefelter syndrome = 47,**XXY**; Turner syndrome = 45,**X**, the 47,**XXX**, and the 47,**XYY** conditions) followed suit from laboratories in Great Britain and the United States, where a rush to study inmates in asylums took place. Controversies broke out over the use of that diagnostic information, especially for long-term studies of the 47,**XYY** condition, which was exaggerated in the press as a criminal syndrome.

17   Neel J.V. 1949. The inheritance of sickle cell anemia. *Science* **110:** 64–66. See also Dronamraju K.R., ed. 2004. *Infectious Disease and Host-Pathogen Evolution*, pp. 1–8. Cambridge University Press, United Kingdom.

18   McKusick V.A. 1978. *Medical Genetic Studies of the Amish.* Johns Hopkins University Press, Baltimore.

19   Følling A. 1934. Über ausscheidung von phenylbrenz-traubensaure in den harn als stoffwechselanomalie in verbindung mit imbezillitat. *Hoppe Seylers Zeitschrift für Physiolgische Chemie* **227:** 169.

20   McKusick V.A. 1998. *Mendelian Inheritance in Man: A Catalog of Human Genes and Genetic Disorders,* twelfth edition. Johns Hopkins University Press, Baltimore.

This immensely useful reference contains a printout of all single-gene disorders in humans with references and a descriptive paragraph. It can also be found online at www.nlm.nih.gov/entrez/query.fcgi?db=omim

CHAPTER 22

The Future and Significance of Classical Genetics

That the nucleus is especially concerned in synthetic metabolism is now becoming more and more clearly recognized by physiological chemists, Kossel concludes that the formation of new organic matter is dependent on the nucleus and that nuclein in some manner plays a leading role in this process.

E.B. WILSON*

IS THERE A FUTURE FOR CLASSICAL GENETICS? In the history of any science, there are fashionable trends when the old looks embarrassing and limiting and the new becomes enormously attractive as the place to be for new discoveries and the excitement of learning something novel. But when the dust settles, these new and old relationships are put into a different perspective. When Harold Plough joined Morgan's laboratory in 1914, he bemoaned that all the worthwhile work was already done. In 1920, entomologist William Morton Wheeler gave advice to young scholars, claiming that "genetics is a cow milked dry." In the 1960s, many molecular biologists wrote off fruit fly genetics as dead. Even as the era of the human genome sequence begins, one can hear murmurs that the new world lies not in the products of the genes and their nucleotide sequences, but in the yet unknown interactions of proteins and their products in the swirl of ignorance known as the living cell. Thus, on the 50th anniversary of the discovery of the Watson–Crick model, Walter Gilbert wrote, "It seems to me that molecular biology is dead. DNA-based thinking has penetrated the whole of biology, and the separate field no longer exists."[1]

Some genetic explanations are best described at a classical level and not a molecular level. Explaining albinism to two pigmented parents with an albino child requires a Mendelian approach if they are to make a decision to have another child. Discussing the significance of an amniocentesis that reveals a trisomy 21 does not require an extensive molecular inventory of the sequences of hundreds of genes present in triple dose. It requires the parent's awareness that having an extra chromosome in every cell of a child's body will have profound effects on health,

---

*1896. *The Cell in Development and Inheritance*, p. 247. Macmillan, New York.

capacity to learn, and even capacity to survive. As long as we live our lives on different levels of awareness, from molecules to populations, we will need classical genetics to understand the distribution of genes, their location on chromosomes, their capacity for recombination, the possibilities of chromosome aberrations, and the differential survival of genetic differences. The technology of molecular engineering relies on a basic understanding of genetic transmission, recombination, mutation, and selection in the design of genetically modified organisms, cells, or vectors.

What is the Significance of Classical Genetics?

Mendel's law of segregation turned out to be closer to universal than even he believed. The unusual cases of *Hieracium* or *Oenothera* proved to have both an underlying basic Mendelism and an obscuring anomaly of how fertilization occurs or how chromosomes are tied together in rearranged complexes. In 1926, Muller recognized "the gene as the basis of life."[2] He was considered a zealot for his views, but his argument was based on a sound inference. Only genes among natural molecules have the capacity to reproduce themselves and their variations (to which Timoféeff-Ressovsky gave the awkward name, "convariant replication"). That fundamental property, which Muller believed to be crystal-like, turned out to be correct. The edifice of molecular biology is built around the reality of the gene as nucleic acid and primarily a double helix of DNA (or, in some instances, RNA).

Classical genetics dethroned vitalist and holist conceptions of life. By reducing the organism or cell to gene-produced and gene-controlled components, geneticists, wittingly or not, discarded centuries of belief in a unique, otherworldly nature of life identified with religious creation mythologies. Classical geneticists also wedded the emerging findings of the chromosome theory, theory of the gene, chromosome aberrations, population genetics, and gene-character relations with Darwinian evolution by natural selection. This merging of classical genetics and evolution reinforced the mechanist and reductionist view of classical genetics. At its broadest level, classical genetics was rightly perceived as threatening to an older worldview of the uniqueness of life, breathed into an inanimate dust by a creator. It is not that classical genetics was ideologically created as an atheistic worldview; it is just easier (and consistent) for classical genetics to be described and used without invoking supernatural properties of life. At best, it permitted an uneasy dualism for those who tried to embrace their knowledge of classical genetics with their spiritual needs.

At this writing, a little over a century has passed since the rediscovery of Mendel's principle of segregation and the use of breeding analysis to dissect genetic traits. During that time, the triumphs of classical genetics have created neither a new warfare of science versus religion nor a withering away of the deep human needs for goals and meaning in life. Rather, the contributions of classical genetics have enriched human life in the arts, humanities, and social sciences. Classical genetics has made humanity aware of the vulnerability of genes to mutation by environmental agents. It has provided a basis for the kinship of all life through descent from distant common ancestors. It has opened up a miniature universe of genes and chromosomes, allowing us to journey in our minds from our nucleotides upward to our genes, proteins, organelles, cells, tissues, organs, individuals, and populations at all stages of our life cycle and throughout the history of humanity and life itself. This is the stuff of meditation as well as a foundation for understanding life.

What is the Role of Ethics in Classical Genetics?

Nowhere is this underlying reductionism of classical genetics more problematic than in human genetic services. The clients of these providers are forced into profound decision making by the availability of prenatal diagnosis, treatment options, and potentials to prolong life without improving the quality of the life of an infant with major impairments imposed by gross genetic or chromosomal abnormalities. However much geneticists understand the disturbances at the level of genes and chromosomes and the limitations imposed on an infant by those disturbances, our human sympathies for life usually override the culling mentality of the practical breeder. It is an essential human trait to be sympathetic and to put ourselves symbolically in other people's places. Throughout history, every generation has contained individuals flooded with caring and healing responses, full of hope, even when there was little that could be done to reverse a bad outcome of nature. But this is by no means a universal response to tragedy. Each generation has also had people willing to cut their losses, to put the needs of the healthy over the needs of the hopeless. This has led to frustration and flawed experiments in eugenics based on oversimplified ideas of sterilization to prevent assumed like-for-like transmission of allegedly degenerate traits and spurious beliefs that social failures are necessarily genetic failures.

We have no consensus among geneticists on the long-term consequences of spontaneous mutation in the human population. There are estimates of genetic load dating back to Muller's contribution on "our load of mutations."[3] The problem with such attempts to predict human genetic futures is that there are many unknowns: We do not know the role of assortative mating for most of our physical illnesses and behavioral traits, assuming (for this is all that can be done at present) that the behavioral traits are primarily inherited in some polygenic or multifactor way with an equally vague environmental component. To advocate eugenic approaches to social and health problems would be seriously flawed without a level of research on gene-character relationships, at least as intense as that devoted to the *Beaded* wing and *Truncate* wing analyses of the 1920s.[4] How much of that analysis will come out of studies of the human genome sequences is equally uncertain without the knowledge of how polygenic traits work at the molecular level. It is not uncommon in science that the most complex and messiest problems are deferred while the basic principles using simpler systems get worked out.

What are the Social Applications of Classical Genetics?

If there are lessons of history, they are rarely obvious because the past repeats itself in new forms that rarely resemble the past. A person voting down a school budget because of economic stressful times does not see the issue as one of depriving the genetic potentials of children of their best expression. Some social policies do have unintended genetic consequences, especially those that allow grossly unequal opportunities based on income, social class, religion, sex, or race. Other social policies may be deliberately based on genetic considerations (frequently false or oversimplified), as is true of almost all racist policies. The revelations of death camps during the Holocaust of World War II did not prevent similar attempts at genocide in other countries in the last half of the twentieth century. Some government policies tried to use genetics as a tool for social progress—the USSR during the Lysenko era of 1935–1965 was such an instance but in this case, the genetics chosen was a pseudoscience that gave the illusion of rapid transformation of crop varieties by the environment.[5]

These past abuses of knowledge have made most geneticists wary of social applications of new knowledge. Whether to promote or ban new applications of knowledge, new legislation is less preferred than a watchdog role for government to assure that abuses do not occur. This has also made the teaching of genetics below the college level a difficult task. The United States experiences recurrent pressure from fundamentalist religious groups to omit or water down the evolutionary aspects of genetics. Many schools avoid much of the controversy associated with genetic screening, prenatal diagnosis, and medical applications of genetics because of the divisive controversy over elective abortion. The philosophic implications of the gene as the basis of life are almost never broached in public school discussions of biology. On many occasions, Muller told his students that genetics is perceived as a subversive science, threatening to governments, because it deals with some of the most fundamental issues of life.

Is Classical Genetics "Milked Dry?"

No one doubts that much can be learned about gene action and the interaction of gene products in working out what we take for granted—the intact living cell and its processes. Our assumption, probably correct, is that most of the insights into how life works will come from the new fields of bioinformatics and proteomics. However, many problems will still require the use of classical genetics for additional insights. Of the millions of species that exist, what portion behaves genetically like fruit flies or humans?

We know that sex determination can be diverse, surprising, and unexpected. How universal is meiosis as a means of generating reproductive cells? Our survey of the genetics (by breeding analysis and cytology) of most of the world's species is limited to a small fraction of all life and phyla.[6] No doubt, such surveys will be done in years to come and in all likelihood they will reveal surprises. We do not have a good explanation for why animals are so "fine tuned" that polyploidy and aneuploidy are usually fatal, whereas in plants, these rarely cause the failure of plant organogenesis. Trans-species gene transfer can be studied in experimental animals. But the effects of polyploidy and aneuploidy cannot be presently predicted by computer models because too many genes and their products, throughout a life cycle, would be involved. We do not know the self-assembly and directed-assembly processes taking place in cells of eukaryotic organisms for the construction of cell organelles. No geneticist doubts, however, that such genes will be isolated for a directed assembly. It is only my limited imagination that limits listing dozens of possible topics for future exploration by classical genetics in eukaryotic systems.

End Notes and References

1. Gilbert W. 2003. Life after the helix: How Jim Watson saw the structure of DNA transform biology (review of McElheny V.K. 2003. *Watson and DNA: Making a Scientific Revolution*. Perseus/Wiley, New York.) *Science* **421:** 315–316; see p. 316.

   Gilbert's point is that new fields will emerge from those fields that presently dominate science because the problems swept under a mental rug will later emerge.

2. Muller H.J. 1929. The gene as the basis of life. *Proceedings of the Fourth International Congress of Plant Sciences, 1926, Ithaca* **1:** 897–921.

   Muller noted how his paper was returned with a request that he change the title to "The gene as *a* basis of life." Muller insisted that the audience and readers of his article, and not the organizers of the sessions, should have the opportunity to interpret his thesis.

3. What makes the estimates difficult are the choices of mutations used for spontaneous rates (e.g., retinoblastoma and Apert syndrome). Among the difficulties are the aging effect in mutation rate in older sperm, changes in reproductive age, changes in family size, and accuracy of diagnosis. What is recognized as a mucopolysaccharide deficiency (e.g., Hurler syndrome) in a large well-staffed municipal hospital may be misclassified as Down syndrome by a physician in a rural area in which such cases are rarely seen. More modern attempts look at changes in nucleotide sequences in selected regions of a chromosome of ethnically close individuals (e.g., Icelanders).

4. In these papers, Muller argued that "Remote as the possibilities of such work may seem to be in the case of such animals as mammals, it is nevertheless difficult to conceive how the genetic bases of the more elusive and complicated characters in them can be determined adequately by any other means." See Altenburg E. and Muller H.J. 1920. The genetic basis of truncate wing: An inconstant and modifiable character in *Drosophila. Genetics* **5:** 1–59; see pp. 1–2.

   Although Altenburg is first author, he insisted, when I interviewed him, that Muller wrote the paper, Altenburg's role, he claimed, was not much more than that of a technician, and he foolishly agreed to a coin flip to decide who should be first author.

5. See Chapter 20 for an account of Lysenkoism during 1935–1965. Although that influence finally abated after its agricultural failures and charges of fraud, the harm done still continued. What was once a great genetics program has never recovered its excellence.

6. There are millions of species (the exact number varies widely) and 96 phyla.

Classical Genetics and the History of Science

If I may throw out a word of counsel to beginners, it is; Treasure your exceptions! When there are none, the work gets so dull that no one cares to carry it further. Keep them always uncovered and in sight. Exceptions are like the rough brick-work of a growing building which tells that there is more to come and shows where the next construction is to be.

WILLIAM BATESON*

But Wilson was well aware that, unlike art, science cannot be satisfied with any construction that achieves only an inner harmony, no matter how appealing. For it has, alas, for many a brilliant mental construction, to match its own creations against the phenomena of the outer world, and for this reason the discipline of its imagination must be greater.

H.J. MULLER†

FOUR COMPONENTS OF A SCIENTIST'S CAREER are legitimate for study by historians of science. I call the first the *scientific component*, i.e., the actual ideas and experiments and the scientific findings associated with a scientist's work. Thus, we would identify with Mendel the concept of hereditary units transmitted from one generation to the next, laws of segregation and independent assortment, and an observation that most of the traits he selected were functionally dominant or recessive. We can examine Mendel's papers and try to determine how he used these ideas and what he discarded or held important among competitive ideas of heredity in the 1860s. We can also see how he presented his data and interpretation and contrast it with the way we see those concepts today. I consider these the primary scientific factors for understanding the science in the history of science.

I call the second component the *social context of science*. This implies that people do not live in a social vacuum. They are constrained by social circumstances (war, epidemics, financial hard times, religious prohibitions on what can be studied, and politics) and these can definitely influence the sup-

*1906. *The Methods and Scope of Genetics*, p. 22. Cambridge University Press, United Kingdom.
†1943. Edmund B. Wilson—An Appreciation, p. 55. *American Naturalist* 77: 5–37; 142–172.

port for science and the opportunities for education and publishing. I am not as convinced that these same factors have a powerful influence on shaping ideas or the design of experiments. I personally believe that they tell us little about the design of experiments or the emergence of scientific theories, but they are powerfully influential on the applications of science to society.

We know much about the third component, which may be called the *historical aspect of science*. Compared to a scientist working in the sixteenth century, one working in the twentieth century has different views of the universe, science itself, the role of the supernatural, and what constitutes science. We also must recognize that certain questions cannot be framed without prior knowledge, and many fields of science are cumulative in their understanding of how things work. The domain of a science and its connections to other fields of science also change over time. Any scientist we study, or whose papers we read, is limited by these historical circumstances. Although science has flourished in past centuries, it is different in many ways from the science of today because science cannot escape the influence of its own era. At issue is not that such historical influences are pertinent and powerful. Instead, we have to ask how much of an influence some of that past era has had on the scientist's work. All centuries have their wars, economically troubled times, natural disasters, and similar hard times. At the same time, science requires financial support, and it may not really matter if that support comes from a prince or a democratic state legislature. It is not easy to reconstruct the society in which the scientists lived and figure out which parts of that culture and its events were pertinent to the type of questions asked and issues available at the time and whether any of these influenced the creativity of any particular scientist.

We know least about the fourth component that I call the *psychological aspect of science*. People are motivated by often unconscious factors associated with their upbringing and personalities. Biographers frequently seek to find these factors in historical figures, whatever their occupation and talents. There are debates about how much of this is driven by hereditary factors, Freudian dynamics, the unhappy circumstances of childhood (divorce, death, financial failure, or psychosis of a parent), sexual conflict, religious conflict, or the imposition of parental will. I do not doubt that these play a part in the way in which scientists treat their colleagues and students, but I am doubtful that any of these could be used at present to explain why a scientist designed an experiment of importance or came up with an important theory. We just do not know enough about these personal factors and their relation to the specific work of a scientist's career. No general theory of creativity is available that can be used to predict the scientific work of a scientist from a knowledge of that scientist's life history.[1]

In this account of classical genetics, I have stressed the scientific aspects because that is what this book is about—how the ideas arose independently and came together to form what we call classical genetics. I am a scientist and when I read a scientific paper, I want to study the experimental design, the tools used, the assumptions made, and those contemporary ideas that were either suggestive or discarded. It is rare that a textbook concept arises in the form seen in a textbook. Much refinement and polishing takes place, a process that I call, using Muller's phrase, "the winning of the facts." This pruning process is very much part of science. I have read accounts of scientists I knew, whose work I knew in intimate detail, that attempted to explain science in sociological (in-groups versus outsiders), political (Marxism versus capitalism), or historical (depression, war, and ideology) contexts, and I found these either false or extraneous.

Muller's Career Is a Challenge for Interpretation

The example I use is that of H.J. Muller. He was a key figure in this history of classical genetics, but many geneticists will argue that I have overblown his importance with respect to the numerous geneticists of recognized talent of those years (1912–1946) during which he was at his zenith. I do not think that to be the case, but I am aware of the enormous influence of his life on my thinking, both as a geneticist and as a historian of science. Readers of this book may reject my interpretation, but I hope that they do so by demonstrating where I have erred. Muller was a flawed individual—insecure, competitive, idealistic, and very talented. He attended Columbia University at a time when only two full-tuition scholarships in all of New York City were available, and he was fortunate to win one. He was also constrained by the death of his father when he was a boy of 10 and the genteel poverty in which he grew up, at a time when there were no social services for widows and their families. Like his father and grandfather, he sympathized with the laboring class, socialism, and the need for reform in a capitalist society that was then largely indifferent to the plight of the poor.[2]

When I began writing the Muller biography, I believed that Muller's basic science work would have remained constant wherever he went and that his applied science (such as eugenics) would have flip-flopped with the country he was in at the time. The reverse was true. Muller dumped old problems and started new ones every time he set up a laboratory in a new institution. It was like watching the multiple moltings of an insect. But the eugenics he proposed—a positive eugenics based on intelligence, social leadership, and a caring personality—was what he pushed in New York, Texas, Berlin, the USSR, Edinburgh, and Indiana. It did not matter whether the government was liberal or conservative,

capitalist or socialist, authoritarian or democratic. Muller's eugenics was like a religious credo.

Some interpret Muller to be a communist who was consistent in his ideological commitment to the Bolshevik cause until his death in 1967 (Willard Libby presented this view when I interviewed him).[3] During the Cold War, Libby and Teller believed that concerns over low doses of radiation were motivated not by scientific inquiry, but by a hidden political agenda that sought to scuttle the development of a nuclear arms program for the United States. In actuality, I have seen no evidence that Muller was a member of the Communist party. His student, Carlos Offermann, told me that Muller had considered that possibility because of his sympathies for the Bolsheviks, but decided that his voice would be more effective for criticizing social policy if he did not join.

Some interpret Muller to be a liberal who was hounded unfairly by the Texas administration in the early 1930s because of his views on racism (he opposed the violations of citizenship rights for blacks), labor (he attacked the inequality of opportunities for those who were exploited or neglected in a system that favored the privileged), and women (he denounced the sexism of the American eugenics movement).[4] This "whitewashing" of Muller omits the facts that he recruited students for the National Student League (a communist-front organization) and helped edit (without signing his name to the articles) an underground newspaper that was known to the FBI as a communist newspaper (called *The Spark*, after Lenin's *Iskra*). Whether or not one agrees with the reactionary views of Texas politics at that time or the administration of the university, Muller did violate his contractual obligations as a faculty member by secretly engaging in these activities.

Table 1. *Students and technicians who worked for T.H. Morgan (other than Sturtevant, Bridges, and Muller).*

Name	Activity with Morgan	Career and comments
J.S. Dexter	variation in *Truncate* wings (1912)	Saskatchewan to study wheat genetics; died 1953
F.N. Duncan	gynandromorphs (1915)	Southern Methodist University; fired for flunking football player; left academics; died 1957
S.R. Safir	white eye series (1913)	Yeshiva University, New York City
R.R. Hyde	fertility (1914); white eye series (1915)	Indiana State Normal School; Johns Hopkins University for immunology
M. Hoge	eyeless (fourth chromosome) (1915)	married Aute Richards and accompanied him to the University of Oklahoma to teach zoology
M. Stark	X-linked lethal factors (1915)	tumor genes in flies
L.S. Quackenbush	unisex broods in *Drosophila* (1910)	left Columbia University in New York City; could not sleep in city (had asthma)
C.R. Plunkett	gene and environment interactions (1926)	New York University; IWW organizer; jailed as pacifist during WWI; fired in 1932 for practicing communism
D.E. Lancefield	autosomal mutations (1918)	Oregon; Queen's College, New York City; studied *D. pseudoobscura*
H.H. Plough	linkage variation data (1915)	Amherst College; environmental effects on *Drosophila* expression
F. Payne	no Lamarckian effects in isolation (1910)	Indiana University; crossover reducers
A. Weinstein	recombination studies	Johns Hopkins University
L.V. Morgan	attached-X chromosomes (1922)	Columbia University; independent scholar
E. Cattell	mapping X-linked lethal (1912)	career unknown
S.C. Tice	Bar eyes (1914)	career unknown
C.J. Lynch	autosomal linkage (1912)	career unknown

Most of Morgan's students did not become as famous as Sturtevant, Bridges, and Muller. I interviewed Sturtevant on May 8, 1968 at his office at the California Institute of Technology and obtained from him some information on Morgan's students. At the time, I did not ask about Cattell, Tice, and Lynch. I already knew about Plough's and Payne's careers. Morgan's female students or technicians are indicated in bold type. In those days, most female students or technicians married and dropped out of research. Cattell published only in 1912 and 1913, Tice in 1914, and Lynch in 1912 and 1920 and thus had some sort of career. One exception is Morgan's own wife, Lillian Vance Morgan, who returned to fly work in the 1920s when her children were older, published articles through 1938, and contributed the attached X stock for *Drosophila* studies. Morgan turned away a few gifted students (such as L.C. Dunn) because his laboratory was overcrowded. Some students who worked for him (such as Fernandus Payne) preferred to do their dissertation work with Wilson. Lancefield also shifted over to Wilson. Most of those listed in this table either shifted to other work and organisms or got trapped in positions that did not foster research. I do not include students who were active with Morgan in the 1930s, such as J. Schultz, C.C. Tan, and T. Dobzhansky; they did most of their work (primarily postdoctoral) with Sturtevant. There is no evidence that Morgan had a consistent standard of excellence or creativity for his selections of co-workers. I also exclude his postdoctoral students of the World War I era, such as Otto Mohr. Morgan liked a busy laboratory with many projects running concurrently.

Ignoring the politics of Muller's social beliefs and focusing on his research, one interpretation (based on my reading of a book on the history of the fly lab) of Muller (and Morgan's school, in general) paints him as an insider who tried to keep others from their share of the glory.[5] Table 1 shows a list of Morgan's students that I discussed with Sturtevant in 1967 to determine what had happened to them. I am very skeptical that Muller, Bridges, or Sturtevant froze out any of those fellow graduate students. Their reasons for failure to become famous include numerous personal factors (health, financial constraints, change of career goals, recognizing that they did not really like the work, and other circumstances). Any scientist who has run a busy laboratory will testify that students come and go for different reasons. That was true in Muller's laboratory when I was there, in Tracy Sonneborn's laboratory also at Indiana University in the 1950s, and in my own laboratory when I was a young scientist in the 1960s at the University of California, Los Angeles. Ironically, the one strong case of a victim of the insider versus outsider theory is Edgar Altenburg. For whatever reason he may have privately held (I suspect it was Altenburg's intense loyalty to Muller), Morgan did not allow Altenburg to work in the fly lab.[6] Yet Altenburg contributed to the discussions and worked with Muller collaboratively both at Columbia and at Rice University. Altenburg was also a solid supporter of the work done by Morgan, Bridges, and Sturtevant, even if he believed that some of that work (experimental design and broad theory) was contributed by Muller.

Morgan, Bridges, and Sturtevant had roots going back to the Revolutionary War, whereas Muller was a third-generation American (not, as one recent book claimed, of recent Jewish immigrant stock with "pushy habits" that offended Morgan).[7] Muller's Jewishness is also doubtful. His father's side came to America as Catholic and became Unitarian. His mother's side was mixed Jewish and Anglican. Muller's maternal uncle, Sam Lyons, was a minister in the Anglican Church. But Morgan and Sturtevant were nonpracticing Protestants who shared a conventional outlook toward American society, and Bridges (a declared atheist of Protestant stock and an impoverished orphan) was a Greenwich Village bohemian with left-wing sympathies, like Muller. Bridges' own idiosyncratic free-love womanizing was viewed as scandalous by Morgan and Sturtevant, who nevertheless protected him. I cannot make any sense of the left- or right-wing values with respect to any of their basic work, nor can I see why nondisjunction should be linked to left-wing sexual promiscuity, whereas chromosome mapping should be linked to a conservative American outlook. I realize that this is simplistic and I do not want to make a mockery of those attempting to find social, economic, historical, or political factors influencing the pure science of these contributors. I do not doubt that these social factors (especially the ideological ones) had a major role in how these scientists applied or reacted to the applications of genetics to society.

One of the dangers of using social explanations for science is that they can be good stories but historically false. Any human being who attempts even modestly to apply Socrates' dictate, "Know thyself," will quickly learn how contradictory our deep assessments of our intentions and past behavior can be and the many different "causes" we can conjure to try to explain our failures and successes. I have been very cautious in invoking such factors in this book, and I leave to future historians, sociologists, and psychologists of science the task of finding such associations.[8]

Was the American Dominance of Classical Genetics Real or an American Bias?

Genetics is international and its roots are almost completely European. What was uniquely American was the synthesis that brought together evolution, cell biology, reproduction, cytology, and breeding analysis. That fusion we call classical genetics. In its formation, American dominance was broken in the period of recovery after World War II when molecular genetics came into being. Molecular genetics was primarily a European synthesis built out of émigré Europeans in the United States (chased out by Hitler) and Europeans who managed to stay alive during the war. Delbrück, Schrödinger, Crick, Wilkins, Franklin, and numerous other Europeans (and one American, Watson, who joined his European colleagues) put together the DNA structure that changed the way genetics was done. I have presented many reasons illustrating why America was dominant in 1900–1930 when the bulk of classical genetics was worked out. This does not mean that Europeans did nothing—they were very active—but my analysis has shown that many of their key contributors in the early years got shunted into work that did not pan out. I believe a major reason that the United States work stole the show was tied to the American graduate school, the agricultural field stations affiliated with college campuses, and the mechanistic outlook of science that pervaded the work habits of American professors.

To demonstrate my point, I list the following topics that would likely be covered in the classical genetics section of a genetics course given to undergraduates. For each topic, I indicate the primary workers who came up with the key ideas or experiments and their nationalities (U.S. in italics; European in Roman type style; multinational with an American component underlined). All of these topics are discussed in this book.

Mendelism. Includes the idea of hereditary units representing traits, the laws of segregation and independent assortment, and the observation that many traits behave functionally as dominant or recessive. Mendel, de Vries, Correns, and Tschermak von Seysenegg. All are continental European.

The chromosome theory of heredity. Includes the relationship of homologous chromosomes to the segregation of traits and the independent alignment of pairs of chromosomes. *Sutton, Cannon,* and *Wilson*

directly made this association as Boveri (Continental) did indirectly.

X and Y chromosomes. The association of specific chromosomes with males or females. Henking, *McClung*, Montgomery, *Wilson, Stevens,* and *Payne*. Henking was Continental and Montgomery was an American who received his Ph.D. in Germany and returned to the U.S. to do his work.

X-linked inheritance. The modified ratios associated with hereditary traits carried on sex chromosomes. *Morgan* (and, without the genetic ratios, Doncaster and Bateson [both British], but Bateson was strongly influenced to shift to heredity after studying with Brooks in the U.S.).

Linkage and crossing-over. The recognition that some genes are associated with the same chromosome and that those farther apart undergo more crossing-over than those closer together. *Morgan* (without the recognition of the distance relation, I also include Bateson and his school).

Mapping. The use of crossover data to compute linear maps of the chromosomes and use these as tools for genetic analysis. *Sturtevant* (also, *Muller* and *Weinstein*, to a lesser extent, for refinements including coincidence and interference, double crossovers, and working out the linearity).

Nondisjunction. The failure of normal separation leading to an excess or deficit of a chromosome; correlation of cytological and genetic factors in nondisjunctional combinations. *Bridges* (with some contributions from *Muller*) and Gates, using *Oenothera* without the genetic association in a less compelling way.

Multiple allelism. The presence of a normal gene and its family of mutant forms, such as the white-eyed series of white, eosin, apricot, etc. Baur (Continental) and *Sturtevant*.

Epistasis. The interaction of two or more different genes leading to modified independent assortment ratios (15:1, 9:7, etc.). Bateson, Hurst (British), *Davenport, Emerson*, and Correns (Continental).

Quantitative inheritance. The additive contribution of two or more genes to a trait producing modified ratios, such as 1:4:6:4:1 running in a graded series, for example, from darkest to lightest. Nilsson-Ehle (Continental) and *East*.

Polygenic traits. The bell curve or Gaussian distribution of a continuous trait, such as height or weight. Johannsen (Continental).

Gene-character relations. The presence of chief genes and their modified expression by modifier genes or environmental factors. *Muller*.

Population genetics. The mathematical description of gene frequencies across generations with effects of detrimental or beneficial genes. Hardy, Fisher, and Haldane (British); Weinberg (Continental); and *Wright*.

Chromosome rearrangements. Structural rearrangements such as inversions, deletions, translocations, and their genetic consequences. *Bridges* and *McClintock*.

Chromosome rearrangements and evolution. Gates, *Bridges, Muller, Painter*, and Dobzhansky (Russian and American), Renner (Continental), and *Cleland*.

Dosage compensation. The equalization in phenotype between a two-dose (**XX**) and one-dose (**X**) state in females and males, respectively. *Muller* and Stern (Continental).

Polyploidy. Multiple sets of chromosomes beyond haploid and diploid, including triploid and tetraploid. Gates, *Davis, Bridges, Blakeslee*, and *Eigsti*.

Lethal factors. Genes that cause death to the embryo or newborn. *Morgan*, Cuénot (Continental), and *Castle*.

Mutation rates. The frequency with which a normal allele mutates to a mutant form. *Muller*.

Radiation-induced mutation. The successful use of X rays to induce mutations. *Muller* and *Stadler*.

Chemically induced mutation. The successful use of chemicals to induce mutations. Auerbach (British) and Rapoport (Russian).

Gene size and gross structure. The dimensions of genes and their compound or unitary state. *Eyster, Demerec, Morgan, Muller*, Agol (Russian), *Sturtevant, Oliver, Lewis*, and Pontecorvo (British).

Note that I have listed 24 European contributions, 45 U.S. contributions, and 8 associated with a multinational work habit. One could consider Muller multinational by counting his work in Europe (1932–1939), which would thereby make him multinational like Gates, Montgomery, Dobzhansky, or Bateson. Even if this were done, a U.S.-associated contribution (italics or underlined) is twice as likely to be associated with a major contribution to classical genetics than is a Continental or British contribution. I say this not out of American bias, but because of the record. I fully acknowledge the European dominance prior to 1902 and the European dominance for working out molecular biology. This book tries to explain why American dominance took place during 1900–1940.

The Relationship of Applied and Pure Science

I stress pure or basic science for most of this history of classical genetics. The separation of applied and basic research, however, is not always clear-cut. Pure science is not exclusively carried out in universities; the U.S. agricultural field stations played a significant role in adding to classical genetics in its formative years. Geneticists were mostly motivated by the excitement of finding something new. Some, like Shull, recognized the commercial possibilities of hybrid corn, but both Shull and East chose not to shift to applied genetics.

I include two chapters on classical genetics and politics and on medical and human genetics because they are important. Both used (applied) classical genetics, but in one case, this led to serious moral issues and monstrous abuse.[9] The other, human genetics, developed many tools for its study and came up with some important principles of classical genetics. Had I omitted those, I believe I would have swept under a mental rug an important motivating aspect of many scientists' lives—the desire to find applications of science for a social good. Human genetics amply demonstrated concepts such as balanced polymorphism for the prevalence of frequencies of sickle-cell alleles in populations. In contrast, eugenics failed whenever it moved from theory to practice, whether in the American eugenics movement or in the Nazi eugenics movement. Similarly, the attempt to destroy classical genetics and replace it with a spurious Lamarckism, by T.D. Lysenko, was one of the tragic chapters in the history of genetics. Not quite as grievous, but sobering when the articles are read with historical hindsight, are the bitter accusations of respected scientists who disagreed over the effects of low doses of radiation on human populations. Although most university-associated scientists tend to think of their work as "pure" and not "applied" science, they or others endorse its usage for society. I do not want students reading this book to conclude that basic scientists do not make mistakes or do not behave badly at times. They can sometimes do evil things, whether or not they are aware of this. That would be the subject of an entirely different book, but I included those chapters on human applications of classical genetics to show that the applications of scientific knowledge are rarely value neutral.

Reflections on Classical Genetics and the History of Science

I conclude this overview of how classical genetics came into being with some reflections on the history, sociology, and philosophy of science. I believe that all three play a part in our appreciation and understanding of how science works. I do not believe that many of the approaches currently being studied are likely to be predictive. Some of the great contributors to classical genetics were easygoing, some were hard taskmasters, some exploited their students, and some neglected their students. Some were hard to get along with as colleagues. Some were apolitical, whereas some had strong political opinions and willingly traveled to march in demonstrations. Some were liberal and some conservative. Some were gay or lesbian, some were conflicted in their sexual identities, and some were conventionally heterosexual. It is not a surprise that an enterprise which has involved thousands of men and women over the past century should include such a broad section of human personalities, psychological disturbances, and different outlooks on life. I just do not believe that anyone can deduce from a reading of their scientific papers much about their sex lives, their generosity or miserliness, their paranoia or ready fellowship, or similar contrasts of human behavior.

The old and the young, male and female, insiders and outsiders, atheists and churchgoers—all have made contributions to classical genetics. A devout Ralph Cleland, deacon of his church, wrote papers on the evolution of *Oenothera* that are as objective as those written by any atheist.[10] I do not doubt that psychological factors are important in determining productivity, in converting laboratory experiments and their results into published articles, or in gaining or barring access to competent students. Some great scientists were ineffective when trying to convey their ideas through scientific meetings. Some scientists sought power and enjoyed accumulating space, grants, and titles. Others were quite modest and shunned a public reputation. Some were slovenly, and others had neat and orderly laboratories and desks. We enjoy reading about these personal habits because it humanizes science for ourselves and for our students.

Was the Morgan School Approach a Paradigm Shift?

There is something akin to a seductive charm to calling events in science paradigm shifts. In *The Structure of Scientific Revolutions*, Thomas Kuhn provided strong evidence that the Copernican revolution was such an event.[11] In his model, paradigm shifts arise primarily by a reorganizing of the language of science. Thus, in the Ptolemaic universe, the sun and the moon were planets because they revolved around the earth. The celestial bodies other than earth were not seen as material objects, but as light-generating (not light-reflecting) entities. In his model, Copernicus placed the sun at the center of the solar system and it thus became a star, with the earth demoted to a planet, rather than the largest, central, and only material body in existence. Galileo's use of the telescope came much later and with it proof for the material composition of the moon and inner planets (Venus had phases like the moon). Jupiter had four prominent moons of its own, whose orbits could be calculated. The moon's mountains and craters cast shadows. This was strong evidence for the Copernican model. But prior to Galileo, in Kuhn's argument, there was no evidence for the Copernican model; it was a shuffling and renaming of the components of the universe through a helio-centric scheme. Does such a model fit the alterations of the chromosome theory (sex-linked inheritance and chromosome mapping) that Morgan and his students developed?

I would argue that they do not. Life science, more often than not, builds on its past rather than replacing it by paradigm shifts. It evolves. In the life sciences, new fields usually emerge through new technologies, such as the microscope for cell theory that slowly evolved from Hooke's buoyant empty boxes to the contents themselves. Even the technology of microscopy evolved with improved achromatic lenses, condensers, and stain technology by the mid-nineteenth century. The evidence of a hereditary role for the nucleus of the cell was not a sudden inspiration, but forced upon Haeckel and others who noted that eggs and sperm contributed equally to offspring. A nucleus of the same size was the one thing both had in common, as Fol's observations of the male and female pronuclei demonstrated. When chromosomes were observed and Flemming, Strassburger, Waldeyer, and others worked out mitosis, this produced considerable speculation about their significance for heredity. It was the new tool of Entwicklungsmechanik that led Boveri, Wilson, and others to favor a chromosomal role for heredity. Similarly, bit by bit, the role of the X element, later the accessory chromosome, then heterochromosomes and idiochromosomes, and finally sex chromosomes sharpened the connection between chromosomes and heredity, in this case, sex itself. Morgan was initially skeptical of his own finding of X-linked inheritance, and he only gave way when he had accumulated experimental evidence (such as the linkage of several X-linked traits). Sutton's attempt to unite Mendelism and meiosis through the chromosome theory, he claimed, was an induction from his cytological work, with Mendelism coming later as a confirmation of his insight. Sutton

relied on a huge body of work that suggested this connection of heredity to the chromosomes. Muller called this collective effort to extend the chromosome theory by the fly lab "the winning of the facts," and I have applied it throughout this history of classical genetics because it demonstrates, in a much more detailed and sequential way, how the main tributaries developed and came together to form classical genetics. Unless one wishes to call all scientific advances revolutions and all scientific theories paradigm shifts (a view Kuhn rejected), I do not see the history of classical genetics as one of one or more paradigm shifts.[12]

Reading and Obtaining
Source Documents Is Important

I believe that the real understanding of science results not from an attempt to force the contributions of science through some social, psychological, philosophic, or political filter, but through a study of the papers themselves and, to the degree that they exist, through discussions in laboratories and meetings, over social interaction, and similar outlets for exchanges of ideas. Because much of the informal ways in which science is done (especially the constant laboratory talk) is not available to historians of science, any reconstruction of how an idea evolves is partial. Not all scientists record the origin of their ideas. Despite those limits on a complete understanding of the evolution of scientific ideas and experimental designs, much can be learned from a reading of primary papers. To Morgan and Bridges, the eosin case, when it first arose, was an example of "bicolorism," a term that meant little more than that the eye color of females was darker than the eye color of males in an eosin stock.[13] By 1932, in

Muller's experiments, bicolorism became "dosage compensation," and he had used deleted X chromosomes carrying extra doses of eosin to show that its expression was additive and not responsive to dosage compensation.[14] Using similar methods, he also isolated other genes to show the existence of dosage compensators—actual genes that equalized the difference between males and females for most X-linked genes. Throughout my overview of classical genetics, I have been impressed by the slow pruning, replacement, and addition of new findings, and the step-by-step process by which an initial finding is purged of misleading or erroneous assumptions to become a self-consistent idea that stands the tests of hundreds of applications and situations.

I have also been overwhelmed by the amount of conflict, personality difference, and other varieties of personal habits of geneticists revealed when I conducted interviews and read their memoirs and correspondence. The scientist who is known through papers is largely a public figure stripped of those idiosyncrasies. What intrigues me is whether the immense productivity and success of Morgan's school had anything to do with that turmoil of jealousy, envy, resentment, self-satisfaction, admiration, hero worship, and similar brew of contrasts one experiences when reading a good novel. I am also puzzled how so able (and careful) a scientist as E.M. East could have said such blatant racist nonsense about African Americans.[15] How does a scientist with rigor applied to plants take so casual a scientific approach to human traits? What limits the carrying over of objective standards from one species to our own? Certainly, we cannot be perfect or consistent beings or anticipate future shifts in public opinion that will occur over one or more generations to come. Yet many of East's contemporaries would never have uttered such words or thoughts.

We assume that it is important to be flexible, to plow through life with bundles of inconsistencies and errors and our capacity for redemption. Art critics, literary critics, and political analysts have struggled with this same problem, and I do not believe that we are even close to understanding the relationship, if any, of scientific imagination, focus, and insight to these very diverse human personality traits.

This is not a reason for despair or a pessimism that I wish to inflict on the history of science. I say it so we do not oversimplify our interpretations of how science is done. For those involved in it, it is an enormous thrill to be a participant of or a witness to great discoveries. I believe that the history of classical genetics is not an anomaly in the history of science. I also believe that the "winning of the facts" accurately depicts how most science is done.[16]

End Notes and References

[1] One of the more interesting books I have read on creativity is Goertzel V. and Goertzel M. 1962. *Cradles of Eminence*. Little, Brown and Co., Boston.

The authors argue that intelligence scores (IQ results) are not sufficient predictors of those who will become outstanding scientists (or contributors to other fields of knowledge). They also argue that some stress is involved in the lives of creative people, including the factors that I cited.

[2] For an account of Muller's life, see my 1981 biography, *Genes, Radiation, and Society: The Life and Work of H.J. Muller*. Cornell University Press, Ithaca.

[3] Willard F. Libby. Interview with E.A. Carlson, Los Angeles. April 8, 1971.

[4] I received a typescript about 20 years ago from a writer in Texas who presented this view in an article for a Texas publication. I do not know if it was ever published.

[5] Kohler R.E. 1994. *Lords of the Fly:* Drosophila *Genetics and the Experimental Life*. Chicago University Press.

Much of this book is useful in discussing the personalities of Morgan and his students and the history of fruit fly research after the 1930s, as it moved to other laboratories. However, it does not explore the individual experiments critically.

[6] In her fine history of the California Institute of Technology, Goodstein J. 1991. *Millikan's School: A History of the California Institute of Technology*. W.W. Norton, New York), Judith Goodstein uses the argument that Morgan liked smart students and Altenburg impressed him as unimaginative (see p. 197). Perhaps, but Hyde, Dexter, and Duncan all received Ph.D.s in genetics with Morgan at about the same time as his more famous students. They were good, but not outstanding. I would argue that Altenburg contributed far more than they in his academic career.

7 Ridley M. 1999. *Genome: The Autobiography of a Species in 23 Chapters*, p. 46. Harper Collins, New York.

8 I recommend that readers check Thomas Mann's 1930 novella, *Death in Venice*. Knopf, New York.

   Mann describes his fictional literary hero, Gustave von Aschenbach, a Nobelist, as driven by homosexual urges and inner turmoil that fueled his great works of literature, without revealing their repressed sources. He goes on to say that it is better that the reader does not know the sources of an artist's creativity, because to reveal them would only cause confusion and alarm. Of course, this is the task that every critic hopes to exploit when writing a biography. Most lives are more complex than the remnants of correspondence left behind at death and this makes a successful analysis difficult or impossible. Most famous people are also reluctant to confess or put into writing their innermost embarrassments and insecurities.

9 See my attempt to present a history of the ideas and events that long preceded the formation of the eugenics movement. Carlson E.A. 2001. *The Unfit: A History of a Bad Idea*. Cold Spring Harbor Laboratory Press, Cold Spring Harbor, New York.

10 Cleland was also a kindly man, who frequently went out of his way to roll up windows on the cars of his colleagues during a rain shower. He was also a monotonic speaker with a nasal drone to his lectures (given right after lunch!). He told our class that during one of his lectures, Watson fell asleep and fell off his chair.

11 Kuhn T.J. 1969. *The Structure of Scientific Revolutions*. Chicago University Press.

12 Dronamraju K. 1989. *The Foundations of Human Genetics*. Charles C. Thomas, Springfield, Illinois.

   Dronamraju provides a table of paradigm categories (over 20) to account for the various activities of science.

My own criticism of this effort is that when a term is used to describe too much, it loses its meaning and effectiveness.

13 Morgan T.H. 1911. The application of the conception of pure lines to sex-limited inheritance and to sexual dimorphism. *American Naturalism* **45**: 65–78.

14 Muller H.J., League B.B., and Offermann C. 1931. Effects of dosage changes of sex linked genes, and the compensatory effects of other gene differences between male and female. *Anatomical Record* (abstract) **51**: 110. The more detailed account is in Muller H.J. 1932. Further studies on the nature and causes of gene mutations. *Proceedings of the Sixth International Congress of Plant Sciences, Ithaca, 1932*. **1**: 213–255.

15 East E.M. Population, p. 621. *Scientific Monthly*, June 1920: 603–624. East expressed his more detailed eugenic concerns in 1929. *Heredity and Human Affairs*. Scribner's, New York.

   I do not single East out as uniquely virulent in his racism; many of his professional contemporaries shared his views. But quite a few did not make such blatant racist comments (Morgan never did). I raise my concern because it is important in understanding how science can be abused through some yet unexplained capacity of an individual to interpret human heredity at a poorly examined level and to interpret the heredity of other plants and animals at a very careful and sophisticated level.

16 Phrases such as "the winning of the facts," "bit by bit," or "pruning process" may not convey the totality of the process that I have described in this account of the origin of classical genetics. If a single word could suffice, I would suggest *incrementalism*. I do not argue that all science progresses by increments, but I believe that it applies to more than classical genetics.

Index